LIBRARY, ST. PATRICK'S COLLEGE, DUBLIN 9
LEABHARLANN, COLÁISTE PHÁDRAIG, B.Á.C. 9

D1582394

GEOGRAPHIES OF SEXUALITIES

ST. PATRICK'S
COLLEGE
LIBRARY

Geographies of Sexualities
Theory, Practices and Politics

Edited by

KATH BROWNE
University of Brighton, UK

JASON LIM
University of Brighton, UK

GAVIN BROWN
University of Leicester, UK

ST. PATRICK'S
COLLEGE
LIBRARY

ASHGATE

© Kath Browne, Jason Lim and Gavin Brown 2007
First published in paperback 2009
Hardback edition reprinted 2009

All rights reserved. No part of this publication may be reproduced, stored in a retrieval system or transmitted in any form or by any means, electronic, mechanical, photocopying, recording or otherwise without the prior permission of the publisher.

Kath Browne, Jason Lim and Gavin Brown have asserted their right under the Copyright, Designs and Patents Act, 1988, to be identified as the editors of this work.

Published by
Ashgate Publishing Limited
Wey Court East
Union Road
Farnham
Surrey, GU9 7PT
England

Ashgate Publishing Company
Suite 420
101 Cherry Street
Burlington
VT 05401-4405
USA

www.ashgate.com

British Library Cataloguing in Publication Data
Geographies of sexualities : theory, practices and politics
1. Homosexuality - Philosophy 2. Homosexuality -
Cross-cultural studies 3. Homosexuality - Social aspects
I. Browne, Kath II. Lim, Jason III. Brown, Gavin
306.7'66'01

Library of Congress Cataloging-in-Publication Data
Geographies of sexualities : theory, practices, and politics / by Kath Browne, Jason Lim, and Gavin Brown [eds.].
      p. cm.
   Includes bibliographical references and index.
   ISBN: 978-0-7546-4761-4
   1. Homosexuality--Philosophy. 2. Homosexuality--Cross-cultural studies. 3.
Homosexuality--Social aspects. I. Browne, Kath. II. Lim, Jason. III. Brown, Gavin.

   HQ76.25.G458 2007
   306.76'601--dc22
                                                                2006103151

ISBN: 978-0-7546-4761-4 (hbk)
ISBN: 978-0-7546-7852-6 (pbk)
ISBN: 978-0-7546-8478-7 (ebk)

Leabharlann Coláiste
000195887
ACC ....
CLASS 306. 766   BRo
DATE 7/9/09
PRICE ....
DROIM CONRACH

S.L.

Mixed Sources
Product group from well-managed
forests and other controlled sources
www.fsc.org Cert no. SA-COC-1565
© 1996 Forest Stewardship Council
FSC

Printed and bound in Great Britain by
MPG Books Group, UK

Contents

Section I: Theories

Section II: Practices

ST. PATRICK'S
COLLEGE
LIBRARY

Section III: Politics

List of Figures

By permission of the Out There! A New Zealand Queer Youth Development Project by Rainbow Youth Inc (Auckland) and the New Zealand AIDS Foundation / Te Tuuaapapa Mate Aaraikore o Aotearoa. The image contains Nathan Brown and his parents Jaye and Russell.

By permission of Heather Renee Russ

Notes on Contributors

Alison L. Bain is an Assistant Professor of Geography at York University, Canada. She is an urban social geographer who studies contemporary culture in Canadian cities. Her current research programme is an empirical study of cultural production and creative practice in the 'inner' and 'outer' suburbs of Toronto and Vancouver.

David Bell is Senior Lecturer in Critical Human Geography and leader of the Urban Cultures & Consumption research cluster in the School of Geography at the University of Leeds, UK. His research interests include sexualities, cultural policy, hospitality and science and technology.

Jon Binnie is Reader in Human Geography and Director of the Institute for the Study of Social and Spatial Transformations at Manchester Metropolitan University, UK. He is the author of *The Globalization of Sexuality* (Sage); co-author of *The Sexual Citizen: Queer Politics and Beyond* (Polity) and *Pleasure Zones: Bodies, Cities, Spaces* (Syracuse University Press). He is also co-editor of *Cosmopolitan Urbanism* (Routledge).

Gavin Brown has recently completed a PhD in the Geography Department at King's College London, UK. His doctoral research examines the social and cultural geographies of gay and queer space in east London, UK. His work on queer geographies, anarcha-queer activism and public sex cultures has been published in *Area, Environment & Planning A* and the *Oral History Journal*, as well as several edited collections.

Michael Brown is Associate Professor of Geography at the University of Washington, USA. His research is on political and cultural geographies of sexuality and the body. He is an editor for *Social & Cultural Geography*.

Kath Browne is a Senior Lecturer in Geography at the University of Brighton, UK. Her work addresses queer geographies, women who are mistaken for men, lesbian geographies, marginalisation within LGBT collectives and spaces of contemporary women's separatism. She has published in various edited collections and journals including *Antipode, Gender, Place and Culture* and *Social and Cultural Geographies*.

Mark E. Casey is a Lecturer in Sociology at Newcastle University, UK. He works in sexuality studies, lesbian and gay studies, gay male travel, the body and sex, urban sociology and has an emergent interest in eastern European gay male sexualities. He also writes for a number of gay male magazines and newspapers.

Vincent J. Del Casino Jr. is an Associate Professor of Geography at California State University, Long Beach, USA. His areas of expertise include social, cultural and health geographies, with an emphasis on HIV/AIDS, sex and sexuality, drug use practices, and community outreach and activism in both the United States and Southeast Asia.

Hanna Hacker is a sociologist and historian. She has widely published on sex/gender transgressions in European history. Her recent research focuses on transnationalities and on New Media cultures. Having mostly worked as an independent researcher, she currently holds a post as Visiting Professor for Gender Studies at the University of Vienna, Austria.

Jin Haritaworn lectures and presents on sexuality and power, Whiteness and gender, race and trans awareness, and the politics of alliance. Jin's publications span topics such as intersectionality, racialised sexualities and gender identities, anti-racist feminism, queer and trans of colour theories and sex radicalism.

RDK (Doug) Herman received his PhD in Geography from the University of Hawai'i at Manoa in 1995. He is currently an Associate Professor at Towson University, part of the University of Maryland system, USA. His interests include indigenous geographic perspectives, Pacific Islands, critical theory, food and culture, and sexuality and space. Among his publications is 'The Aloha State: Place Names and the Anti-Conquest of Hawai'i' in *Annals, Association of American Geographers*, 1999.

Phil Hubbard is Professor in Urban Social Geography at Loughborough University, UK. He has wide-ranging interests in the city and its ability to accommodate difference and has written widely on the geographies of female sex work. His recent books include *The City* (Routledge, 2006) and the forthcoming *Key Texts in Geography* (edited with Rob Kitchin and Gill Valentine, Sage, 2007).

Larry Knopp is Associate Dean of the Graduate School and Professor in the Department of Geography at the University of Minnesota-Duluth, MN, USA. His research interests include feminist and queer geographies, sexuality and space studies, and urban, social, cultural and political geographies.

Jason Lim is a Lecturer in Human Geography in the School of Geography at the University of Southampton, UK. His research interests include considerations of race, ethnicity and heterosexualities; theories of affect and everyday practice; and the cultural politics of the so-called War on Terror.

Tatiana Matejskova is a PhD student in the Department of Geography at the University of Minnesota, USA with research interests in the urban and cultural geographies of (im)migration, citizenship, state and sexuality. Her dissertation research examines the spatial politics and practices of immigrant 'integration' and citizenship in Berlin, Germany.

Catherine Jean Nash teaches urban social and cultural geography in the Department of Geography, Brock University, Ontario, Canada. Her research interests focus on the constitution and regulation of gender and sexuality in urban spaces and draws on feminist theory and practise, queer theory and LGBTQ and trans studies.

Farhang Rouhani teaches geography at the University of Mary Washington in Fredericksburg, Virginia, USA. His interests include globalisation and state formation in Iran, migration politics, queer anarchism and slow living.

Matthew Sothern is Lecturer in Human Geography at the University of St. Andrews, UK. His research interests centre on the cultural politics of (neo)liberalism with a focus on those bodies that cannot be rendered safe, normal or even intelligible within contemporary liberal political formations.

Acknowledgements

This collection owes its existence to a Philly cheese steak sandwich during a break from the Philadelphia AAG conference in 2004. It has been a long and interesting journey since then and we owe a number of people thanks for ensuring the existence of this edited collection. We would like to thank all the authors for their exciting and innovative contributions. The support we have received from 'senior' academics has been invaluable. We are indebted to the academics who initially discussed sexualities within Geographies. It is their bravery and efforts that have enabled not only this collection, but also wider discussions of queer and sexualities within this once hostile discipline.

We would like to thank Ashgate and, particularly, Valerie Rose for her patience, support and work on this collection.

We each, individually, would like to make the following acknowledgements:

Kath: I think the majority of my gratitude goes to the other two editors of this book. Our various skills and time commitments have all be used in producing the volume. This work would not have been possible without the support of my department, and I am indebted to the Hove women's rugby team and close friends for their personal and social support.

Gavin: I owe Loretta Lees and Tim Butler a debt of gratitude for their inspiration and encouragement, as well as for their patience with this book when it distracted me from my doctoral studies. Juggling this book, a part-time PhD and a demanding job would not have been possible without the love and spirited support of JoJo, Lauren and Tallulah.

Jason: I would like to thank my co-editors for their hard work, support and inspiration. I would also like to thank Dawn Robins for her help in getting the typescript ready. I owe a debt of gratitude to all those friends and colleagues who have lent their support during the process of putting this book together, but most of all, I would like to thank Ruth for all her unstinting love, understanding and patience.

Introduction, or Why Have a Book on Geographies of Sexualities?

Gavin Brown, Kath Browne and Jason Lim

Geographies of Sexualities: Theory, Practices and Politics seeks to address some of the most recent developments in geographies of sexualities and to gesture towards a number of initial questions, directions and tensions that are currently emerging from this exciting and proliferating field of study. The book considers these ongoing developments in light of the continuing influence of 'queer theories' upon geographies of sexualities, and it asks what possibilities are offered by this intersection of ideas. It explores a host of themes – sexualised difference, social relations, institutions, desires, spaces – but does so through the framework of three interlinking perspectives: theory, practices and politics. Sexual geographers work in a diverse theoretical and political terrain, often dominated by different aspects of queer theory. Their eagerness to engage with concrete social relations and practices in their research means that their insights are materialistic, spatialised and affective. It is through this focus on the materialistic, spatialised and affective that sexual geographers help to contribute something both distinctive and innovative to broader thinking on sexual difference, relations and desires. One of the aims of this book is to create a space for a consideration of the emerging implications that are raised by such contributions, theoretically, empirically and politically.

This book has come into being because we feel that the relatively young field of geographies of sexualities has blossomed over the past decade or so. This proliferation of ideas, insights and knowledges across an ever-widening range of concerns warrants concentrated attention. *Mapping Desire* – the first (and still most significant) major edited collection to address the breadth of geographies of sexualities – was published in 1995. Since then, the importance and influence of geographies of sexualities has spread steadily, although not without resistance, throughout the discipline of geography and beyond. Within geography, geographies of sexualities are now taught on many, if not most, undergraduate courses, at least in the UK and US; the amount of research being carried out has witnessed an explosion; and ideas from geographies of sexualities are becoming increasingly influential elsewhere in the discipline.

In this introductory chapter the groundwork for understanding these developments in geographies of sexualities will be laid out. For students, it will provide an overview of the major strands of thought comprising the field and will explore the development of its most important concepts. For researchers and teachers, this chapter, in particular the latter parts, will also highlight important points of debate and conceptual tensions that exist between different theoretical approaches and between different writers. The next section of this introductory chapter will address the mutually constitutive relationships between sexualities, spaces and places. It will also explore how geography is important in attempting to understand contemporary sexualities and will consider how geographers and others have considered the spatialities of sexualised relations. The third section of the introduction will provide an overview of the development of geographies of sexualities since the 1980s, starting with earlier geographies of prostitution, running through geographical investigations of gay ghettos and lesbian appropriations of urban space, and moving on to the continually developing encounter with queer theory, including its effect of reinvigorating concerns with heterosexual geographies. The fourth section of the introduction will outline the organisation and themes of the contributors' chapters that comprise this collection.

Sexualities and Space, or What Exactly Have Geographies got to do With Sexualities?

While geographies of sexualities emerged in the 1980s and 1990s, the past decade or so has witnessed an increase and diversification of writings on sexualities, space and place. This proliferation has not only occurred within the boundaries of the discipline of geography, but it has also taken place outside it, as sociologists, anthropologists, cultural theorists and others have started recognising the importance of spatial and temporal context and fluidity (Hunt 2002; Kuntsman 2003; Conlon 2004; Eves 2004; Jacobs 2004; Skeggs et al. 2004; Taylor 2004; Grosz 2005a; 2005b). The central theme of this explosion of work has been the exploration of the relationship between sexualities, space and place – questions about the ways in which sexualities are geographical, or the question of how spaces and places are sexualised. There are, of course, many such ways in which sexualities are geographical, but here we will discuss a few examples that show a number of important themes and that highlight the importance of thinking about both politics and practice in order to understand these themes. The first set of examples draw attention to how, in various ways, everyday spaces are produced through embodied social practices – that is, we are starting with bodies and what they do. In turn, it is often through these practices that the norms regulating such spaces – and the sexualised relations between bodies, selves and others that constitute these spaces – become enacted. The second set of examples suggest how sexualities can usefully be understood through the institutionalisation of spaces at a variety of scales, from the national to the transnational.

The norms regulating acceptable sexual behaviour in public or shared spaces are an example of how everyday spaces are sexualised. Public spaces are normally governed by unspoken understandings, enforceable by both official authority (for example, the police) and by the verbal interventions or looks of passers-by. These constrain displays of sexual desire: a kiss might be acceptable on a busy street, but rolling around on the ground with a lover might not be. As Gavin Brown (forthcoming; Chapter 16, this volume) considers, however, participants in sex parties at Queeruption[1] gatherings attempt to constitute different understandings of what is acceptable in shared and public spaces.

---

1   Queeruption is the name adopted by a series of international radical queer DIY gatherings that have taken place (almost) annually since 1998. Every gathering is a little

The sexualisation of space, however, does not only apply to spaces where people might expect to engage in explicitly sexual activity of some kind. Many other banal and everyday spaces are structured by sexuality. Many geographers have written about the ways in which home (Valentine 1993a; Johnson and Valentine 1995; Elwood 2000; Gorman-Murray 2006) and work (Valentine 1993b; McDowell 1995; 1997; Kitchin and Lysaght 2003; Kawale 2004) are sexualised. Home, for many people, is taken for granted as a place of comfort, a retreat from the world, a place to be oneself. For many lesbians, gay men, bisexuals and trans identified people, however, home can be uncomfortable and alienating, shaped by the assumptions of heterosexuality that are present in their social relations with parents, siblings, neighbours and others in and around the home. Such heterosexual assumptions often structure familial relations at home. Hiding lesbian or gay sexual desires and practices from parents is an understandable strategy when 'coming out' as lesbian, gay or bisexual might turn the home into a space of violence – violence meted out by parents and other family members. Even without the prospect of such physical violence, the everyday regulation of normal behaviour, identity and practice may still have to be negotiated. Such regulation might take the form of direct political and social injunctions against lesbians, gay men, bisexuals and other sexual minorities (for example, overt social disapproval in housing policies, social security assumptions etc.). Or it might take place through indirect means: for example, assumptions about 'normal' sexuality that structure conversation (how do you refer to 'a partner'?); discussions of life aspirations that presume a heterosexual relationship and marriage ('are you not married yet?'); disparaging comments about 'gays'; and jokes that presume all present share a common distaste for those who do not conform to the heterosexual norm, backed up by labelling those who do not want to 'get the joke' as 'spoilsports'. These everyday contexts, discussions and practices not only create an 'Other' to heterosexuality, they also constitute spaces as heterosexual and, indeed, constitute heterosexuality itself.

The spaces that we can understand as structured by sexuality, however, are not just these kinds of everyday spaces. Rather, these everyday spaces intersect with various other scales of spatiality, including national, international and transnational spaces. Luibhéid (2005) has reflected upon the history of the regulation of international migration to the US. Through various legislative acts and practices of interpretation and enforcement of those acts, the US has sought to exclude what Luibhéid calls 'queer' migrants. It is through this exclusion of 'queer' migrants that a normative construction of nation, citizenry and citizenship is produced. Even though explicit discrimination against lesbian and gay immigrants to the US was removed in 1990, lesbian and gay relationships are not recognised as a legitimate basis for legal permanent resident status. Such implicit exclusion forces these immigrants to resort to other means. Turning to other kinds of visas – student, tourist or work – offers less stability of immigration status and nothing of a secure basis for a long term relationship. Still others turn to heterosexual marriage as a way to gain entry, raising the possible irony of a lesbian leaving a home country to avoid marriage,

---

different to the rest, responding to the possibilities offered by its location and to the creativity of the community of people who plan it and who turn up to make it happen.

only to end up in a sham marriage in order to stay in the US and other 'safe' countries (Poore 1996).

What these examples demonstrate is that sexuality – its regulation, norms, institutions, pleasures and desires – cannot be understood without understanding the spaces through which it is constituted, practised and lived. Sexuality manifests itself through relations that are specific to particular spaces and through the space-specific practices by which these relations become enacted. The example of 'queer' immigration to the US suggests how sexualities are both lived and regulated through a contested moral economy that becomes expressed as an imagined geography: a centring of heterosexuality as 'American' and a positioning of 'queer' subjects as a moral threat from abroad. Such a sexualised imagination of the US and abroad becomes instituted not only in national and international laws, but also in how it is practised and enacted. It is this institutionalisation of sexualised imagined geographies (the centring of heterosexuality and the 'moral threat' of queer) that becomes contested politically. What is at stake in such political contestations is the power to define who belongs and to define what bodies are allowed to do, when and where. This example, then, suggests how imaginative, representational and figurative spaces become related to material effects that make a difference to people's lives (e.g. whether people can enter a country or not; whether they can be safe in particular spaces; what they can and cannot do in safe or unsafe spaces; what channels of immigration are open to them; what tactics they might employ to negotiate the policing they are subject to). This relationship between, on the one hand, the figurative, representational and imaginative and, on the other hand, the material also pertains to other kinds of sexualised spaces. On one level, 'closet space' might be thought of as a metaphor for the concealment and denial of lesbian, gay, bisexual and/or transsexual (LGBT) lives and desires, yet 'closet space' is also lived in very material ways through countless practical and political acts and through experiences of threat and marginalisation (Brown 2000).

The consideration of the 'everyday' spaces of sex parties and of the home also points to the ways that participants' experiences of spaces are regulated by norms regarding what is acceptable or expected sexual practice. These norms and expectations are not set in stone, but can be challenged and renegotiated. The spaces, whether sexualised, heterosexualised or even homosexualised, are constituted through the enactment, negotiation and contestation of norms of appropriate sexual conduct, even where the sex act itself may seem to be 'irrelevant'. What we do makes the spaces and places we inhabit, just as the spaces we inhabit provide an active and constitutive context that shapes our actions, interactions and identities. A consequence of this set of ideas is that we can never take a given space or set of practices for granted or assume that they are fixed. A home, a nation, a bathroom, a workplace – no space exists in a timeless state. Each is created in particular ways, often associated with sexualised and gendered norms and conventions that are historically and geographically specific. Not only are the places we inhabit made through our repeated actions such that we take their normality for granted, but these places produce us precisely because we so often do what we are *supposed to* do – what is 'common sense' in a given place. Nonetheless, although, for example, there are a set of hegemonic gendered and sexualised norms about what to do in a public lavatory, activities such as cottaging (that is public sex in toilet spaces) suggest that it is always possible to follow the desire to do something differently. In (repeatedly) doing something differently, this can become established as 'the norm', even if temporally.

Sexualities in Geography: A Historical Overview, or How Did We Get to Here?

Geographers have commented on the discipline's approach and attitude towards human sexualities (and homosexualities in particular) since the mid 1980s. Initially, some geographers addressed the absence of discussions of sexuality within geography by contesting the discipline as a heterosexist institution (McNee 1984). Over the last two decades, the range of geographical engagements with human sexualities has proliferated to the point where many recognise the geographical study of sexuality to be an emerging sub-discipline within geography with its own internal tensions and identifications of hierarchies. Nonetheless, the growth of geographies of sexualities is still an uneven process both within the discipline as a whole and amongst national academies. Most work by sexual geographers is still located within the broader sub-discipline of social and cultural geography, although this book also includes work by scholars who locate their research primarily within urban, political, economic and health geographies. There are also those who could be defined as 'sexual geographers' who do not work within the discipline of geography.

This section of the introduction seeks to chart the development of geographical studies of sexuality and the expansion of geographical knowledge about the mutually formative relationship between sexualities, space and place. For the sake of simplicity, we present the history of geographies of sexuality through a broadly chronological sequence of thematic concerns, moving from studies of gentrification in the gay 'ghetto' through to the most recent engagements with poststructuralist queer theory. It is not our intention that this sequence should be read as implying that each new development was either an inevitable progression from that which preceded it, or that earlier theoretical models and objects of study do not continue to have valence for geographers of sexualities. Although we believe that a certain theoretical orthodoxy is beginning to develop within the sub-discipline at the expense of earlier approaches, we still believe that there exists a healthy theoretical pluralism within the work produced by geographers of sexualities.

Neither academic theories nor the spatial practices that constitute sexual identities and spaces develop in a vacuum. Both influence and are influenced by politics. By 'politics' we mean not just the formal political power and practices of state institutions, but also broader contestations of power and, in this context, heteronormative power relations in particular. We understand power as an amalgam of forces, practices, processes and relations (Sharp et al. 2000, 20). Power might be understood as myriad entanglements of resistance and domination that are mutually constitutive of one another. Power is not just something that happens to us; we are always engaged in these entanglements. Power operates through how we interact with one another, how we regulate each other's behaviour and consequently make the spaces that we inhabit. Rather than an ethereal force, power is something that we are all located within. In this way, power can be productive and can take many forms. Of course, political beliefs and practices are historically and geographically located. In the remainder of this section, we analyse developments in the field of sexual geographies within the political contexts in which the research was undertaken. We believe that these broader contexts, as much as academic trends, explain some of the developments within geographies of sexuality and the specific concerns that have arisen amongst geographers in different countries at different times. Of course, these connections are never that straightforward or complete, but we still believe there are connections between (sexual) politics, theory and practices. In many ways, this is particularly clear in the case of early work on the gay ghetto.

Geographies of the Gay Ghetto

Some of the earliest work that sought to understand the relationship between sexualities and the creation and uses of space were studies of the residential concentration of gay men in inner city areas of American cities (Castells 1983; Lauria and Knopp 1985). This work attempted to map these gay residential and commercial clusters in the tradition of American urban geography at the time and, by so doing, legitimise the study of gay lives as an appropriate topic for geographical research (Knopp 1987). Although clusters of gay bars and other businesses had existed in many major European and Australian cities for some time, by the mid 1980s the co-presence of residential concentrations of gay men with these businesses was most visible in the United States. Initially, this was read as a result of gay men moving from rural and small town America in search of more liberal environments that would offer some respite from the pressures of heterosexist society. As geographers explained at the time, this appropriation of territory in major urban centres served as a defensive base where gay men could feel safe (Castells 1983; Knopp 1987; 1990; Warren 1974). However, the structures of the American political system also encouraged the residential concentration of gay men as a means by which they could exercise political power in pursuit of civil rights (Lauria and Knopp 1985; Knopp 1987; 1990). The urban geography of gay space in British cities throughout the 1980s was more uneven (Bell and Binnie 2000), although the dominance of 'municipal socialism' in some British cities throughout the period (Cooper 1994) offered a limited political voice to gay people against the vicious homophobia of the Thatcher government. It was not until British local governments began to take a more entrepreneurial and competitive approach to place-marketing in the 1990s, most notably in Manchester (Quilley 1997; Whittle 1994a), that significant gay centres in Britain's major urban centres began to be studied by British urban geographers.

The majority of the early work on gay space was focused on urban enclaves (in North American cities) and primarily analysed fixed territorialisations in the form of gay bars and other businesses as well as (gentrified) residential clusters. Podmore (2001) has argued that this focus is problematic for the study of lesbians' uses of urban spaces because although specific places might be identified as 'lesbian', these sites are less likely to utilise the visible signifiers that usually mark an area as 'gay'.

(Re)placing Lesbians in Geography

The approach to lesbians within gay male geography can perhaps be summarised by Quilley (1995, 49) who states "[b]y gay community I refer mainly to men." Castells (1983) and Knopp (1990), amongst others, argued that lesbians were less likely to be able to afford to concentrate their homes in a given neighbourhood and were less likely to achieve local political power (again, in an American context). Castells (1983) believed that the power relations between women and men visible in society were reproduced within gay spaces. He believed that these were due to men's essential need to claim space, an innate need that women did not possess. Similarly, although they acknowledged that it was easier to live as 'gay' if you were white, male and middle class, Lauria and Knopp (1985) contended that gay men appropriated urban space more than lesbians did because gay men were more oppressed *as men* in relation to heterosexual men and consequently had a greater need for 'safe' spaces in the city. There are, however, an increasing number of explorations of lesbian appropriations of urban space in North American cities that contest these assumptions. As early as 1978, Ettorre's work challenged the assumption that lesbians were not involved in urban politics. Geographers have also contested the assumption that lesbians could not or did not wish to appropriate urban space at a neighbourhood level (Rothenburg 1995; Winchester and White 1988).

In a British context, Valentine (1993a; 1993b; 1993c; 1995b) demonstrated how women living in a small English town created lesbian spaces, ranging from more materially grounded spaces at a neighbourhood level to the temporary appropriation of heterosexual spaces such as bars and clubs. For Valentine, 'lesbian landscapes' incorporated more than just appropriations of space; these geographies consisted of complex time-space relations where different places took on different meanings over time. This work recognised that space could not easily be categorised as either 'gay' or 'straight', but that particular groups made differential uses of space (for example, when an occasional lesbian night is held in an otherwise straight bar). Building on ideas about the temporal constitution of spaces, lesbian geographies broadened the focus of the geographical study of sexuality and space beyond the inner city and incorporated discussions of home, work and street (Johnston and Valentine 1995). These geographies recognised the sexualised construction of particular spaces, such as home, as well as considering the complex negotiations of public spaces by lesbians. Moreover, this work opened up a discursive space for the study of rural sexualities (Bell and Holliday 2000; Kramer 1995; Phillips et al. 2000), although this is an area of work that is still under-developed.

More recently, Podmore (2001) and Peace (2002) have contended that it is how urban space is analysed and understood that excludes explorations of lesbian uses and appropriations. They critique (gay) urban geography for its narrow focus on territoriality and on singular identities and for an over-investment in the importance of 'visibility'. Podmore suggests that lesbians have very different means of making themselves visible (to each other) than gay men and that to properly explore these practices, geographers need to (re-)integrate the domestic sphere into their interpretations of urban space (see Jay 1997). Furthermore, both Podmore and Peace both argue that the artificial separation of public and private space from each other masks the multiplicity of different identities that can be found within a given neighbourhood. Instead of the concentration on residential clusters and commercial premises that has typified much gay geography, they propose that geographies of lesbian space can only be advanced through an attention to women's social networks and their daily circulation through quotidian urban space. We believe such a project would not just enable a better understanding of lesbian space, but would also reveal the complexities of the everyday geographies of queers of colour, gay men who do not participate in the commercial gay scene, and others who are rendered invisible through the focus on fixed territories (see Casey, Chapter 10, Nash and Bain, Chapter 13, this volume).

Early Encounters with Queer

Although what we now call 'queer theory' was largely developed by academics working in humanities departments in US universities, this diverse set of ideas was first taken up within geography by British rather than American geographers of sexualities (Bell et al. 1994; Binnie 1997). Knopp (1998) has suggested that this was partly a reflection of the more conservative outlook of American geography departments, but also a measure of the different political and cultural traditions of gay 'communities' on either side of the Atlantic – a British 'cultural politics of resistance' versus American attempts to spatially consolidate gay economic and political power. As Binnie (1997) noted, the work of Knopp, Valentine and others had successfully added lesbian and gay concerns to the pot of geographical analysis, but there was still a considerable amount of 'stirring' needed in order to challenge the heteronormativity of space and the many ways in which everyday spaces reinforce the invisibility, marginalisation and social oppression of queer folk. Queer is a highly contested term, one that has a variety of uses, applications and, some would argue, misuses. Throughout this introduction and in the conclusion, we (the editors) will take the position that queer is not just synonymous for lesbian and gay. We therefore challenge one use of queer, which is as an umbrella term for LGBT. Instead, we consider queer to question the supposedly stable relationship between sex, gender, sexual desire and sexual practice. This challenge to the supposed correspondence between desires, identities and practices consequently disrupts the stability of heterosexuality that this correspondence shores up. Queer theories do not only understand sexuality as varying historically and culturally, an insight arising from historical studies of the invention of the categories of heterosexuality and homosexuality in late nineteenth-century Western societies. Queer theories also challenge heteronormativity – "the set of norms that make heterosexuality seem natural or right and that organize homosexuality as its binary opposite" (Corber and Valocchi 2003b, 4). Heteronormativity allows heterosexuality to go unmarked and unremarked upon – to be thought of as normal – by making homosexuality operate as heterosexuality's binary opposite. Homosexuality is made to function as the marked, abnormal Other of heterosexuality. The categories of 'heterosexuality' and 'homosexuality' are, thus, mutually constitutive and cannot be understood autonomously. The intelligibility of the categories of 'heterosexuality' and 'homosexuality' is also reliant upon the opposition between 'male' and 'female' and upon the supposedly natural sexual desire between these two sexes.

Throughout this book it is clear that the term 'queer' is not used homogenously. Here, in focusing on how 'early' geographies of sexualities have developed into 'queer geographies', one can see how this diversity begins to emerge. These initial queer geographies initiated a discussion about how sexed and gendered performances produce space and, conversely, how spatial formations shape the ways in which sexual dissidents present and perform their sexualities in public spaces. This discussion drew on theories of performativity developed by Judith Butler (1990; 1993a; 1997), which suggested that identities, for example the gender identities 'male' and 'female', are not simply 'there', always-already existing as an expression of natural sexual difference. Rather, it is through the reiteration of social and discursive conventions that our actions (speech, practice etc.) transform bodies so that they become recognisable as male or female. Building on this performative approach, geographers have contended that space is not simply the vessel in which things happen, but is actively constituted through the actions that take place. Work on sexualities in this mode challenged how the everyday repetition of heterosexual relations becomes normalised such that quotidian space is not assumed to be sexual at all (Bell et al. 1994; Binnie 1997). Geographers have sought to explore not only how spaces come to be hierarchically sexualised, but also how racialised, classed and other forms of social hierarchies come to structure seemingly unitary categories of sexuality such as 'gay' or 'lesbian' (Nast 2002). This most recent queer geographical work looks not just at the hegemonies of heterosexuality, but also at the multiple diversities between those who identify as 'gay' (see Haritaworn, Chapter 8, this volume).

One of the first geographers to publish an extended consideration of an appropriate epistemology for the analysis and understanding of queer geographies was Jon Binnie (1997). He argued that the production of geographical knowledge had excluded sexual dissidents and that the (then) 'new cultural geography' was as marked by heterosexism as earlier positivist strains of geography within the discipline. Binnie advocated that sexual geographers should place a greater emphasis on the lived experience of sexual dissidents, but that, in doing so, their work should also "include a greater critical awareness of the material conditions for the production of 'knowledge' about sexuality" (1997, 224). To achieve this awareness, Binnie proposed that a queer geographical epistemology would need to have at its centre a renewed commitment to an honest acknowledgement of the embodied positionality of the author-researcher and "a recognition of the value of camp" (1997, 228). 'Camp' works because it is simultaneously knowing and innocent. It resists fixed and definite interpretations. In this respect, camp operates in similar ways to much queer space, where the delineation of identity boundaries and gender roles is constantly blurred and in flux. In Chapter 2 of this volume, Binnie revisits, extends and interrogates some of the ideas from his earlier paper.

A related approach to the fuzzy distinction between gay and straight space has been offered by Hemmings's (1995; 1997; 2002) work on bisexual spaces. To her, bisexuals occupy both lesbian/gay and straight spaces and have played a role in the construction of both. Bisexuals may think of either gay or straight spaces (or both) as 'home', but a bisexual identity is never predominant in either set of spaces, and although the presence of bisexuals may be acknowledged, it is seldom fully included.

For Hemmings, this posed several theoretical challenges as, by implication, bisexual space is neither gay nor straight space, and the presence of bisexual identities in space is always partial. Her solution to this problem was to focus on witnessing the spatial enactment of embodied acts and desires, although she accepted that these seldom easily correlate to a self-identified bisexual identity. She examined the production of bisexuality and the negotiation of bisexual desires in relation to other queer bodies in particular queer spaces. By taking this approach, it was Hemmings's intention to re-emphasise that desire is "enacted through our bodies" (Hemmings 1997, 149) and is site-specific.

In attempting to write queer geographies, these authors drew both on queer theory and on broader social theory. In doing so, they produced an interpretation of queer that was quite distinct from that in other disciplines. In particular, their concern with the production of space, everyday social relations and the materiality of embodied queer performances differed from other forms of queer theory that relied heavily on more discursive analyses and metaphorical understandings of space (Brown 2000). Nevertheless, this body of geographical work still remained primarily concerned with the performance of lesbian, gay and bisexual identities. It seldom followed through the logic of the concern with the site-specific embodiment of desires to offer a more thoroughly queer critique of the production and performance of all sexualities.

Geographies of Sexualities: Taking Heterosexuality Seriously

Although feminist geographers have long examined the ways in which patriarchal social relations are seen to reinforce and be reinforced by heterosexist relations within the home, the work-place (McDowell 1997; Gregson and Lowe 1994; WGSG 1997; Domosh and Seager 2001) and elsewhere, it took geographers of sexuality some time to turn their attention to the spatial production of heterosexual identities and desires (Nast 1998; Hubbard 2000). Latterly, however, these geographies of heterosexuality have highlighted the heterogeneity of different forms of heterosexuality and have demonstrated that these too are contextually specific. This work has recognised that heterosexual space is variously sexualised or desexualised by and for different people at specific times, with heterosexuals caught up in various modes of self-production and self-surveillance. In using queer theory to deconstruct normative heterosexuality (see Hubbard, Chapter 12, this volume), there has been a recognition that some heterosexualities are 'queerer' or more dissident than others and can themselves pose a challenge to established heteronormative power relations.

'Prostitution' and the vilification of 'red light districts' have been used to highlight the diversity of heterosexualised spaces, the ways in which these spaces are regulated and the moral panics that play a pivotal role in these processes of regulation (Hart 1995; Howell 2000a; 2000b; 2003; 2004; Hubbard 1998; 1999; 2002; Hubbard and Saunders 2003; Tani 2002). The internationalisation of spaces of sex work, including sex tourism, points to the geographical complexity and multiple forms of power that constitute the 'sex trade' (Brown 2000; Law 2000; Hall 1994; Nagle 1997). Central to these discussions have been the complexities of agency, coercion and processes of regulation by which 'moral' and 'immoral' heterosexual geographies become differentiated (Hubbard and Saunders 2003). However, the focus on more dissident forms of heterosexuality, such as sex work and prostitution (Hubbard 1998; 1999), sex tourism (Law 1997), BDSM (Herman, Chapter 7, this volume) and 'dogging' (Bell 2006), means that much work still needs to be done to understand the mundane processes by which everyday expressions of heterosexuality are (re)produced in social space. This is particularly important for the study of all sexualities because, as Blum and Nast (1996) have suggested, the construction of heterosexuality is central to the construction of all forms of alterity and difference.

The 'unremarkable' hegemonic status of heterosexuality is beginning to be remarked upon and deconstructed. Like lesbian and gay geographies, most of this work focuses on urban and suburban space, with little attention paid to the construction of rural heterosexualities (although see Little 2003). As the cracks of heterosexuality begin to become investigated, exposed and considered, Hubbard (Chapter 12, this volume) argues that there has been a proliferation of work on heterosex and he argues that "straight geographies have gone queer too".

Taking Queer Further

The social and political terrain on which geographical critiques of sexualities take place has changed significantly over the last fifteen years. There have been significant advances in civil rights for some lesbians and gay men in Britain, many other European states, Canada and elsewhere (with, some might pointedly argue, a decline of such civil rights in the US). The commercial gay scene has grown in size, scope and location in these countries and others too (for example, see Matejskova, Chapter 11, this volume). Positive representations of gay people are now far more common in the mainstream media as well. However, this progress has not been uniform or universal; it remains uneven within and across national borders. Queer theorists and activists have interrogated such apparent achievements. They have suggested that such rights are only granted on the condition that lesbians and gay men conform to the normative model of a monogamous, long-term, consumerist (and, more often than not, white and middle class) relationship. The price paid for such rights is the reproduction of these norms, hence delegitimising those whose sexual lives do not conform. In this context, geographers have recently engaged critically with these changes in sexual citizenship (Bell and Binnie 2000) and the uneven spread of a 'global gay' identity across the world (Binnie 2004). Such work has questioned who benefits from these changes and at what cost.

As Boellstorff (2003) demonstrates in relation to the Indonesian context, this questioning considers the specific articulations in local and national spaces of the transnational circulation of 'queer' discourses and discourses about 'lesbian' and 'gay' identities. Boellstorff notes that many men he encountered during his research in Indonesia identified as 'gay', their initial exposure to such understandings reportedly occurring through globalised mass media. Despite identifying as gay, most of these men retained an expectation of getting married. Unlike in many contemporary Western contexts, this expectation of marriage was not understood to conflict with a 'gay' identity or lifestyle. Indeed, it was seen as a source of potential fulfilment. The expectations held by 'gay' men in an Indonesia context need to be understood as a specific articulation of various *contemporary* globalised and local norms. Indeed, these discussions, along with others that explore nationalities and sexual practices (Lambevski 1999), illustrate that although the terms 'lesbian', 'gay' and 'bisexual' often connote 'common sense' ideas about desire, relationships and gender, the links between actions, identities and labelling processes are far from homogenous and are not easily definable. Questioning the transnationalism of identities such as 'gay' or 'lesbian' (see Hacker, Chapter 5, this volume) exposes the construction of these sexual identities and illustrates their historical (Faderman 1981; Weeks 1985) and geographical specificity.

Whereas 'queer' is used by some as a short-hand, umbrella term for all lesbians, bisexuals, gay men and transgendered people, it is increasingly being used as an appellation for sexual positionalities that contest not just heteronormativity, but also homonormativity. As with 'queer', the concept of 'homonormativity' has a range of meanings and is used to understand how homosexuality is constructed within class, racial and ethnic norms. As Nast (2002) has stressed, some white, middle class gay men have achieved a certain degree of 'liberation' because of their inclusion into more mainstream capitalist social relations, whilst many working class gays and queers of colour are still denied access to these privileges. Homonormativity, in this sense, has been used to extend the queer analysis and contestation of the practices and privileges of those gays and lesbians (in the main) who are prepared to assimilate on the basis of largely capitalist and heteronormative values.

Queer activisms can explore a distinct set of politics, and these can conflict with those who seek to advance lesbian and gay rights claims (as they are currently formulated). Queer's emphasis on deconstruction and the desire to question heteronormativity calls into question the tenets of these rights arguments. Whereas mainstream gay politicians may seek formal equality with heterosexual institutions (such as marriage), queer questions the uneven application (even amongst straight people) of the 'rights' that gay rights activists want equal access to. It asks 'equality to do what and with whom'? In contrast, queer political projects can be productive, experimental and utopian. They are often associated with creativity, fun, playfulness and the contesting of gendered, sexualised and racial norms in order to produce new forms of sociality, practice, desire and affect (see Gavin Brown, Chapter 16, this volume). This does not, however, mean that contemporary queer activism never deploys rights claims nor ever utilises more established identity categories to advance certain strategic causes (for example, to enable 'queers' from the Global South to claim political asylum in the European Union by asserting 'gay' identities that may have little currency or meaning in their countries of origin (Luibhéid and Cantú 2005)). It does, however, mean that, in contesting normativity, queer politics can be strategically deployed and that certain activities once seen as 'queer' can become mainstreamed.

In parallel with queer activism's turn towards a more productive and utopian politics that transcends straightforward civil rights claims, queer geographers and others have begun to engage with queer theorisations of becoming. Knopp (2004) has re-evaluated conceptions of gay migrations through the ontological lens of actor-network theory and more-than-representational geographies. He considers queer movements and placelessness to be part of an on-going quest for belonging and identity, which offers the opportunity to continually experiment with alternative modes of being and to engage in active processes of reinvention (see also Knopp, Chapter 1, this volume). Such queer movements and placelessness reveal the perpetual incompleteness of a queer identity. It is this continual process of becoming that challenges essential or pre-determined bodies, identities or spaces. It also prompts questions about how things come to be materialised and about the regulation of such materialisations. More broadly, then, these theoretical engagements with ideas of becoming explore how bodies come to take shape and the importance of emotions and affects in these constitutions (see Lim, Chapter 4, this volume). In challenging the rigidity of gender and sexuality, academics have explored how categories of gender are enacted and reiterated in a plethora of ways that move, for example, between pleasurable playing with drag through to the painful policing of gender ambiguous bodies in toilet spaces (Browne 2004).

The queer interrogation of gender and sexual difference not only problematises the idea that one's gender identity should match up with one's biological sex (male or female), but problematises the very idea there are (only two) immutable and natural biological sexes. This interrogation of gender and sexual difference is in contrast to the rigidity of dichotomous conceptions of gender within earlier lesbian and gay geographies. The queer problematisation of the relationship between sex, gender and sexuality is important, for instance, in understanding the relationships 'trans' subjectivities, bodies and practices have to prevailing ideas of sexuality. Queer understandings of gender in such contexts have been complex and far from homogeneous. Trans theorists and activists have argued for the right to be recognised as their chosen gender (McCloskey 1999; West 2004) and have simultaneously deconstructed the idea that gender must be embodied within a man/woman binary (Halberstam 1998; Hird 2000).

Here is Queer?

The narrative so far might suggest that all sexual geographers have been enthusiastic adopters of queer theory. This is not the case. Many scholars, both inside and outside geography, have pointed out the limits of queer or have affirmed the value of other political and theoretical stances (Jackson 1999; 2003; Jeffreys 2003; Witz 2000). The value placed on these other political and theoretical stances often arises in the context of a broader appreciation of the longer histories of sexual dissidence and the movements that have agitated for rights for those who are othered by the dominant norms of heterosexuality (Nash 2005; see Gavin Brown, Chapter 16, this volume). Given that queer theory is a body of work that has largely, although not exclusively, been developed within the humanities, its adoption within a predominantly social science oriented geographical arena has presented both problems and interesting points of departure. A focus on discourse and an inheritance of poststructuralist conceptions of power has added depth to the analytical tools available to geographers, but has by no means supplanted social science concerns with how institutions regulate social relations in material ways. Some of the major challenges, then, faced by sexual geographers over the past decade or so have been how to materialise and spatialise the insights of queer theory and how to combine the insights offered by queer theories with considerations of power that focus on institutions, practices and material social relations.

Indeed, the development of queer theory and its adoption and furtherance by geographers can only be understood within a much wider political context. Queer activism arose from older traditions of sexual dissidence, traditions that continue to be important today. Elsewhere around the world, the globalised circulation of queer understandings of sexualities and of other Western discourses by which lesbian, gay, bisexual and trans identities become comprehensible has intersected with many other local, national and regional understandings of non-normative sexual desires, practices and identities (Boellstorff 2003; Blackwood 2005; Collins 2005). To say we are all now 'queer' would be to dismiss the complexity of queer, its uneven use across geography and those geographies of sexualities that contest and challenge assumptions about the wholesale adoption of specific discourses and concepts. Throughout the book the diverse uses of 'queer' are apparent, as is the non-linearity of geographies of sexualities, which continues to engage with concerns that were originally brought to the fore in the 1980s, albeit now with a wider range of theoretical tools at its disposal.

Geographies of Sexualities

This book encompasses both queer geographies and other approaches to the geographies of sexualities. The work presented here not only explores lesbian and gay identities, lifestyles and embodied practices, but also seeks to question heteronormativity and other modes of sexualised power relations. What is also apparent is that the authors' research and insights cut across the distinctions we have elaborated in this introduction. We think that this crisscrossing is important because it illustrates that categories are not finite but fluid and that there are potentially important intersections, overlaps and crossovers between distinctions such as heterosexual and homosexual, gay and lesbian, or gay/straight and queer.

In reflecting upon and intervening in current theoretical debates, as well as examining contemporary issues and historical manifestations of sexuality, this book offers accessible insights into 'desire' and the spatiality of both heterosexualities and homosexualities; yet, it also seeks to move understandings of sexualities beyond these tropes. One of the goals of this book is to show how geographers' efforts to address the challenges and potentials offered by queer theories have modified the theoretical terrain on which studies of sexualities take place. Not only have geographers shown how queer theory can be applied to social scientific questions, but they are starting to show how queer theory might be taken forward, enriching the theoretical terrain shared with others outside of the discipline. There are four chapters (those by Bell, Knopp, Hubbard, and Michael Brown) within this book in which the contributor has responded to our request to reflect upon developments over the past decade or so in the parts of geographies of sexualities in which they work.[2] Not only do these

---

2   It is with regret that, despite our efforts, there are no contributions by female writers amongst these reflective pieces. Please see the Conclusion for our discussion of this problem.

chapters provide a succinct overview of these parts of the sub-discipline, but they also provide a number of contrasting insights into the impact that queer theory has had on geographies of sexualities during the past decade. They offer a view of how queer theories have posed new questions and new problems for geographers and of how geographers have contributed to the development of new queer ideas and of ideas at a tangent to queer. Most interestingly, these chapters ground these insights in the context of particular sets of research questions and concerns.

Defining the structure of a book is a necessary evil. We have chosen to divide the contributions to this book into three categories – theory, practices and politics. As editors, we are conscious of the artificiality of these categories and recognise the overlaps and intersections between them. We chose to define this book in terms of these three broad areas because they offer specific foci that enable an exploration of the breadth and diversity of contemporary geographies of sexualities. The themes also intersect with other divisions that might have been chosen to structure the contributions to this book, such as spatial scale or specific substantive issues. However, the division into theory, practices and politics allows a flexibility enabling the authors to work with and through different spatial scales and enabling us to include diverse issues within each section.

Theories

The 'Theories' section of the book comprises two reflective chapters that engage with a wide breadth of theoretical developments within geographies of sexualities and four chapters that offer more specific and focused explorations of theoretical debates both within geography and in cognate disciplines.

Larry Knopp's chapter starts off this section by suggesting some nascent theoretical developments that might be pursued by sexual geographers. He calls for a further queering of the geographical imagination, envisioning such a queering to involve a move beyond the divides that oppose materialist view of the world to discursive ones and that oppose emotion, affect and desire against the rationality of the mind. Knopp also explores the possibilities for queer spatial ontologies. Rather than offering fixed ideas of place, such ontologies stress ephemeral connections, movements and gatherings.

Jon Binnie's chapter explores the political and erotic relationships between sex, sexuality and knowledge. In the chapter, Binnie raises questions about the relationship between sexuality and 'the field'. He revisits his 1997 paper on queer epistemologies, problematising the earlier interrogation of heteronormativity and also extending his analysis to considerations of homonormative relations to the field. The chapter also challenges the ethnocentricity of queer knowledges and debates, and the sexualised assumptions that surround teaching and other pedagogic practices.

In his chapter, Vincent Del Casino Jr. seeks to bring together ideas from sexuality studies and those from health and medical geographies. By doing so, he attempts to open up space for questions about how heteronormativities shape and are shaped by issues of morality and constructions of various disabilities. Del Casino attends to the construction of HIV+ bodies and the effects such constructions have on HIV policies.

He also explores how biomedical discourses construct abilities and disabilities in relation to the pregnant body and to erectile dysfunction.

Jason Lim attempts to bring into conversation two bodies of theoretical thought: queer theories and (Deleuzian) theories of affect. He examines ways in which theories of affect might offer insights into the politics of events and encounters, and might offer insights into embodied memories of how to desire. He also considers the implications of theorising affect for thinking about the nature of political change and for fostering ethical and reparative stances alongside our critiques of heteronormative practices and institutions.

Hanna Hacker's chapter problematises the ways in which development discourses frame what counts as acceptable desires, a framing that has the effect of globalising Western ideas about sexuality. In the context of development funding for community organisations or HIV projects, for example, the desires of those who are deemed to be 'less developed' are compelled to orient themselves to the identities and categories offered by Western development discourse. Against this, Hacker explores the potentials for transnational enjoyment or *jouissance* that resists such development discourse. Such *jouissance* offers a way of thinking about how to form transgressive zones and networks of connections between cultures, while also acknowledging, tapping into and unsettling the potential for violence and domination that arises in the context of such transnational connections.

David Bell finishes off the 'Theory' section of the book with his retrospective look at how geographies of sexualities have fared over the past decade and a half and at how the influence of queer theories have spread within geographies of sexualities in that time. In particular, he writes of the continued need to queer the discipline of geography and to transgress the sexualised norms of enquiry. These moves are of particular importance given the ongoing difficulties in getting material on sex and sexualities published.

Practices

This section focuses upon empirical research regarding how sexualities are practised and policed, and how these practices and processes of regulation (re)make sexualised spaces and spaces of sexualities.

Douglas Herman's chapter explores BDSM (bondage and discipline, sadism/ masochism, although Herman explores the problems of this term) in the United States, arguing that these practices can be considered queer in the sense that many of these practices lie outside of the mainstream. He follows Hubbard in arguing that queer is not solely located within LGBT lives, experiences and identities. In exploring the spatialities of BDSM, Herman explores the regulation and performance of the 'perverse'.

Jin Haritaworn offers a critique of Butler's (1990) 'Gender is Burning' – specifically Butler's exploration of interracial desire – and uses this critique in an exploration of the experiences of people of Thai decent who had non-Thai social or biological parentage. Jin shows how for racialised queer bodies being identified as (or mistaken for) heterosexual does not offer privilege and how queer is not necessarily a 'safe' site. Jin's examination of the complexities of Thai multiracialities in the UK and Germany shows how specific hegemonies are created that belie the possibility of a single 'queer' space.

Kath Browne explores the (re)constitution of femininities, arguing that rendering sex fluid contests the very tenets of sexualities. She uses an incident at Dublin Pride 2003 (in the Republic of Ireland) to examine the hierarchical deployment and deploring of diverse forms of femininities across sexed embodiments amongst those who define as lesbian and gay.

Mark Casey examines the differentiation and hierarchies within the use of 'the scene' in Newcastle, UK. His contribution complements Matejskova's by arguing that other forms of difference than de-sexualisation work to produce, marginalise and exclude those who might be termed the 'queer unwanted' from increasingly visible and hegemonic 'gay' scenes. In this way, Casey offers insights into the effects of operationalised power relations for those who are deemed undesirable with respect to strategies of capital accumulation.

In examining the negotiation of heterosexualities within the Groover bar in Bratislava, Tatiana Matejskova offers a critique of the literature that solely problematises the presence of straight women in these gay bars. She argues instead for a more nuanced conceptualisation of the negotiation of such presence and for a diverse understanding of how heteronormativity can be contested.

Phil Hubbard's reflections conclude this section by examining how queer has become 'a game for all the family'. He argues that using queer to explore heterosexualities renders the boundaries between straight and LGBT problematic and contestable.

Politics

Building upon the previous considerations of sexualised spatial theory and of empirical engagements with the performance of sexualities in space, this section considers the politics of geographies of sexualities. It attempts to both theorise the impact of political power on the production of sexualised spaces and to consider the activist engagements of sexual geographers in their academic and affective communities.

Catherine Nash and Alison Bain explore the grassroots political response to a police raid on the Pussy Palace, a women's bathhouse event in Toronto. They explore how, through the defence campaign, the 'queerness' of the event and the practices it sought to foster were 'normalised' by mainstream gay (male) activists who offered practical support to the women defendants. Nash and Bain offer a study of the differently gendered geographies of queerness and homonormativity in Toronto in the early 2000s.

Farhang Rouhani utilises Oswin's (2005) concept of 'queer complicities' within the context of his study of the recent development of a transnational queer Muslim movement. Rouhani's chapter seeks to tease out the complex politics of assimilation and differentiation, capitulation and resistance, within sections of this transnational social movement. He positions the emergence of queer Muslim identities in relation to Islamic scholarship on sexuality and to forms of radical-progressive Islamic activism and spiritual practice.

In his chapter, Matthew Sothern interrogates the rhetorical figure of The Child as it haunts queer and, more particularly, HIV+ bodyspace. For him, The Child represents everything that queer is not. He examines homonormative evocations of the family through a close reading of a series of advertisements produced by the New Zealand AIDS Foundation.

Gavin Brown's chapter examines queer autonomous spaces created by radical queer activists as a model of alternative means of being queer beyond the reach of the commodity. Brown proposes an alternative understanding of 'queer' on the basis of these ethico-political practices. He suggests that participants in these autonomous spaces *become* queer precisely through their active participation in the creation of prefigurative experiments in post-capitalist means of engaging with sexual and gender difference.

This section concludes with Michael Brown's reflections on the development of political geographies of (homo)sexuality. He questions whether queer theory and broader poststructuralist critiques now operate as restrictive orthodoxies within sexual geography. His concern is to explore the potential for a tactical use of quantitative methods, GIS and other 'scientific' methods by critical sexual geographers in order to better intervene in policy debates from which we too often exclude ourselves. Brown's chapter offers a timely reflection on the politics of doing research on sexuality that is relevant beyond the academy.

Conclusions

The conclusion offers a polyvocal exploration of how this book has come together in the way that it has. It also offers some thoughts about possible future directions for geographies of sexualities. It comprises three sections each written by one of the three editors, and it represents an attempt to highlight the diverse writing styles, opinions and theoretical orientations that are often hidden in co-authored pieces of work. In offering divergent and potentially conflicting narratives, the conclusion finishes a book that does not seek to prescribe or dictate but rather attempts to embrace the possibilities of multiple ways of thinking about and doing geographies of sexualities.

SECTION I
Theories

Chapter 1

From Lesbian and Gay to Queer Geographies: Pasts, Prospects and Possibilities[1]

Larry Knopp

Introduction

Queer[2] geography is often conflated with sexuality and space studies and/or lesbian and gay geographies. In fact, these are distinct (though overlapping) fields. While their subject matter and approaches do often overlap, sexuality and space studies examine a wide range of issues pertaining to sexuality and space from a variety of theoretical, philosophical and even disciplinary perspectives, while lesbian and gay geographies traditionally focus on issues pertaining to LGBT experiences from a geographical perspective.

Queer geography did not develop independently of these other areas, however. Indeed it could never have existed without the theoretical and methodological tools that both of them provided, including, notably, a focus on gender and sexual relations at multiple spatial scales, from the body and family to the state and beyond. Lesbian and gay geography and sexuality and space studies grew out of engagements with a variety of critical social theories and analyses, including, especially, feminist but also postmodernist and poststructuralist theories and analyses. This opened up a world of theoretical and analytical possibilities into which first LGBT and later queer geographies stepped.

Queer geography initially carved out a very particular niche within these fields and has since expanded beyond them. Its initial distinctive contribution was the

---

1   Portions of this article have been reprinted from *The Professional Geographer* with permission from the Association of American Geographers (AAG), copyright AAG, 2007.

2   'Queer' is quite obviously a contested term with multiple and contradictory roots. My use of it here refers primarily to a family of intellectual and political developments growing out of the academy (specifically, the humanities – especially literary criticism – but also, more recently, the social sciences) and radical sexual politics (Jagose 1996). Its hallmarks are anti-essentialism and inclusivity, goals that are sometimes in tension. As an umbrella term for non-normative sexual (and other) subjectivities, 'queer' can be quite expansive and inclusive, but at the expense, sometimes, of making certain oppressed subjectivities (e.g. 'lesbians', 'queers of colour') invisible and/or of rendering *all* subjectivities 'queer', hence potentially weakening its anti-status quo political efficacy. See Butler (1990; 1993), DeLauretis (1991) and Fuss (1991) for examples of early inspirations, and Bell et al. (1994), Binnie (1997), Knopp (1999) and Nast (2002) for more recent ones situated within the discipline of Geography.

application of certain postmodern and/or poststructuralist perspectives to sexuality and space studies, generally, and lesbian and gay geographies, in particular. This meant rethinking those studies in a way that critiqued the taken-for-granted categories and essentialisms of structuralist and modernist thought – for example, identity categories – as these have been applied to the fields of sexuality and space and LGBT geographies. As a project, queer geography has been deconstructive and critical, and suspicious of certainties, universal truths, and ontological imaginations about the way the world works that are mechanistic or instrumental. It is also, arguably, part of a larger project, shared by some strains of contemporary feminist geography, of drawing on those humanistic philosophies and epistemologies that are consistent with poststructuralist and/or postmodern thought and of redefining human geography in ways that seek to bridge the division between social science and the humanities.

As a self-conscious intellectual movement, queer geography emerged in the 1990s. It highlighted the hybrid and fluid nature of sexual subjectivities, and it reimagined the geographical dimensions of these accordingly. In lesbian and gay geographies, for example, the focus shifted from reifying and celebrating 'gay spaces' to understanding the always multiply fluid, ambiguous and contingent sexualised spatialities that are constants in human experience (c.f. Ketteringham 1979; 1983; Weightman 1981; Lauria and Knopp 1985; and Bell et al. 1994; Phillips et al. 2000).

Queer geography has not stopped there, however. Its adherents have taken some of its most powerful insights – about the significance of sexuality and desire in human social life and institutions, and about multiple, hybrid, and fluid sexual spatialities – and applied these to other realms of geography. One of the most ambitious and persuasive of these is Gibson-Graham's work on applying queer notions to economic geography (Gibson-Graham 1996), but, as Bell in Chapter 6 of this volume also makes clear, queer geographers these days seem to be queering just about everything, from the body to the state to cyberspace (Bell et al. 1994; Brown 1994; 1997; 2000; Longhurst 1995; Johnston 1996; Wincapaw 2000; Knopp and Brown 2003).

Still, there is much more that queer geography could do. What follows is a discussion of some areas into which queer geography might expand. Their intimate connections to, and dependence upon, a relationship with various forms of critical and humanistic geography should be evident in the descriptions of these areas.

Queer Spatial Ontologies

Queer geography has tremendous potential for helping us rethink spatial ontologies. Feminist geographers have provided the foundation for much of this work through their considerations of gendered spaces and the spatiality of gender (for example, important deconstructive work on spatial dualisms such as 'public-private' and 'home-workplace', but also reconceptualisations of the city, commuting fields, and links between these spatialities and broadly-based exercises of power – see Mackenzie and Rose 1983; England 1993; 1994b; Peake 1993; Kobayashi 1994; Kobayashi and Peake 1994; Pratt and Hanson 1994; Stiell and England 1999); considerations of the control of women's movements and bodies through coercion, fear and violence (Valentine 1989); and reconceptualisations of boundaries, borders and other spatial demarcations in terms of their roles in constructing socially meaningful group differences and categories (see Pratt 1998 for an excellent example). A queer perspective, informed by embodied and lived queer experiences, can similarly help us to rethink some additional spatial ontologies, including place, placelessness and movement. Many queers' lived experiences, for example, entail a radically different relationship to these notions than that of more sedentary non-queers. The visibility that placement brings can make us vulnerable to violence as well as facilitate our marginalisation and exclusion from the security and pleasures that placement typically brings members of dominant social groups. It is no surprise, then, that queers are frequently suspicious, fearful and unable to relate easily to the fixity and certainty inhering in most dominant ontologies of 'place'. Indeed, many queers find a certain amount of solace, safety and pleasure being in motion or nowhere at all. Social and sexual encounters with other queers can feel safer in such contexts – on the move, passing through, inhabiting a space for a short amount of time – and a certain erotic (or just social) solidarity can, ironically, emerge from the transient and semi-anonymous nature of such experiences (see Bech 1997).

Whatmore (1999) and Thrift (1996; 1999) use actor-network and non-representational theories in a way that is particularly helpful in reconceptualising place from this kind of a queer perspective (see Knopp 2004 for a full discussion). Their radical notions of agency, in which non-human as well as human 'actants' can have causal powers; their associative, topological imaginations, in which connections and flows are as important as sites and nodes; and their humble theoretical aspirations and ethic speak directly to many queers' experiences. Thrift (1999), for example, conceptualises 'places' not as fixed ontologies but as 'passings' that are elusive, ephemeral and always in the process of becoming *and* disintegrating. These passings are always becoming and disintegrating due to their emergence from both material practices and meaning-making processes that are themselves always evolving and incomplete. Such ideas lend themselves particularly well to the analysis of many queer experiences, for example 'coming out' (itself a heretofore ontologically problematic notion for the vast majority of geographers). If we think about cruising (the arguably masculinist practice of circulating through space in search of anonymous or semi-anonymous erotic or sexual encounters), clubbing and other forms of queer congregation, we can see how Thrift's ideas can help avoid conceiving of queer congregation as belonging to, or occurring in, places that are static or fixed. Cruising is not so much tied to a fixed site but is all about the flows of movement and passings. Clubbing, meanwhile, is an ephemeral experience – both the ephemerality of erotic encounters with others on and off a dance floor, but also the ephemerality of the club night itself, coming together to use a space for one night a week or one night a month, congregating and then dispersing. Such queer experiences and practices might be understood in terms of Thrift's idea of an 'ecology of place' as a 'spectral gathering', in which various practices, meanings and 'actants' are examined for their myriad and ephemeral interconnections.

Similarly, 'placelessness' might be productively reconceptualised as an embodied and material practice, one that offers certain pleasures and other benefits (such as security) through its various perceived qualities (heterogeneity, temporariness, anonymity, cosmopolitanism), rather than just as a lack. Cybersex and other online encounters, but also social and sexual experiences in fleeting spaces and moments, such as festivals, parades and transit stations, are examples. Movement, meanwhile, might be thought of as a set of highly contingent and fluid practices involving personal reflection and reinvention, rather than in terms primarily of autonomous actors and objects circulating amongst and through vectors and nodes. The key here is rethinking spatial ontologies in ways that address their emotional and sentimental meaning and significance, not just their materiality or abstract intellectual utility.

Queer Spatialities of Gender

Queer geography is already contributing to expanded understandings of spatialities of gender. Namaste (1996a) and Browne (2004), for example, have looked at transgender spatialities in the context of the violent regulation of 'gender outlaws' in public space, through what Namaste calls 'gender-bashing', and in the context of the anxieties and hostilities that gender-ambiguous bodies provoke in certain other kinds of spaces, such as sex-segregated bathrooms. Some related spatialities that queer geography can help explain include the impacts on personal relationships, home and community when individuals undergo sex-changes; the different spatialities associated with essentialised versus non-essentialised transgender practices (e.g. androgyny versus sex-changes); and the implications of transgender practice for citizenship, border-crossing and immigration.

Queer geography can also extend understandings of the spatialities of drag and gender performativity. In their famous paper on gay skinheads and lipstick lesbians in public space, Bell et al. (1994) raised important questions about intentionality, authorship, authenticity, consumption and spatial context when it comes to the relationships between gender performance, erotic desire and social power. One very important question that could be explored in detail is under what spatial conditions drag and performativity are recognised as such, rather than as simply the norm. Another has to do with the implications of drag for sexual subjectivity: how, when and where do heterosexual and homosexual subjectivities survive gender transgressions, and where do they not?

Finally, queer geography can help with understandings of spatialities of resistance to gender regimes. What forms of activism and resistance might challenge the hegemony of current gender regimes? What kinds of spatial strategies, in particular, might be effective? Is cyberspace a potentially productive field for transgender activism? And what might a less trans-phobic world look like spatially?

Queering Homophobias and Heterosexisms Spatially

In a similar vein, queer geographies might help us to better understand (and resist) homophobias and heterosexisms. Nast (1998; 2002) and Muller (2007 a; 2007 b) are exploring this through examinations of the ways in which socially constructed *and deconstructed* desire are policed in different kinds of spaces. Muller is doing this through her study of homophobia and heterosexism in the spaces of women's sport, while Nast has looked at how male desire – even 'deconstructed' desire – in both gay male consumer spaces and spaces of academia, is policed through deep-seated material practices and cultures of patriarchal power (for example, the gay male travel and tourism industry's ageist, racist, patriarchal and implicitly heterosexist body aesthetics). There's much more that could be done in this regard, such as examinations of spaces of pornography, including the gendered and sexed urban (and cyber) political-economy of the pornography industry; spatialised representations of desire in media more generally; the role of travel and movement in policing erotic desire, rather than unleashing it; and the policing of reconfigured desire in carceral spaces. In addition, analyses and inquiries into the dynamics of deconstructed desire in women's and feminist spaces – particularly forays into the links between homophobias, heterosexisms and feminist spatial practices – would be tremendously valuable (though they would probably best be done by queer feminists).

Getting beyond the closet-ghetto polarity is another way of queering our spatialised notions of homophobia and heterosexism. One of the most insidious manifestations of both homophobia and heterosexism is the persistence of a closet-ghetto polarity to describe the world of lesbian and gay (and even, to some extent, queer) spatialities. The two terms' dominance in discussions of LGBT and queer studies implies that lesbian and gay (and even queer) lives and spaces are inevitably separate and distinct from 'straight' lives and spaces. Queer mapping projects (e.g. Northwest Lesbian and Gay History Museum Project 2004; Brown and Knopp 2006) and other queer projects of representation can show how enmeshed with 'straight' environments (and each other) closets and ghettos actually often are, and in the process how homophobias and heterosexisms operate through the closet-ghetto polarity.

Related to this, overcoming the materialist-discursive divide in social (and spatial) theory will be central to queering our notions of homophobia and heterosexism. Homophobias and heterosexisms are quite patently discursive and material at the same time. We need to do much more by way of studying them as integrated practices, for example by looking at connections between representations and the material policing of 'perverse' spaces. Here Thrift's non-representational theory may again be particularly useful, since its ontological imagination begins from the premise that meanings and materiality are inseparable.

Generational Cultures as Queer Sites

Generational cultures are among the chief destabilisers of social categories, including gender and sexuality. One cohort's radical social change is its successor's mundane, taken-for-granted reality. How does this happen, and under what spatial conditions? Why, for example, are attitudes of American young people about gay sex and gay rights so different from those of their parents? How does this vary across space? How does spatial context intersect with time to produce these changes?

In light of the crisis of relevance that feminism, arguably, now faces, at least in the minds of many young people (including many young women), one must also ask about generational dialectics within (and beyond) feminist and queer geographies. How are new generations of scholars positioning themselves in relation to earlier cohorts of feminist and queer geographers?

We might also ask how generational cultures diffuse. Generational cultures and generational changes embody, in many ways, the kind of hybridity and fluidity that is at the heart of queer experience. They are products and producers of what Thrift (1999) calls 'lingering legacies' and, at the same time, of powerful and passionate forms of agency, which can fruitfully be conceptualised, a la actor-network theory, as 'relational achievements' involving human and non-human discursive, material, semiotic, technological, institutional and other forces (Whatmore 1999, 26). They reproduce, reinvent and transgress almost all social and theoretical boundaries, from bodies to beliefs. It is clear, then, that complex discursive-material regimes shape and define generational world-views and cultures.

Cultural Politics

The so-called culture wars[3] are also plainly queer. They are conflicts in which the stakes are nothing less than the social construction of reality.[4] A queer geographical approach to the culture wars might, as a start, look at their spatialities in terms of the framing of debates (see, for example, Brown, Knopp and Morrill 2005). How are space, place and environment implicated in the framing of so-called 'cultural issues'? What topics get placed 'on the table', what ones don't, and why? And how are the issues framed differently from place to place?

Secondly, how might we understand, using the tools of queer geography, 'moral landscapes'? That is, how do some moralities come to dominate some spaces and places, what are their contradictions and dialectics, and how are they related to other moralities in other places? Furthermore, how do these landscapes change?

On a somewhat different note, we might also look at specifically queer forms of politics – that is, politics that seek simultaneously to deconstruct essentialism and inject a sexual-minority perspective into political analyses and agendas – from a spatial perspective. In what kinds of places are a politics of 'queering' likely to be practised, in what kinds of places are they likely to be successful, and what might a queer spatial practice look like? For instance, is there a place for the strategic denial of 'lesbian' and 'gay' identities in patently homophobic cultural contexts, such as some conservative rural environments, or would such strategies just validate local cultures' homophobia? Similarly, what are the limits, if any, of 'queering' public

---

3 'Culture wars' is a term that has gained particular currency in the U.S. Many of the conflicts to which it refers are, in fact, increasingly common throughout the West, if not beyond. These include debates about the place of religion in civil society, political struggles over abortion, reproductive freedom generally, sexuality, 'gay rights', assimilation vs multiculturalism in immigration policies, 'values' instruction in state-run schools, public funding of the arts (especially controversial art and artists), and popular culture's 'values' (among other things).

4 From this perspective, it should be no surprise that the producers of popular entertainment (e.g. 'Hollywood'), lawyers (who trade in multiple and competing 'truths'), and political operatives are so reviled in much popular culture. All three lay bare the constructedness of our everyday realities, and most consumers of the realities they produce feel powerless in the face of these elites' perceived hegemonic control over those realities.

policy debates in already 'queer-friendly' environments? And what particular forms of 'queering' (e.g. challenging constructions such as 'patriarchy' or 'capitalism') are most (and least) likely to succeed in what kinds of contexts?

Finally, given how balkanised many societies around the world are becoming culturally and politically, what might a queer approach be able to do for us in terms of understanding the various 'camps' or 'fortresses' that have come to characterise so much of the cultural and political landscape? How might we use the tools of queer geography to make sense, for example, of gated communities, gay resorts or Promise Keeper[5] rallies?

Conclusion: Queering the Geographical Imagination

What I am advocating here is a radical project of queering the geographical imagination in general. This is an argument as much about ethics as it is about theory, and about epistemology as much as ontology, even though I have concentrated more on theory and ontology here. At the heart of this new imagination are the following goals or values:

The first is getting beyond, once and for all, the idea that materialist and discursive approaches to knowledge and understanding are incompatible. From a queer perspective the material is always discursive and the discursive always material. The second has to do with the epistemological implications of the first. Abandoning the materialist-discursive dichotomy means taking seriously the body, sentiment, emotion and desire as co-equal sources of knowledge, and ones that are not only equal in value to reason, rationality and the mind, but *integral parts of them*. While some might see this as naïve romanticism or clever rhetoric, I see it as part of an epistemology and ethic that is grounded in the messy realities of the human experience, including a critical awareness that all knowledge is situated, partial and incomplete – and that that is something to celebrate rather than bemoan.

Third, geography needs to expand its empirical terrain to include more of these messy realities, including fluidity, hybridity, incompleteness, moralities, desire and embodiment. Again, Thrift (1999) and Whatmore (1999) offer inspiring articulations of how such a project might begin, while I have attempted elsewhere to flesh out explicitly the implications of their call (Knopp 2004). Fourth, we need to temper our intellectual (and other) ambitions with humility and, in particular, with a recognition that our practices and the knowledges we produce are themselves always in the process of becoming, rather than complete.

Finally, we need to modify our geographical-ontological imaginations such that our objects of study are viewed more relationally and topologically than autonomously and discrete, more reflexively than objectively, and more humbly than ambitiously. The result, I think, will be a geography that is less arrogant and

---

5   Promise Keepers is a US-based conservative Christian men's organisation, founded in the early 1990s, that has now become international. The organisation holds large male-only 'conferences' designed to promote fellowship and solidarity among men as they work to keep certain 'promises' derived from Bible verses.

elitist, more hopeful than fear-driven, more possibilistic than deterministic, and more human (and humane) than inhuman (and inhumane).

Chapter 2

Sexuality, the Erotic and Geography: Epistemology, Methodology and Pedagogy

Jon Binnie

What I want to do in this chapter is to consider the place of sexuality and the erotic within geographical knowledge. In doing so I revisit an earlier paper I wrote on this subject (Binnie 1997) that focussed on the relationship of sexuality to 'the field' – the sexual and erotic politics of fieldwork and the sexualised construction of geographical knowledge. I consider whether a queer epistemology can come any closer to fruition. I wish to reconsider the difficulties of embodying geography and question the mainstreaming of sexual geographies within the discipline. This discussion seeks to contribute towards theoretical understandings of the geographies of sexualities particularly in relation to the erotic/knowledge divide. A re-thinking of the reasons for the original paper and the re-evaluation of its main arguments are necessary for a number of reasons. There has been an explosion of work on sexuality and space within human geography in recent years, which has meant that epistemological and methodological issues can no longer be sidelined but instead need to be foregrounded. Within critical human geography and elsewhere there has been a reaction against what have been seen as the excesses of the cultural turn. In this context it would be easy to see a concern with sexual transgression and the erotic as trivial or apolitical in the context of the moral turn (Lee and Smith 2004; Smith 2000) and the policy turn (Bell forthcoming; Martin 2001) within human geography. Moreover, within queer/sexuality studies we appear to be witnessing a particular turn away from considering the erotic in favour of wider political concerns such as citizenship, globalisation and economics. While this turn should be welcomed as a corrective to the early 1990s neglect of the social and economic components in sexuality studies, there is a danger that the erotic is again marginalised. In developing this discussion the chapter is constructed in the following way. The first part of the chapter provides an overview of debates on the relationship between sex, sexuality and geographical knowledge. A key concern in this section is the notion of a queer epistemology, and whether we can use the insights of queer theory to re-think how we conduct research on sexuality in human geography. This discussion leads to a consideration of the place of the erotic within queer geographies and methodologies. Having examined the relationship between queer theory, the erotic and the politics of knowledge, I go on to question the stability of the term queer, reflecting criticisms from a postcolonial perspective and from the 'new queer studies' that queer theory suffers from ethnocentrism. Having discussed the desirability and the possibilities for a queer epistemology, the final part of the chapter examines these issues in relation to pedagogy and the disciplinary specificity of geography, arguing that field trips are particularly sexualised spaces – the study of which can offer us insights into the distinctively sexualised nature of geographical knowledge.

Sex and Geographical Knowledge

In 1997 I wrote a paper entitled 'Coming Out of Geography', which sought to examine how sexuality was implicated in the research process. This intervention sought to problematise the relation between the researcher and the researched in terms of desire, erotics, politics and knowledge. At the same time it attempted to challenge the various ways in which the geographical academy sought to limit what could and could not be spoken about embodied sexual practices. David Bell's paper '[Screw]ing geography' (revisited in Chapter 6 of this volume) was another intervention that sought to make visible the institutional constraints on doing work in geography (Bell 1995). Central to both of these interventions was the call for geography to reflect on its own squeamishness when it came to sex and when it came to bodily matters more generally. In my essay, I argued that this squeamishness needed to be confronted if we were to produce meaningful discussions of the relationships between erotics, communities and identities. I sought to put on the agenda the relationship between the construction of geographical knowledge and the exclusion of sexual dissidents from within it. I therefore wanted to examine how epistemological and methodological issues could explain the marginalisation of sexual dissidents within human geography. I argued that the mainstream geographical community often displayed squeamishness about sex:

> Heterosexuals seem to cope with queer theory in its most abstract, intellectualised, disembodied form, but then to run scared when confronted with the materiality of lesbian, bisexual and gay lives, experiences and embodiments.   (Binnie 1997, 227)

Re-reading this statement some ten years later, I feel rather uncomfortable about the level of generalisation about straight geographers and the degree of certainty about the homosexual/heterosexual binary as the principal divide between queer and non queer. The issue of squeamishness around embodiment is, of course, not simply a question of the homo/hetero divide. The notion of homosexual respectability is one that has come to the fore in debates within lesbian and gay studies and within queer theory. We therefore see that concerns about respectability and the concomitant squeamishness about embodiment applies on both sides of the homo/hetero divide, challenging the centrality and stability of that divide as the primary one in studies of sexuality. We must recognise that race, gender, class and disability are highly salient to squeamishness around sex. Beverley Skeggs' work on class, gender and respectability reminds us that notions of respectability and middle-class morality have been fundamental in drawing distinctions between class formations.

Since the blossoming of work on sexuality within geography in the early 1990s, we have witnessed many transformations in the sexual political landscape. These in turn have informed changes in the terrain of queer theory and sexuality studies. In the queer theory of the early 1990s there was a preoccupation with *transgressive* sexual practices and performativities – for instance reflected in the Institute of Contemporary Arts (ICA) conference 'Preaching to the Perverted' and in the *New Formations* special issue on perversions (Squires 1993). In the UK, this concern with transgressive sexual practice reflected specific events on the ground, for instance the effective criminalisation of sadomasochism following the Operation Spanner trial (Bell 1995; Bibbings and Alldridge 1993). Since the mid 1990s, there has certainly been a turn away from this focus on transgression and a turn towards a focus on topics that are seen as perhaps being more worthy. This is despite the fact that many of the concerns about the marginality and criminalisation of sadomasochism are still salient today (see Dandridge 2006) – witness the proposed legislation in the UK to criminalise the possession of violent sexual images.

In critical human geography we have witnessed a non-representational turn, partly inspired by Judith Butler's work on performativity, examining all manner of topics. As Butler's work on performativity has become assimilated within mainstream critical human geography, it would appear that the body has become more abstract than ever. We see, for instance, Butler's work critiqued for rendering the discussion of the physicality and guts of embodied sex as abstract and safe as ever. Yet, to ask for attentiveness to the physicality of sex is not to call for a deployment of the body in essentialist terms: the body is always social. Nor does it mean a simple celebration of embodiment in which we are obligated to provide confessional discussions of how our own bodies and sexual practices are implicated in our own research. Kim England's (1994a) essay is exemplary in this regard in providing a discussion of the limits and constraints upon who does work on what group or identity. When I read it originally, I felt a strong sense of relief that at last someone had had the guts to confront these issues head on – and that straight researchers were at last admitting to some sense of humility in recognising the limits in their ability to represent the voice of the Other. Some years later, I feel perhaps much more ambivalent in this regard. While, certainly, researchers of sexuality need to show considerable sensitivity to difference, does this not also lead to silencing and to absences?

In the context of the rise of non-representational theory and the prominence of discussions of desire rooted in psychoanalysis, the ever more abstract rendering of the sexualised body means that the question of sex and the relationship between erotics, sex and sexuality remain as troubling as ever for geographical enquiry. This is despite earlier pioneering attempts to make these visible as objects of study. To speak of sex acts can often appear clumsy or embarrassing – and certainly not very sexy – see, for instance, the paper on public sex by Michael Warner and Lauren Berlant (1998) in which they narrate an evening at a club devoted to the erotic margins. More generally, certain sexualities are seen as having an erotic excess: gay men have been pathologised as hypersexual and promiscuous, and further discussion of the erotic nature of gay male identities may in effect reinforce stereotyping.

In the mid-1990s a number of interventions were made that sought to examine the sexual politics of ethnography (for instance Kulick and Willson 1995; Lewin and Leap 1996; Newton 1993). These were significant in raising the issue of the connection between academic authority, status, power and sexuality. Following these interventions, we can see that discussions of power in the research process have to acknowledge the relationship between sexuality and power/knowledge. However, speaking of sex can sometimes be seen or interpreted as a form of subcultural capital – an affirmation of the authority of experience. We should also recognise that attempts to be reflexive about one's own power and authority have their limitations. As Beverley Skeggs (2002) argues, such attempts can often end up being anything but, and can simply make visible the researcher's own insensitivity or lack of reflexivity.

Erotic Politics and Queer Practice

There is a need to understand a complex shifting terrain and cartography of politically marked acts and identities. We need to understand the connection between acts, practices and identities. Thus there is a need for a turn to consider erotic practices. Focussing on the erotic rather than the category of sexuality may be fruitful in helping us to get to the materiality and physicality of sex as opposed to identities and communities, which have been much studied within human geography (although this is, of course, an artificial distinction as erotic connection may be seen as the foundation of sociality – even community – within some sexual dissident communities). Thus far, we have seen that a queer epistemology has to take on board the relationship between knowledge and the erotic. However, any attempts to articulate such an epistemology risk advancing essentialism. Moreover, given the challenges to the notion of queer epistemology that I will outline in the next section, we need to tread carefully in thinking about and even legitimating what might be an appropriate way of working through this question. I must reiterate that I am not simply calling for a celebration of the erotic for eroticism's sake, but rather to acknowledge *the politics of erotics in knowledge creation*.

Samuel Delaney's (1999) work on the purification of space in Times Square in Manhattan is exemplary in terms of the erotic geography I am calling for. It examines the erotic politics of zoning legislation in New York (see Dangerous Bedfellows 1996; Papayanis 2000), which has had the consequence of purifying the space and reducing opportunities for eroticised contact and encounters with difference within this part of the city. Delaney's partly autobiographical account of the disappearing streetscape of porn cinemas and of disappearing opportunities for erotic encounters on 42nd Street is marked by a matter-of-factness and candour in its discussions of sex that are far from voyeuristic. This comfort in discussing sex is refreshing but rare in academia. That such frankness is rare is unsurprising given the costs associated with discussing the erotic in such a direct manner in academic writing. As Ralph Bolton (1995, 158) argues: "The few who have written most explicitly about such matters are precisely those on the margins who, not by choice but because of discrimination, have been excluded from academically respectable positions in major universities – in other words, individuals with nothing to lose by being honest and forthcoming".

In highlighting the work of Delaney, I do not want to appear prescriptive in stating that some forms of methodology are preferable to others in terms of a progressive politics of sexuality. Can there be an essential queer method or technique? I think here we need to learn much from discussions of epistemology and methodology within feminism. The notion that some methods and techniques are more feminist than others seems to be increasingly questionable. In relation to matters queer, can a specific technique be deemed queerer than others? I am not so sure. Perhaps rather than articulating a queer standpoint theory, we should more modestly pay attention to how techniques are used and to what end? Stephen Valocchi (2005) suggests that ethnographic methods are best suited for advancing a queer understanding of identities and practices. While I would not necessarily share Valocchi's faith in ethnography, it is often assumed that qualitative techniques may be more appropriate for research in this area given that they are deemed to be more suitable for the investigations of emotions, feelings and opinions. However, can we make a straightforward link between methodological approach and the queer effects of a piece of research? Ethnographic approaches may well do considerably more harm than a questionnaire (see, for instance, the debates around Laud Humphrey's notorious research on sex in public toilets). Perhaps rather than trying to prescribe certain methods as being queerer than others, we should pay attention to the queering potentialities of different types of research. For instance, the generation of a data set on poverty among lesbians and gay men would have queer effects by challenging stereotypes of gay affluence and by making visible an issue long neglected in different research communities (see Chapter 17 for Michael Brown's reflections upon similar questions).

Challenges to a Queer Epistemology

While there are a number of challenges to the notion of a single, stable entity of queer and an associated queer epistemology, I wish to discuss two in particular, namely challenges surrounding the idea of 'normativity' and the relationship between nation and epistemology.

Heteronormativity and the Homonormative

Heteronormativity has been a powerful concept in challenging the way society is structured along the two gender model – norms that enshrine heterosexuality as normal and therefore lesbians, gay men, bisexuals and transgendered people as Other and marginal. However, I am not so sure about its usefulness now. The notion of heteronormativity tends to lump all heterosexuals together in the same box, and can mask or obscure the differences between and within sexual dissident identities and communities. It can sometimes appear that heterosexual identities are uniformly normative and socially conservative, while non-heterosexuals or sexual dissidents are constructed as radical, progressive or outside of social norms. However, are these norms as stable, immutable and static as these assumptions would suggest? As straight men are disciplined in their lifestyle choices by gay men in *Queer Eye for the Straight Guy,* are we not witnessing subtle changes in the sexual landscape whereby what is valued as normal or normative is increasingly confused and up-for-grabs? Moreover, we are also witnessing the media attention devoted to pathologising and 'outing' non-normative heterosexual practices, for instance dogging, which involves straights meeting for public sex in car parks (Bell 2006).

At the same time, there has been considerable discussion about the notion of homonormativity (Bell and Binnie 2004; Duggan 2002; 2003) and the development of the Gay Right in the United States (Goldstein 2002; Robinson 2005). The notion of homonormativity refers to the mainstreaming of lesbian and gay politics and the associated focus on gay marriage and consumption. The increasing visibility and power of affluent white gay men has been accompanied by the marginalisation of the politics of both lesbian feminism and sex radicalism, and has highlighted the exclusions within queer communities on the basis of race, class, gender and disability. In this context, Diane Richardson has written about what she sees as the professionalisation of lesbian and gay subjectivity under a neoliberal politics of normalisation (Richardson 2005). We need, therefore, to think about what homonormativity and the normative mean in relation to queer studies.

Whose Queer Epistemology?

The second major challenge to the notion of a single coherent queer epistemology comes from those writing from a postcolonial and/or transnational perspective. In this context, David Eng et al. (2005, 15) call for a certain 'epistemological humility' in that US researchers should be less ethnocentric in their academic practices: "Our attention to queer epistemology, queer diasporas, and queer liberalism might be considered one modest attempt to frame queer studies more insistently and productively within a politics of epistemological humility". They go on to argue that:

> [s]uch a politics must also recognise that much of contemporary queer scholarship emerges from U.S. institutions and is largely written in English. This fact indicates a problematic dynamic between U.S. scholars whose work in queer studies is read in numerous sites around the world. Scholars writing in other languages and from other political and cultural perspectives read but are not, in turn, read. These uneven exchanges replicate, in uncomfortable ways, the rise and consolidation of U.S. empire, as well as the insistent positing of a U.S. nationalist identity and political agenda globally. We propose epistemological humility as one form of knowledge production that recognises these dangers.   (Eng et al. 2005, 15)

This statement is a well-meaning attempt to challenge ethnocentricity within US queer scholarship. It is a call for much greater engagement with the writing of scholars working in queer studies outside of the United States and with work written in languages other than English. This is entirely laudable. This demand for epistemological humility comes from an introduction to a special issue of *Social Text* entitled 'What's Queer About Queer Studies Now?' It is significant that all the sixteen essays in this special issue are by US or US-based scholars. In this context, calls for epistemological humility can be seen to be doing little to address American ethnocentricity. We therefore see the irony of the conversation about the American-centric nature of this debate as it is played out among American scholars. Thus, we need to be aware of the national question in framing queer epistemology. As geographers, we should be more attuned to this, but this is not necessarily the case.

Eng et al. (2005, 12) argue that "[m]uch of queer theory nowadays sounds like a metanarrative about the domestic affairs of white homosexuals. Surely, queer studies promises more than a history of gay men, a sociology of gay male sex clubs, an anthropology of gay male tourism, a survey of gay male aesthetics". They are surely right to argue that queer studies needs to be more than this, but I do think it is a bit worrying that Eng et al. pick on the study of gay male sex clubs as a target for their criticism. Surely a sociology of gay male sex clubs is something that is valuable and necessary, given rising rates of HIV transmission? We have yet to see many studies of gay male sex clubs in Geography, although Michael Brown's work on political responsibility and media discourses around the transmission of sexually transmitted infections between men in Seattle is significant in drawing attention to the spatial politics of the erotic (Brown forthcoming). Queer studies may certainly need a clearer theoretical direction and a more solid basis on which to deal with some of the big questions that are raised in Eng et al.'s very broad brush introduction, such as globalisation, citizenship, migration and racism. But what about the disciplinary background of the authors? It is very hard to escape the conclusion that queer studies (new or old) is still overwhelmingly located in the humanities – and the US humanities at that.

A Queer Epistemology of Geography: The Disciplinary Specificity of Geography

In assessing the relationship between sexuality, erotics and knowledge in human geography, it is very important that we do not overlook pedagogical concerns. I think this is important given the disjuncture that sometimes occurs between research and teaching. For instance, we need to recognise the particular socialisation of geographical knowledge that takes place via the field trip, a particularly sexualised space. We are all meant to recognise the value of field trips within geography. They are a valuable recruitment tool – a way of branding the discipline, making it distinctive from other natural and social sciences. Field trips are also a significant way of distinguishing one geography degree programme from another. We all know the value of field trips on open days and the demand to offer ever more exotic destinations to prospective students as one means of differentiating our courses from those of our competitors.

The relationship between pedagogy and geographical research on sexuality is the key focus of Karen Nairn (2003) in her valuable and challenging paper on the sexual politics of field trips. Discussing the geography of sleeping arrangements on residential field trips, she examines the politics of gender, race and sexuality in the production of geographical knowledge on such trips. A focus on embodiment can fruitfully interrogate the way these are intertwined with notions of squeamishness and comfort. As Nairn argues, the discomfort felt by some of her informants reflects uneven power relations in terms of whose subjectivities are valued and whose are rendered marginal.

> Hegemonic spaces and discourses favouring particular sexual and cultural identities have the potential to acts as gatekeeping mechanisms to readily include geography students and

staff most comfortable with those spaces and discourses while those who do not conform
are implicitly provided with the message that their subjectivities and needs for particular
kinds of spaces do not matter.   (Nairn 2003, 77)

In the conclusion to her article, Nairn goes on to question "whether fieldwork needs
to take place away from universities and schools to count as 'real' fieldwork and
whether indeed it offers the best means of relating theory and practice" (2003, 78).
So, perhaps the centrality of fieldwork within the discipline – or rather the specific
privileged status accorded to it – could partially explain the specific attitude towards
sexuality within the discipline. However, is there not a certain over-generalisation
being articulated here, too? Do all queer students and staff experience the kind of
dislocation, marginalisation and sense of exclusion that Nairn speaks about? Nairn
makes the valid point that "geographic knowledge is not only the formal and explicit
curriculum but also the informal and implicit curriculum that is often referred to as
the hidden curriculum". So, we are speaking about the values that are communicated
down through fieldwork but also through the social environment, particularly on
residential fieldtrips. Gill Valentine (1999, 418) speaks of the blurring of home and
work, and of public and private spheres that takes place on fieldtrips: "Nowhere is
this more so than in the ubiquitous geography field class, where part of the *raison
d'etre* is often to encourage informality and a 'get to know you' relationship between
staff and students ... Even if personal information is not volunteered, it is certainly
sought out by students." As Nairn and Valentine argue, field trips can be sites of
considerable anxiety and discomfort for both students and staff.

  While field trips may present specific challenges, what about the mundane
space of the everyday geography classroom? Valentine argues that because of the
assumption that mundane environments are coded as straight, lesbian, gay and
bisexual staff are faced with the issue of whether to be out with their students. There
is an anxiety that simply because a lecturer teaches about sexuality, students will
make assumptions about that teacher's sexuality. I think this has certainly changed
in my experience: with the mainstreaming of work on the geography of sexualities
and the proliferation of undergraduate courses on which sexuality is taught, this is
now perhaps becoming less of an issue. Coming out is only the beginning, though,
as there are many forms of coming out and many kinds of closet. For instance,
Glen Elder (1999, 87), discussing coming out in the geography classroom, notes
that "the politics of nation, immigration, 'race', and sexuality in contemporary US
society require that as a white South African national I 'come out' on several fronts:
anti-racist, foreigner, African and gay". Whatever forms of closet or coming out,
Valentine (1999, 421) suggests that "in the small and incestuous world of academia,
once you are 'out' there are few possibilities to retreat into the closet if the going gets
tough". The personal risks associated with coming out should be balanced against
the notion that being a certain kind of gay or lesbian is fashionable and has cultural
capital associated with it. In an era marked by subjects such as the metrosexual,
can being out always be seen as such a disadvantage? Perhaps we should examine
the way in which professions are sexualised – so that having the right kind of self-
presentation, speech, clothing and performance matters. So, rather than focussing on
coming out per se, we should be more concerned with which public performances

of queer sexuality are acceptable and which are not; which are acceptable within mainstream Geography and which are not? However, we must still acknowledge the fact that, as Lee Badgett (2001) has reminded us, there is an economic cost for people being out. That cost is not borne evenly. For instance, financial or economic independence is often a prerequisite for coming out. Here I am thinking about class issues. So, there may be significant differences in economic power and how this relates to performances of sexuality. While those with less may have less financial independence to be out, others with more may also have more to lose by behaving inappropriately.

Valentine (1999, 421) comments that:

> Visibility across the curriculum is important for dissident sexualities because it normalises them. Otherwise the danger is that lesbian, gay and bisexualities are missed out altogether or become relegated to the one-off 'exotic other' lecture, which only serves to reinforce stereotypes and perpetuate the taken-for-granted heterosexuality of the university environment.

But is this visibility always so desirable for all sexual dissidents? Can visibility do harm? Is it always a positive experience? Likewise, is *normalisation* so desirable either? For many, sure, normalisation is preferable to enfreakment or to being marked out as an exotic other. I guess the question is whose kind of normal? I think what Valentine really means to emphasise is the mundanising aspect of normalisation. There is also the potential embarrassment that queer students may feel regarding being lectured about queer sexuality by an older lecturer. The possibility that student and teacher may share very little other than a common disposition means that classes on sexuality and space may well produce dis-identification as easily as identification, in the same way that students may be embarrassed by pedagogic attempts to relate to popular and youth culture.

Concluding Thoughts

In a paper on value and geographical knowledge, David Bell (forthcoming) provocatively discusses the institutional politics of contemporary geographical research. Research in the contemporary university must have a visible use value otherwise it is seen as worthless. The utility of research will obviously depend on the specific institutional context and matrix of power/knowledge. This question of utility and the value of research has been one that has mattered a lot to those researching sexuality – wanting to claim a place at the table of human geography. Research on sexuality, like any other area of research within human geography, is seen as valid if it furthers the goals of the institution or discipline – in the UK context, whether it is 'RAE-able'. Is this always compatible with other goals such as the development of research that is concerned with erotic practices and sexual transgression? Who is research on the geographies of sexualities for? To what purpose is such research conducted? Who should be the subject and object of such research and to what ends? I think that research should challenge norms that regulate and control – that put people into boxes. Research should counter homophobia, somatophobia, erotophobia and transphobia, and it should develop critical thinking. What, then, is the purpose of studying the erotic in geography? It is surely to understand how sexuality is deployed as a force and for what purposes. While welcoming the increasing institutionalisation of geographies of sexualities, we must be attentive to normalising tendencies that accompany such transformations and not lose sight of the sometimes troubling relationship between erotics and politics in the production of geographical knowledge.

Chapter 3

Health/Sexuality/Geography

Vincent J. Del Casino Jr.

Introduction

> Sex acts, what sex itself entails in its many possible permutations, but also the specifics
> of how sex is translated into discourses of medicine, health, and public policy; all of
> these components of the box marked 'sex', and many more ... remain very much on the
> sidelines of population geography ... even when expressly concerned with fertility, birth
> control, sexual health and the like.
> (Philo 2005, 328)

> More recently, health care geographers have begun to express more concerns for an
> alternative kind of difference, cultural difference, which recognizes that individuals and
> groups of people characterized by ethnicity, gender, sexuality and physical or mental
> disability have different health beliefs, practices and experiences that must be taken into
> account in research.
> (Gesler and Kearns 2002, 96)

As this publication suggests, there has been an emergence of critical theory related to
the geographies of sexualities since *Mapping Desire* (Bell and Valentine 1995) was
published over ten years ago. In similar ways, medical and health geographies have
undergone rapid and dynamic change over the last fifteen or so years. In the 1990s,
medical geographers and so-called post-medical or health geographers debated the
efficacy of a purely biomedical-centric geography of health and health care, illness
and disease. As they did, theorisations of health, the body, illness, experience and
performativity became more nuanced, paralleling to a certain degree the important
debates taking place in social and cultural geography. The inclusion, for example, of
a chapter by Pamela Moss and Isabel Dyck (2002) – two feminist health geographers
– in *The Handbook of Cultural Geography* implies that questions of healthy and ill
bodies need to be the concern of geographers more broadly.

Yet, despite the parallel in theoretical developments in health and medical
geography and in social and cultural geography, there is strikingly little work being
done at the intersections of health and sexualities studies in geography. Thus there
remains the possibility to queer the subject of health and medical geographic inquiry
by challenging the presumptive heteronormativities implicit in medical geography.
As Philo suggests in the aforementioned quotation population geography, like health
geography, eschews discussions of both sexuality and sex. A quick examination
of English-language medical geography textbooks illustrates that most medical
geography remains only marginally interested in current social theoretical questions
related to socio-spatial identities, particularly those related to the multiplicity of
sexual/sexed identities. Gatrell's (2002) *Geographies of Health* does not address

sexuality at all, although he does examine the structural conditions for sexual violence. Meade and Erickson (2000), whose textbook *Medical Geography* is still one of the most noted in the field, fail to mention sexuality either, although they do spend some time on an analysis of HIV/AIDS diffusions. In both cases, these volumes allude to a linkage between sex/sexuality and health. Yet neither suggests that sexualities may be performed, constructed and relationally constituted through, for example, engagements of desire or through sex acts themselves. Even recent so-called 'geographers of health' such as Gesler and Kearns (2002) only briefly touch on sexuality and geography in their book *Culture//Place//Health* where they dedicate three pages to a discussion of the 'geographies of sexual orientation' by focusing mostly on the literature on HIV/AIDS. They fail to examine the complexity of sex as an aspect of the performative nature of sexualities. In the end, questions related to the intersections of sexuality and health studies remain fairly marginal to the overall cultural turn that has 'hit' medical and health geography in recent years.

Reciprocally, the study of sexuality in geography, particularly as it is informed by queer theory, has, for the most part, marginally examined questions related to either health or the medical. When geographers, including queer theorists in geography, have turned their attention toward questions of the intersections between health and the medical, it is has often been in the context of HIV/AIDS (e.g. Brown 1995, 1997; Craddock 2000b; Wilton 1996, see also Sothern, Chapter 15, this volume). This is not to suggest that this is not a significant or important body of work. Rather it is to argue that there might be a much more lively conversation about healthy (and ill) bodies within the context of queer geographies. It is critical to make connections between health (broadly conceived), the medical (pluralistically imagined) and sexualities (in all their complications) more explicit. After all, discussions of neighborhood safety and analyses of the powerful representations that try to suture certain practices to particular sexualities are, at their core, about maintaining or constructing healthy and (un)healthy spaces. As Brown and Knopp (2003, 320) argue, there is a need "for geographers to investigate in further detail the dangerous spaces of abuse, harm and bashing that threaten the queer body quite materially". It is equally critical that we consider how certain medicalised discourses narrate sex and sexualities in ways that reproduce what Sothern (e.g. Chapter 15, this volume) suggests constitutes a narrowly articulated liberal political subject. Returning to Philo (2005, 328), we might also begin to think about how "sex [and sexuality] is translated into discourses of medicine, health, and public policy".

In the next section, I want to examine how medical geography's own relative marginalisation within the context of critical human geography has left it largely bereft of theorisations of sexuality. In so doing, I briefly trace how and why sexuality studies have been marginalised within the context of health and medical geography. I follow this section with a discussion of how queer spatial theories might help us rethink health and medical geography, bringing the latter set of subdisciplinary practices into conversation more explicitly with critical human and queer geography. I do this through an examination of the intersections of: (1) health, morality and sexuality; and (2) sexuality and disability studies. I conclude this chapter by suggesting ways in which queer geographers and others interested in sexuality and space studies might more fully appreciate the theoretical insights offered by a re-reformed medical and health geography.

Medical/Health Geography and Critical Human Geography

Medical geography, at least in its own historical rendering, often suggests its origins lie in the work of Jacques May (1958; 1961). His work on disease ecology gave rise to a significant body of research on tropical diseases and other ecological studies of the distribution and diffusion of illnesses. During the 1960s and 1970s in the United States, geographers interested in medical geographic questions found themselves influenced by both the social relevancy movement and the quantitative revolution in the discipline. Many became focused on the siting of health care facilities in a way that made 'spatial sense' (e.g. Shannon and Dever 1974; Smith 1986). Both traditions found themselves strongly influenced by positivist approaches to studying the 'medical'. Some of this work became particularly important in the study of disease distribution and diffusion. Medical geography, as its title suggests, took as its starting point biomedical definitions of health and illness, and some went as far as to suggest that medical geography has been an 'arm' of biomedicine (and not necessarily in a positive way). Over the course of the 1970s and 1980s, and into the 1990s, medical geography became more insular, mired, in some regard, in an exclusively positivist framework, one that tended to eschew cultural politics or social theoretical innovations (see the critique raised by Kearns and Gesler 1998, 2–4; also see Brown 1995). As a result of the rather insular nature of medical geography, it became, at least in a US context, a fairly marginal subdiscipline leaving those critical human geographers interested in medical or health-related questions centring their own identity outside of the subdscipline of medical geography (e.g. Philo 1996).

In the 1980s and 1990s, a debate broke out within the medical geography community as some began to question the dominance of 'the medical' and exclusively quantitative approaches to studying questions of health care and illness. Medical geographers questioned the biomedical-centric nature of the subdiscipline. Kearns and Gesler (1998; 2002), in particular, began to push medical geography toward what they called a 'geography of health' that engaged with humanistic and structuralist approaches to studying health (e.g. Kearns 1998). Their work echoed the structure-agency debate that raged in the broader discipline at the time (Giddens 1984; Gregory 1981; Pred 1984). The new geography of health, they argued, was more in tune with cultural geographic approaches, particularly critical landscape studies (Cosgrove 1984; Duncan and Ley 1993). In a series of academic volleys published in *The Professional Geographer*, medical geographers argued about the future direction of the subdiscipline, including its name as medical geography or health geography (Dorn and Laws 1994; Kearns 1993; 1994a; 1994b; Mayer and Meade 1994; Paul 1994). Within the context of those debates, Dorn and Laws (1994) suggested that medical geographers needed to engage with theorisations of the body, particularly as these related to psychoanalytic theories of abjection. Unfortunately, even the new self-proclaimed 'health geographers' seemed to resist this turn toward poststructural theorisations of the body and the development of a psychoanalytic medical geography. This resistance might be tied to the theoretical emphasis on structuration theory, critical realism and symbolic interactions (Litva and Eyles 1995) that seemed to dominate this new health geography (see critique by Philo 1996). Kearns (1994a), while agreeing in principle with Dorn and Laws, did little to engage their invitation to consider the work of Julie Kristeva (1982), Elizabeth Grosz (1994) or Michel Foucault (1982). In its programmatic statements, at least, health geography took a 'cultural turn' and became dominated, to some degree, by Gesler's (1992) notion of 'therapeutic landscapes' (see Smyth 2005 for recent discussions). For Gesler and others interested in merging health geography with new cultural geography, landscapes as places invested in cultural notions of health and healing became important sites of investigation. The focus on landscape (as both a site of humanistic feeling and a political economic structure) shifted the focus in medical geography away from applied spatial science and toward landscape studies. At the same time, this turned health geography's attention away from corporeal experiences and the body as key sites of theorisation and practice. From a disciplinary standpoint, a reformed medical geography as a geography of health created a new disciplinary space for the concern with cultural and social geographies of health. The emergence of the journals such as *Health and Place* announced this new found passion for cultural geographic questions as they relate to health and shifted the subdiscipline, while partially re-ghettoising medical geographic inquiry.[1]

The cultural turn in medical geography toward questions related to health and healthy spaces meant that bodies, as both material and theoretical subjects/objects, were marginal to discussions within medical and health geography. Ostensibly, the discussion launched by Dorn and Laws remained on the edge of the cultural reformation in health geography. Perhaps, new health geographers, like many geographers, remained reticent to address the materiality of bodies. As Longhurst argues in her book *Bodies*:

> The fluid volatile flesh of bodies…tends not to be discussed. There is little in the discipline [of geography] that attests to the runny, gaseous, flowing, watery nature of bodies. The messy surfaces, depths of bodies, their insecure boundaries, the fluids that seep and leak from them … [G]eographers … often fail to talk about a body that breaks its boundaries – urinates, bleeds, vomits, farts, engulfs tampons, objects of sexual desire, ejaculates and gives birth. (Longhurst 2001, 23, as cited in Parr 2002, 247)

Geographers, health geographers included, remain somewhat 'squeamish', to borrow Brown and Knopp's (2003) phrasing, about the materiality of bodies themselves. This may have impacted discussions of sex and sexuality within medical and health geography.

Hester Parr (2002) thus calls on medical geographers to be more critical of their own subject by calling attention to Longhurst's (2001) critique of how geographers have discussed bodies. She does so while also calling on medical geographers to continue to build their own critical analyses of bodies and body spaces. "While

---

1   I would also argue that *Health and Place*, while setting out an agenda for a health geography that engages with critical social theory, is also an important site for classic medical geographies.

using words like 'fart, vomit, and tampon' may serve to shock us out of corporeal complacency, we need to think beyond these tactics to envisage how attention to such bodily processes might be part of a critical medical geography" (Parr 2002, 247). What Parr is suggesting is that we refocus our attention on the medical not simply as a 'way of seeing' or 'doing' medical geography but as an object of analysis itself. Turning our attention to how spaces and bodies become medicalised should be a critical question of medical geographic inquiry; it is also an important insight for thinking through the complex ways in which sexed and sexualised bodies and spaces intersect. Following on the heels of Foucault's valuable work (1978; 1985; 1986), we should remain interested in the socio-spatial and historical contexts for the organisations of sexed and sexualised bodies as both subject and object. We must also engage with the real messiness of the medical and with the historical process of normatively medicalising various bodily acts and practices as healthy or ill, abling or disabling. This turns medical and health geographers attention toward how certain sexualities and sexed practices are constituted as pathological and deviant, othered within their broader socio-spatial contexts (Hubbard 2000). And, I would argue, medical geographers could begin to challenge the heteronormative assumptions of the medical through a more thorough engagement with sexuality and space studies, particularly as these are informed by queer and poststructuralist identity theories. To do so means that we must refuse to "isolate the practices of sex from other systems of power – as if it were a space of transcendent nirvana … [Instead, we should] analyze their interconnections and theorize the potentialities of 'sex' both to consolidate and disrupt the co-constitutive systems of heteronormativities, sexisms, racisms, capitalisms, and abelisms" (Sothern 2004, 183).

Now, it is unfair to say that medical geographers have completely ignored studies of the body, sex or sexuality. In fact, there have been a number of important studies that interrogate the intersections of health, health care and body politics. The work of Moss and Dyck (2002) comes to mind almost immediately, as does the work of Anderson and Kitchin (2000). Much of this work, as Parr (2002) recognises in her review of the field, focuses on women's bodies and examines the intersections between women's lives and their experiences of illnesses, particularly chronic illnesses and disabilities. And, these studies have tended to focus more on gender, race and class identities rather than on sexuality *per se*, although Gill Valentine's (1999)[2] recent work on masculinity and disability is an exception. As Mona Domosh (1999, 429) has argued, early feminists' attempts to bracket sexuality and sex makes sense because of fears that a focus on sexuality might "reduce women to the erotic". Because of this fear, feminists' social constructions of gender, as they first appeared in the 1960s and 1970s, tended to theoretically construct 'sex' as the underlying 'real' thing, different from gender. Importantly, though, "recent work has revealed the complex interwoveness of cultural sexual practices and gender identities, [which is] fracturing our simple gender/sex dichotomy into hundreds of pieces" (Domosh 1999, 430). The challenge of a medical/health geography focused on sexualities is to resist re-constituting bodies and spaces as deviant others, while simultaneously

---

2   Again, however, Valentine would probably be reticent to classify herself as a medical or health geographer.

maintaining a rigorous commitment to difference and multiplicity. So there is real value in turning our attention toward queer theory and queer geographies if we are going to begin to untangle the complicated sets of relationships between, say, heternormativity, sex, sexuality and health.

On the Moral Geographies of Health and Sexuality

> Through the various discourses, legal sanctions against minor perversions were multiplied; sexual irregularity was annexed to mental illness; from childhood to old age, a norm of sexual development was defined and all the possible deviations were carefully described; pedagogical controls and medical treatments were organized; around the least fantasies, moralists, especially doctors, brandished the whole emphatic vocabulary of abomination. Were these anything more than means employed to absorb, for the benefit of a genitally centered sexuality, all the fruitless pleasures? All this garrulous attention which has us in a stew over sexuality, is it not motivated by one basic concern: to ensure population, to reproduce labor capacity, to perpetuate the form of social relations: in short, to constitute a sexuality that is economically useful and politically conservative? (Foucault 1978, 36–37)

In the early 1980s, an illness, now known as HIV disease, was identified among self-identified gay men in the United States and France.[3] At the same time, Ronald Reagan was inaugurated President of the United States. Reagan would not even publicly speak of HIV or AIDS until 1987, long after this particular illness and syndrome had spread to all continents of the world (see Cerullo and Hammonds 1987; Watney 1988; Whippen 1987). The diffusion of this particular virus transcended the attempts by clinicians and other medicalised practitioners to suture HIV disease to particular risk groups and spaces, but the powerful discourses of blame that were part and parcel of the 1980s discourse of HIV perpetuated the demonisation of certain sexualities and sexualised practices, while artificially protecting others (Hammonds 1990). Reagan's rather narrow vision of sexual citizenship prohibited the federal government from providing budgets dedicated to fighting the spread of HIV or to finding medications to either stem the transmission or reduce the long-term effects of HIV disease. In the logics of the Reagan administration it was not important to teach people how they might fuck or use drugs safely; rather it was only important to know whom you fucked or with whom you shared needles. And, if that sex or drug (or sex/drug) partner did not meet the norms of a narrowly defined heterosexuality, then it became rather easy to link that particular identity to a deviant, aberrant sexuality that need not be engaged by the state. Since Reagan, the US government has maintained a love/hate relationship with safer sex campaigns, particularly those they see as 'sexually charged' and tied to so-called aberrant sexualities – those of sex workers, intravenous drug users, or queer subjects – that operate outside the norms of 'decent society' (see Patton 1994 on how certain people become 'victims' and others 'complicit'

---

3   It is important to note that the geography of how HIV 'came to be' is critically important to the history of the epidemic. Marginalised (and ignored) in early HIV-related rhetoric were those peoples and places, such as intravenous drug users and peoples of the developing world, for whom advanced diagnostic biomedical care was not available. So, the epidemic was always multiplicitous even if the epidemiological discourse was not.

in the spread of HIV). In the era of George W. Bush, safer sex campaigns are once again under attack because of their portrayal of 'deviant' sexualities that stray (or rather run away screaming) from a centred heteronormativity based in monogamy, for example (e.g. GMHC 2003).[4] At the same time, such campaigns might reinforce other normative and exclusionary practices and politics. As Woodhead (1995, 242) argues we have to be careful not to ascribe safer sex programs solely as sites of resistance because "such projects ultimately rest on the singularizing moves that assume that *out* gay men can somehow reflect *en masse* the worries of *all* 'gay' men, or at least those at risk". Safer sex campaigns might therefore be tied in both discourse and practice to a normalised 'out gay' man who artificially stands in for the experience of all 'gay' men. This masks the complexity of gay experiences and identities and how that complexity is tied to various HIV-related risks.

The complicated process of constructing HIV, its transmission and how to stem its diffusion is tied to a larger biomedical imperative to trace and isolate illness origins and diffusions (see, for example, in geography Gould 1993; Shannon et al. 1991). As Michael Brown (1995; 1997) has argued quite effectively, early representations of the HIV epidemic by medical geographers did little to acknowledge the social constructedness of the epidemic or the ways in which the moralised discourses of biomedicine managed to reduce HIV to a series of vectors whose bodies were gay men. HIV was seen as operating within a linear teleological trajectory (see also Craddock 2000b). Recent theorisations of globalisation suggest, however, that these discourses of disease vectors and sexualities play a powerful role in articulating how health, illness and health care are produced in ways that disrupt the neat linearity of disease diffusions and distributions (Altman 2001; Murray 2001; Treichler 1999). In the context of Thailand, for example, the global identification of HIV with a monolithic and narrowly defined 'homosexuality' (in the form of a 'gay related immuno-deficiency syndrome' or GRID) served to marginalise this illness from the geographical imagination of most Thais, including government officials (Lyttleton 1994a; 1996; 2000). Linking gay identity with the 'West' and with the 'urban' allowed the Thai government firstly to deny the possibility of the spread of HIV in Thailand (Del Casino 2006); and it secondly afforded rural Thais with the comfort that somehow HIV existed outside *khwampenthai* (or Thainess), the feeling of being Thai (c.f. Thongchai 1994). HIV was not a part of a normalised heterosexuality, and other sexually transmitted diseases (STDs), such as syphilis and gonorrhea, which have 'easy' biomedical fixes, were not seen in day-to-day discourse as co-constitutive of the spread of HIV. In this case biomedical practices based in presumptive notions of a universal and simplistic cure created space for a Thai heterosexual masculinity linked to drinking, gambling and commercial sex work (Singhanetra-Renard et al. 1996; Sweat et al. 1995; VanLandingham et al. 1993). The circulation of certain epidemiological models of how HIV is transmitted and by whom meant that HIV received little attention in countries like Thailand which denied its own 'gay community' and thus HIV (Del Casino 2006). The results were that self-identified

---

4   Also see the discussion of the Bush administration's 'Abstinence-only' campaign by Human Rights Watch (http://hrw.org/english/docs/2002/09/18/usdom4291.htm), accessed 8 February 2006.

heterosexuals believe themselves to be safe from HIV. Ironically this exacerbated the epidemic and its spread (Lyttleton 2000).

In the case of HIV, risk is not simply a process of practising certain 'behaviors'; risk has been constituted by the very discursive practices of biomedicine itself which identifies certain spaces and identities as risky or safe (e.g. Farmer 1992). Returning to the Thai context, HIV risk has been tied to particular 'practices,' such as paying for sex, which means that other sexual acts, such as having sex with a non-commercial partner, are deemed to be perfectly safe (Lyttleton 2000). HIV is thus sutured to a biomedically and politically defined promiscuity that is only constituted through certain forms of heterosexuality. By ignoring other heterosexually 'safe' practices, such as sex with one's wife or girlfriend, HIV transmission becomes geographically situated in certain spaces (Del Casino 2006; Lyttleton 1994b). Most sexual decision making, at least in discourse in Thailand, is left in the hands of men, which allows them discretion in their own sexual practices, particularly in relation to marriage or dating. This creates the possibility of the spread of certain epidemic illnesses, many of which are sexually transmitted. Parallel discourses emerged around HIV in Sub-Saharan Africa.

> While sophisticates could joke about Europeans' notions, the emphasis in public health advice and in Western mass media on promiscuity and on sex with prostitutes reinforced the perception that Westerners continued to stigmatize Africans' sexuality as 'excessive,' 'diseased,' and 'dirty.' Depiction of prostitutes as 'a reservoir of infection' fueled local constructions of AIDS as 'a disease of women,' or of the 'lower orders,' from whom the 'morally pure' required protection.   (Grundfest Schoepf 2004, 21)

Sexualities are not simply things 'out there' but are tied to complicated localised and globalised sets of medicalised discursive practices that operate differently and differentially (e.g. Patton 1990; 1994). Sexualities are also tied in very complex ways to socio-cultural and political-economic processes that have real spatial effects and affects (Brown 2000). HIV, as an example, has a complicated sexualised history that has dramatically impacted the diffusion and distribution of this particular virus.

While HIV remains an important focus for geographic inquiry,[5] perhaps because it has so dramatically impacted politically active queer communities, other STDs are more marginally examined. And, yet, the spread of illnesses such as syphilis, gonorrhea, and chlamydia, whose rates are once again on the rise in a variety of global contexts (World Health Organization 2001), are important co-factors in the spread of HIV (as the presence of these diseases may facilitate the transmission of HIV more easily) and are potentially serious diseases in and of themselves. Chlamydia rates, for example, among college students in the United States continue to increase (CDC 2001). This suggests that unprotected vaginal, anal and oral sex is likely also increasing (Koumans et al. 2005). Neither medical nor queer geographers have examined these particular processes which are linked to discourses of sexuality, sexual identity and intimacy. As others in the field of medical anthropology have argued (e.g. Kochems forthcoming; Kochems and Patti 1993), the emergence of new sexualised discourses,

---

5   It is interesting to note, however, that there has not been one publication in the *Annals of the Association of American Geographers* on HIV/AIDS since 1991.

such as bare backing (having anal sex without a condom), may be tied to the shifting meaning of what it means to be intimately 'gay', for example (also see Del Casino et al. 2004; Kochems and Del Casino 2004). Similarly, in certain communities in the United States, anal sex among male-female sex partners may serve as a way to protect virginity or maintain the boundaries of the heterosexed body (see Mosher et al. 2005 for a recent study of sexual behavior trends in the US), perhaps reinforcing compulsory heterosexuality and an emergent hegemonic Christianity (Simon 2006).[6] These practices may also be a way to rethink the 'normative' boundaries of heterosexuality, suggesting that geographers consider more seriously how "sex acts, sex work, sex workers, sexual diseases, sexual health, and sexual policies ... [are all] framed by prevailing attitudes about the where, when, and how of acceptable and non-acceptable intimacy, procreation, and reproduction" (Philo 2005, 330). Constructions of sexuality that are complicated by normative notions of what it means to be 'gay' or 'straight' – and these norming processes happen within any number of so-called 'communities' – are also sometimes linked in important ways to the diffusion and distribution of certain diseases. Following Kesby (2005), this suggests that medical and queer geographers might consider thinking through adolescents' constructs of sexuality and sexual practices as well as those of adults.

Mired in a politics and economics of access not only to information but also to condoms and other safer sex practices, STDs, such as syphilis, will continue to diffuse through various socio-sexual networks that are themselves situated within discursively constituted readings and spaces of how sexualities are to be performed and practiced. STDs are also important to examine because they are politically linked to so-called 'deviances' (Craddock 2000a, 89–95). Unlike cancer which is not our 'fault', STDs are normatively considered to be illnesses of 'choice'. For liberal subjects embedded in a growing autonomous world of individual human rights, STDs are not the responsibility of 'others'. What this suggests is that there will continue to be a narrowing of what constitutes 'appropriate heterosexual identities' in a globalised political economy of difference as certain societies, such as the US, choose to largely ignore the diffusion of HIV and other STDs in the marginalised economies of, for example, Sub-Saharan Africa. This is justified within powerful discourses in countries such as the US because of the ways in which sexualities in these other spaces are constructed as 'abnormal' or 'deviant'. Thus Hubbard (2000, 197, following Richardson 1996) suggests the idea of

> 'queer' heterosexuality to distinguish between heterosexual norms of procreative sex and forms of sexual practice that might include promiscuous sex or anal and oral intercourse, with these latter 'queer' forms of heterosexuality potentially including what are commonly regarded as 'homosexual' acts. Conversely, individuals who fail to conform to heterosexual expectations by remaining celibate or living alone may also be described as possessing 'queer' heterosexual identities.

---

6 This would be a very interesting site of geographic inquiry, examining the spatialities of 'safe sex', 'virginity' and 'intimacy' among self-identified heterosexual teens. It should also be noted that there is very little work on how individuals link identity to sexual practice in geography (as an exception see Sothern 2004, for an excellent analysis of queer radical geographies).

What we have to do in examining these complex processes is avoid the easy reductionism of biomedical discourse that wants to create rather simplistic definitions of how sexuality and illnesses (and health) are related (Craddock 2000a; 2000b). Instead, we must pay "[a]ttention to the social organization which surrounds urination and bleeding, for example, [which] may help us to understand how bodies which do this (in)correctly are categorized and managed by a medicalized society" (Parr 2002, 247). We must also examine the corporeal experiences of sex, the exchanges of fluids, the mashing together of bodies, the eroticisms and subtleties of intimacies, the 'virtual' possibilities of sex, which are socially co-constitutive, part and parcel of the shifting dynamics of the discursive constructs that constitute certain healthy (and ill) sexualities as they shift in definition temporally and spatially. We must also challenge the moralised assumptions of biomedicine as they are deployed in the political discourses of conservative forces globally, which reinforce abstinence over safer sex education, population expansion over responsible family planning, and compulsory heterosexuality over a dynamic and ever-changing continuum of sexualities and sexualised identities.

Rethinking (Dis)Abilities and Sexualities

> Despite long running telephone sex lines and more recently the development of cybersex, sexual activity is still seen as a very physical, embodied experience which 'deviant' bodies cannot participate in satisfactorily.   (Butler 1999, 209)

> In relation to disability and sexuality, the common cultural representations are ones of asexuality, with disabled people uninterested or unable to take part in sexual activity, or as sexual 'monsters', unable to control their sexual drives and feelings.
>   (Anderson and Kitchin 2000, 1164)

Queer theorists and geographers have long argued that sexualities are contingent, socially constituted and materially experienced. In similar ways, the meaning assigned to the 'ability' and 'disability' of people is also contingently constructed in relation to shifting socio-cultural and political-economic dynamics across a myriad number of spaces. As social constructionists have long argued, being disabled has shifted and changed (Parr and Butler 1999) particularly in relation to capitalist and neoliberal socio-spatial relations that mark certain bodies as 'productive' and others as 'unproductive' and, as the aforementioned quotations aver, in relation to various discourses of sexuality. As bodies have been assigned certain designations in relation to capitalism and changing political economies, the medicalisation of certain bodies as unproductive and therefore dis-abled has also intensified (Gleeson 1999). This is not to suggest that people living with disabilities do not experience "'true' pain and inconvenience" (Parr and Butler 1999, 4) or that people living with various disabilities (both mental and physical) are not "victims of sexual abuse or violence in institutional or community based residential facilities" (Gleeson 1999, 131). Rather, it is to argue that how we constitute what it means to be disabled is epistemologically similar to the ways in which we go about de- or re-sexualising certain bodies (c.f.

Butler 1999; Moss and Dyck 2002; Valentine 1999).[7] And, more importantly, perhaps, in the context of our own queer geographies is the notion that "[c]rossing the divide and entering gay space [for example] involves a recognition of identity and can be a big step for any individual … If the welcome is less than supportive due to ableism the experience can be all the more traumatic" (Butler, 1999, 206).

It is also interesting to consider how certain illnesses, such as HIV, which do not always manifest a visibly outward and tangible marking on the body, intersect with how we practice and perform our sexualities: one need look no further than the literature on HIV+/- discordant relationships to consider how this particular disability opens up and closes off certain sets of sexualised identities and practices (e.g. Beckerman 2002). HIV can be both enabling and disabling, productive of new sexual possibilities and limiting in other ways, particularly as antiretrovirals at least in Western economies shift the temporality of HIV and re-create it as a chronic illness.[8] HIV *could* enable the emergence of new communities of self-identified HIV positive individuals engaging in sexualised practices that are not tied to concerns about contracting HIV or even other sexually transmitted diseases. Once HIV manifests itself physically in outward appearance the relationships to sex and sexuality may once again shift as those HIV positive individuals with outward illnesses are marginalised in their sexuo-social communities as diseased. It is possible that a person living with HIV disease will move in and out of the regulatory frameworks of normative constructions of health and illness and therefore move between being constituted as sexual and asexual. Importantly, this suggests that the embodied practices of various disabilities challenge the presumptions of a singular unitary and bounded body. This is why Shildrick (2004, 2) argues that "there is no distinction to be made between one corporeal [bodily] element and the next" (see also Shildrick 2002); we must consider the ways in which bodily practices and performances express inter-corporealities and stretch bodies beyond the boundaries ascribed to them discursively by various regimes of authority such as biomedicine. It is the 'promise of connectivity' that enables people occupying normatively marginal subject positions to create new possibilities and sexualised identities that may help us in "rethinking disability" (Shildrick 2002, 3) and in rethinking the intersections between various disabilities and sexualities.

There are thus important intersections to be interrogated between the shifting terrains of sexualities and disabilities by both medical and queer geographers who are particularly interested in how certain discursive and embodied practices can both enable and disable certain sexualities. Take, for example, the pregnant body. On one hand, this body is the quintessential heterosexual object, fulfilling what Hubbard

---

7   It is important to note that Moss and Dyck are interested in chronic illnesses, particularly as these impact women. Chronic illness, more generally, is another important area of inquiry for medical/health geography and sexuality studies.

8   As we enter a period of 'post-AIDS' in western economies where antiretroviral treatments are becoming more pronounced, it is possible to think of HIV disease as constructed as a chronic illness (e.g. Sothern, 2006, for a longer examination of this issue). Of course, at least in the United States, a diagnosis of having 'AIDS' means that one is eligible for disability compensations from the federal government. In this way, AIDS is very much still a disability within the political economy of health care in the US.

(2000) identified above as the normalised routine of reproduction, a project that is supportive of both capitalism and heteronormativity. In the same instance, however, pregnancy is dis-abling, as the body becomes an asexual object of potential revulsion and abjection. "Constructions of the pregnant body as ugly, alien and not 'sexy' or sexual help to explain why the pregnant body is so often considered to be private and in need of concealment" (Longhurst 2001, 53). Biomedical discourse constitutes pregnancy as a pathology and as a temporary state of disability that demands vitamins, strict diets and regulatory frameworks, and reduced physical activity. Moreover, the pregnant body is sutured to home within the confines of medicalised practices (Fannin 2003), further spacing women's bodies and heterosexual identities as reproductive. The reduction of women's bodies to the core of a normative heterosexuality through pregnancy is reinforced by a hegemonic biomedical set of discourses and practices that makes these bodies readable as reproductive. This move toward reifying women's heterosexual bodies illustrates how such moves are embedded within a set of complex contradictions: at another point in time and space these same bodies may be sexed and sexualised in relation to heterosexual male desire. Of course, as in vitro fertilisation allows for a destabilisation of the heterosexual norm of the 'straight nuclear family', while allowing others to rethink their reproductive role in both the family and society more generally, the discourses of who can and cannot constitute a parent are being challenged. The queering of the family in this way may serve to undermine certain heteronormative assumptions that suture parenting to heterosexuality. At the same time, such 'deviant' practices unleash governmentality effects that seek to discipline these new socio-spatial arrangements and families as well (c.f. Foucault 1991).

Biomedicine also plays other important roles in reinforcing certain heterosexual identity positions, particularly those related to virility and masculinity, universalised within the norms of what it is to be a heterosexual man (e.g. Potts et al. 2004). One of the more recent 'disabilities' to be identified and medicalised within the confines of a set of heteronormative discourses is erectile dysfunction, which has been brought into the spotlight with the advent of drugs like Viagra™ and other sexuopharmaceuticals (Del Casino 2007). As Potts et. al. (2004, 490) argue, "[i]n medical constructions of 'erectile dysfunction' there is little reference to the significance of factors outside or beyond 'the body' … Importantly, the medical domain now advocates drugs as the most effective and efficient means of treating erectile dysfunction, *regardless of aetiology*[9]" (emphasis in the original). The assumptions underlying the production of such drugs are that they reproduce certain universal practices within the private spaces of sexual relationships between a man and woman. Marshall (2002, 134) thus suggests that "[t]he 'function' is 'successful' intercourse, which is 'functional' for the [heterosexual] couple, which is 'functional' for society" (also see Katz and Marshall 2004). The advertising blitz that has followed the introduction of Viagra and other drugs like it has focused almost exclusively on a heterosexual market. There are interesting geographies to the distribution and diffusion of such biomedical 'cure alls' and important ramifications for how the biomedicalisation of erectile dysfunction operates in and across different spaces – within the confines, for example, of sex

---

9   This term is defined as a cause or set of causes of a particular disease.

tourism, and outside the bounds of the normative heterosexuality upon which this particular drug is based. The use of Viagra among self-identified gay and bisexual men, for example, suggests that this drug can be used in a variety of socio-spatial contexts, such as party-and-play environments[10] to mitigate against other drugs such as methamphetamines which can sometimes cause impotence (Kochems and Del Casino 2004). In another space, male sex workers might use such sexuopharmaceuticals to work longer hours. Understanding the spatialities of how erectile dysfunctions are constituted and how the 'cure' is then practiced would be a valuable area of inquiry for geographers interested in investigating the intersections of sexuality and health. As Potts *et al.* (2003, 706) argue quite effectively, the introduction of drugs like Viagra creates "a powerful kind of pressure to have sex against all other odds – which women would arguably experience if they are positioned as 'good' feminine sexual subjects within a male sex drive discourse – [a pressure that] is potentially intensified when a male experiences erectile dysfunction". The presumptive norms of a masculine heterosexuality are thus reinforced within the spaces of health by a biomedical discourse that sees erectile dysfunction as a pathology and a problem (as opposed to a part of the natural progression of life), albeit one that has a potential cure.[11]

Conclusions

> Radical geography will be impoverished, politically as well as intellectually, if we cannot find ways of balancing concerns with diverse experiences of oppression and exclusion, with continued struggles to advance our understandings of the causal processes helping to perpetuate sexism, heterosexism, ableism, racism and ageism.
>
> (Chouinard and Grant 1996, 196)

Chouinard and Grant offer a provocative analysis and call important attention to the limits of 'radical geography' when they suggested certain (dis)abilities and sexualities are marginalised within the academy. Over the last ten years or so, queer geographers and certain medical geographers have been very interested in extending this project, pushing the boundaries of what geographers see as 'important' or even 'relevant' sites of geographic inquiry. I have argued in this chapter that working across the boundaries of sexuality studies and health/medical geographies may help to open up new questions and foreground new possibilities for understanding how heteronormativities are set in place and constituted in relation to, for example, issues of morality or constructions of various disabilities. There remains more to be done, however, if the project of creating space for the multiplicity of sexualities and abilities manifest in everyday practice is to be realised. At the same time we should not shy away from the important project set forth by Foucault and others that asks

---

10 Party-and-play means situations where individuals integrate drugs and sex.

11 There is an important political economy to the production of such drugs, and the investment in research and development in the area of drugs such as sexuopharmaceuticals creates greater disparities, as less time and energy are invested in addressing other life-threatening diseases such as malaria, tuberculosis, cholera, and HIV/AIDS.

us to interrogate the bounding and naturalising of various spatialities and identities that can be harmful to the overall health and well-being of both individuals and broader communities. Only through our continued interrogation of how power and knowledge operate through the institutionalised spaces of health, illness and health care can we fully appreciate the emergence of the multiplicity of sexualities, sexual practices and other (re)enabling identities that are part and parcel of the everyday practices of various sexualised subjects.

Acknowledgements

I would like to thank all those who have supported me in the research that informs this paper, including Michael Brown, Dennis G. Fisher, Lee M. Kochems and John Paul Jones III. In addition, I have to thank the Universitywide AIDS Research Program (Grant # ID02-CSULB-042) who provided support for the research that inspired sections of this paper. The ideas and all the associated errors included in this paper are, of course, my own.

Chapter 4

Queer Critique and the Politics of Affect

Jason Lim

Introduction

Recent years have seen an explosion of work influenced by theories of 'affect' take place within geography (Anderson 2004; 2005; 2006; Carter and McCormack 2006; Dewsbury 2000; McCormack 2002; 2003; 2005; Saldanha 2005; Thrift 2004). These ideas have sought to open up a space for thinking about how bodies may affect other bodies and be affected by them in ways that are not prescribed or proscribed in advance. Such thinking has sought to enable the envisioning of new modes of practice and new ways of relating to other people, places, situations and material things. It has sought to offer insights into the creativity involved in the processes of making community and of negotiating (social) encounters. In offering such potentialities, much of this work also raises questions about how best to conceptualise social change and about the role of critique in advancing such change (see Anderson 2004; 2006; Massumi 2002).

Importantly, a very similar set of questions has been posed by those working with queer theories. These questions explore the possibilities for desiring in ways that move beyond the politicised regulation and representation of bodies and desire. Queer theory seeks to facilitate new practices and new ways of relating that are beyond normative understandings of what bodies should and should not be able to do. Because these normative understandings rest on ascribing a putative identity to these bodies, queer theories attempt to destabilise or sometimes even do without identities. Such a questioning of identity might be thought of as a movement away from the fixed co-ordinates of a 'politics of recognition' (Fraser 1995), where particular identities (of social groups and their political interests) are already staked out and recognised, which underlies many forms of critique. So, some queer theorists have also questioned the place of critique, suggesting that it might need to be augmented with more careful attention to the question of how to act creatively. Indeed, some queer theorists, prominent among them Eve Kosofsky Sedgwick (2003), have turned to ideas about 'affect' for just such inspiration.

Surprisingly little has been written within geography on the potential for an exchange of ideas between theories of affect and queer theories, although Gavin Brown's chapter later in this book is an exception.[1] In this chapter, I seek to bring

---

1 Knopp (2004; Chapter 1, this volume) and Binnie (Chapter 2, this volume) have interrogated some of the opportunities and some of the problems raised by thinking about the closely-related field of 'non-representational theory' in the context of geographies of

together specific strands of both these approaches and to explore some of the possibilities and some of the pitfalls that a conversation between these two bodies of thought would open up. This discussion takes place, firstly, in the context of the progressive adoption of queer theories and its insights by geographers of sexualities; and, secondly, in the context of the aforementioned spread of theories of affect within geography, in particular the Spinozan-Deleuzian conceptions of affect that I shall be concentrating on here.

This chapter tackles questions about the nature of ethics and politics in thinking about geographies of sexualities. It explores what different ways of thinking about social and political change can offer to those seeking to create space for new forms of sexual desire and sexualised relationships and spaces. It asks whether the very notion of critique, which underlies much of contemporary thinking on how sexuality is regulated in both an institutional sense and in everyday settings, requires rethinking. The next section of the chapter provides a brief sketch of Spinozan-Deleuzian theories of affect. The third section of the chapter suggests what such theories of affect might have to offer to considerations of geographies of sexualities that work with queer theories. This section also asks why, given the similarities of some of these ideas, theories of affect have not been more widely taken up by geographers of sexualities. I pose three responses to this problem, each of which is tackled in a subsequent section. The fourth section of this chapter, then, discusses how we might think of a politics of affect and how this task might require us to think about politics in unconventional ways. Not only does this section suggest how affect might prompt a reassessment of how 'the political' is understood, but it also suggests a reassessment of how affect is conceived in relation to politics. The fifth section relates these discussions to different views as to the nature of social, political and cultural change. The final section considers how we might think about queer theories and practices as already welcoming the ethical potentials to act and relate in new ways that are presented by constantly unfolding events. Such a line of thought presages a discussion of how far queer critique is open to such potentials, as against taking up a purely oppositional stance.

Affect and Deleuze

The conception of affect developed by Gilles Deleuze (Deleuze 1988; 1990a; Deleuze and Guattari 1988; Buchanan 1997; Connolly 2002; Massumi 2002), inspired by the ideas of Baruch Spinoza, helps us to think about bodies and what they can do. Affect can be thought of, very simply, as the capacity or power of bodies to affect other bodies and be affected by other bodies. 'Bodies', here, are not to be understood only as human bodies. Rather, 'bodies' can be defined as a site where forces are 'actualised' (an idea that will be discussed later). Practically, this means that bodies might take many different configurations, for example, material composites of any

---

sexualities. Hacker (Chapter 5, this volume) brings together some Deleuzian ideas with ideas from queer theory in ways that are suggestive of links to the themes of this chapter.

kind, discourses, ideas, social collectives, or human bodies in conjunction with tools, technologies, other objects and/or other humans (see Deleuze 1988, 127).

Deleuze's take on the idea of affect directs our attention towards the question of what bodies can do and how they can be affected (Deleuze 1988, 123–128), demanding that we understand this question as an open one. This approach asks about the body's potential for action, its abilities to connect, to enter into relations with other bodies. This is where Deleuzian thinking about affect becomes political because it demands a change in the ways that bodies are defined and understood.[2] If the question of what a body can do is an open one, then this approach attempts to avoid starting with taken for granted ideas about a body's function or properties. While Deleuzian theories of affect recognise the historically specific political processes that ascribe to bodies the functions or properties by which they become identified, they also seek to open up a space for thinking beyond this specification of bodies. In other words, thinking about affect can help one to step outside of moral economies that specify what bodies are and are not supposed to be able to do, and that repudiate the bodies that affect each other in supposedly improper ways.

It might be thought that the Deleuzian approach to thinking about affect seeks to redress the potential injustices that might arise from the enforcement of such moral economies. If it does so, however, then it does not primarily do this through a validation of modes of desire or modes of entering into relations that are normatively but unjustly negated. Rather, thinking about affect allows one to think about what might take place beyond the limits of normative modes of regulating life and relationships. By focusing on the way that bodies' capacities to enter into differing relationships is, as we shall see, constantly shifting and amenable to augmentation, an attentiveness to affect can be enormously empowering. It can foster a sense of exploration by gesturing towards how bodies are able to forge new connections and new alliances, to forge new pleasures and new ways of being productive. An attentiveness to affect also brings with it a new kind of ethics. Such an ethics urges us to make judgements about what is right and what is wrong, what is safe and what is dangerous, or what is harmful and what is beneficial from within and on the basis of the particular (social) situations and ways of living that we are constantly forging. This is not only a disavowal of thinking of moral frameworks as if they are frozen relics that apply to all times and all places, but it is also a validation of the vitality of ethics, of the fact that ethical questions are always in the process of being negotiated as the character of 'the social' changes.

The question of what a body can do is an open question because of the indeterminacy of affect: "no one knows ahead of time the affects one is capable of ... [Y]ou do not know beforehand what good or bad you are capable of; you do not know beforehand what a body or mind can do, in a given encounter, a given arrangement, a given combination" (Deleuze 1988, 125). A body's affects – its capacities for action and relation – are indeterminate, in part, because these capacities are constantly changing. They change because they are the outcome of the interactions and encounters the body has with other bodies. As bodies are perpetually

---

2   Thanks to Paul Harrison and his post to CRIT-GEOG-FORUM at www.jiscmail.ac.uk 29 June 2004 for the inspiration to follow up this point.

doing different things – entering into different situations and interactions – so their affects are continually modified as a result. Here, two other Deleuzian concepts are important for understanding this indeterminacy: firstly, the difference between 'the virtual' and 'the actual'; and secondly, the idea of the 'event'.

The difference between the virtual and the actual is coincidental with the difference between the body's affects – what it can *potentially* do – and the things the body *actually* does. The virtual field is made up of all the things that might happen in any given encounter. Alternatively, from the point of view of something that has already happened, it is all the things that might have happened. The virtual field is always multiple – there are always myriad things that might be done, out of which only one becomes actualised. The virtual, then, underwrites the indeterminacy of life – the way that there are always different things that can happen, always different things that bodies can do or different relations that they can establish with one another (see Massumi 2002). It is out of the multiplicity of things that might happen that something *actually* comes to pass. So, we can talk about a process of 'actualisation' as the selection of what occurs from all the things that might have occurred.

The idea that the virtual is a field is indicative of how affect is not just a capacity within the body, but is to some degree outside the body, too. What a body can do is always to some extent determined by its multiple relations with other bodies in the various contexts and environments it inhabits. The indeterminacy of the virtual arises in part because of the continual changes to the actual relations between bodies. These actual changes, for instance in something as simple as bodies' relative proximity or speed to one another (Deleuze 1988; Deleuze and Guattari 1988, 260–265), not to mention changes in the qualities of a host of other, often social, relations between bodies, reconfigure what any one body could potentially do. These actual changes, in other words, alter bodies' affects, presenting these bodies with new opportunities for connection, relation or movement.

The difference between the virtual and the actual is important for understanding the nature of events and how events always present different potentials for action and relation, for forging connections and (new) ways of living. For Deleuze (1990b), the 'event' can be thought of as virtual (and hence as enveloping a multitude of virtual worlds), as distinct from the actual 'states of affairs' by which events come to pass. Events are always (just) past or just about to come, but are never the present (state of affairs), although they are contemporaneous with the present. Events are intimate to bodies (Deleuze 1990b, 5) but not actually corporeal themselves. Rather, they are incorporeal, an effect of the mixture or interaction of bodies. They can be thought of as a sense or logical attribute that shadows or hovers over actual states of affairs (1990b, 100). This sense neither originates in the subjectivities of individuals nor is fixed in place by an objective reality. In Deleuze's (1990b, 100–101) example of a military battle, the event of the battle is grasped differently by different participants, and it is the event in its multiplicity (all the things in the battle that have just happened and all the potential things that might be done) that causes each of these different participants to do different things – to flee or charge or shoot when they do.

More generally, actualisation, when thought of from the point of view of the event, is the process by which the virtual (the event) is repeated, but is repeated differently from itself. Memory, for example, can be thought of as an event, a virtual collection of percepts that are not organized in any particular way, but that are ready for recollection – actualisation – in a particularly organised way (Colwell 1997). While memory as virtual is the past contained in the present, ready to be brought back to life at any moment, the process of actualisation repeats the percepts of memory in a different way from memory itself. The actual, in other words, does not resemble the virtual that it differentially repeats. A recollection repeats a memory, but this repetition is different from memory itself, as it is arranged in a way that has a bearing on the present (Colwell 1997), used in the context of a body's action or interaction.

Affect can be thought of as an event. It straddles the virtual and the actual (Massumi 2002) at the intersection of the demands of the present and the virtuality of the past. It is worth reiterating that the process of actualisation involves repetition but also, importantly, difference. And it is also worth remembering that this difference is underpinned by the virtuality of the event – the multiplicity of different things that might be done or brought back to life. This does not mean that anything could happen, but only that we cannot say in advance what will, at least not with any certainty.

Affect and Queer Theory

It might be thought, then, that these theories of affect could be brought together with queer theories within geographical research and thinking regarding sexualities. Both queer theory and Deleuzian approaches to affect relate to questions of how to desire, who to desire, and which bodies desire and are desirable by which other bodies. Both sets of ideas seek to open up the question of desire. Queer theory problematizes these questions, in part, by posing a challenge to heteronormativity (and, more recently, homonormativity) insofar as these practices and institutions effectively circumscribe desire. Indeed, in a curious circularity, heteronormativity operates precisely by, on the one hand, identifying bodies as sexed (male or female) by reference to their putatively essential functions – their procreative functions – and, on the other hand, by normalising heterosexuality on the basis of a supposedly natural relationship between male and female sexed bodies. Desire – presumed to proceed from a sexed body – becomes a function of bodies defined by their supposedly innate properties. The heteronormative distinction between 'normal' and 'abnormal' (e.g. lesbian or gay) desires also rests on this putative functionality.

Queer challenges to heteronormativity proceed along many fronts, not least by pointing out the way that it requires performative repetition, institutionalisation, discipline and policing in order to shore up the normality of the correspondence between sexed bodies and sexual desire (see, for example, Butler 1993a; 1997; 1999; 2004). Queer theories and practices also seek to liberate desire from these constraints so that desire can proceed beyond not just procreation but also potentially beyond other heteronormative strictures such as marriage, monogamy, class and race expectations, and gender expectations. In seeking to multiply the possibilities for desire and for different kinds of action, queer theories and practices attempt not to prejudge the question of what a body can do. Indeed, queer thinkers have pointed out how heteronormativity and homonormativity are belied by the diversity of desires and relationships that *already* characterise many people's lives, and that transcend the ideological identities and proprieties that heteronormative and homonormative institutions attempt to enforce.

Desire can, thus, no longer be thought of as having a fixed functionality; rather, both the qualities of the desiring body and the object of desire are open questions. Such a move has clear resonances with Deleuzian approaches to affect. Indeed, desire not only plays a major part in Deleuze's philosophy, but can be understood through affect. For Deleuze, desire is the process of entering into relation. As such, desire is productive – it produces reality because it is the forging of the social, material, biological, physical etc. relations by which bodies connect in order to work, to do things, and to exist (Deleuze 1997; Deleuze and Guattari 1988). Yet desire, for Deleuze, must be understood as affect and event, as *processes* of assembling various heterogeneous elements – social, biological, political. These heterogeneous assemblages, because they are always still in the process of being assembled, cannot be pinned down once and for all as the result of the work, agency or drives of identifiable subjects, social structures, or bodies understood organically (Deleuze 1997, 189). Desire as a process of relating always implies that there is some kind of 'line of flight' – a way of doing things otherwise, an escape from the 'stratified' organisation of social (material, biological, political etc.) relations, even as these strata are being organised in the same processes of assemblage.[3]

Another interesting resonance between theories of affect and some strands of queer theorising is in their respective conceptions of temporality. While affect can be thought in terms of its eventfulness, 'queer temporality' (Halberstam 2005; Sedgwick 2003) also suggests ways of living that welcome the potential to affect and be affected in ways not scripted in advance. The development of the idea of queer temporalities can be placed in the context of an anti-Oedipal project – again, something shared with Deleuze (Deleuze and Guattari 1984; 1988). Sedgwick (2003, 147), in particular, writes about queer temporalities as disrupting an Oedipal generational narrative ("it happened to my father's father, it happened to my father, it is happening to me, it will happen to my son, and it will happen to my son's son"). Generational relations do not always proceed with this level of regularity and repetition. In a move that is echoed in geographical research on affect (Anderson 2004; 2006), Sedgwick (2003, 146) mobilises *hope* to create room to realise that the future *may* be different from the present. In doing so, she gestures towards a quasi-Deleuzian notion of potential, distinguishing between that which actually happened and the wider field of the things that might have happened but did not. This distinction – between the virtual and the actual, in Deleuzian terms – is crucial,

---

3   Deleuze's thinking on desire is hence in contrast to that of Foucault, for whom 'Sex-Desire' is a discursive, technological and institutional regime that produces bodies as sexed and constrains desire in the service of these idealised bodies (Foucault 1979; Butler 1999). Deleuze's 'desire', however, alongside his concept of 'bodies without organs', might be thought to converge with Foucault's (1979) idea of not yet determined 'bodies and pleasures' as ways in which we might think about how to forge new kinds of social and sexual relations and pleasures (Deleuze 1997, 190; Robinson 2003).

Sedgwick asserts, to being able to think about the possibility of things happening differently (2003, 151 n5).

Judith Halberstam (2005) takes up the theme of queer temporalities to think about life unscripted by the conventions of family, inheritance and child rearing. She counterposes queer temporality to 'repro-time' (the idea of women's 'biological clock', and scheduling for married couples) and 'family time' (the scheduling of daily life, especially with relation to beliefs about children's needs) (2005, 5). Queer time, by contrast, often operates outside of the privileged and respectable temporalities of maturity, adulthood, marriage and parenthood. One of the forces that has deroutinised life for gay communities in particular has been the AIDS epidemic since the 1980s (see also Sothern, Chapter 15, this volume). The prospect of a constantly diminishing future because of foreshortened life expectances has created an emphasis on the here and now, of living life in its immediacy. These ways of affecting and being affected have broken free from the functionality of family, reproductive and generational lifestyles. The focus on the potential offered by the moment has meant an attentiveness to squeezing new possibilities out of the time at hand (Halberstam 2005, 2). One implication of such queer temporalities is the emphasis on creating friendships that are not based on a sense of identity. So, rather than forming connections only with people with whom one shares an identity, or with whom one's life narrative overlaps, queer friendships and communities become based on the intimacy of intensity and the immediacy of continuing to live now (Sedgwick 2003, 149). In the sense that queer temporality seeks to maximise the potential of intensive encounters and seeks to map ethical paths of connection, desire and friendship, it might be thought that theories of affect might help in the thinking of such projects.

It would seem, then, that there is plenty of scope for fruitful conversation between those who work with Deleuzian theories of affect and those working with queer theories. This scope for an exchange of ideas prompts the question: why have theories of affect not been more widely adopted by geographers of sexualities who are influenced by queer theory?[4] There are several reasons for this. Firstly, I think that some geographers who work with Deleuzian theories of affect have attempted to think beyond some of the conventions of critical thinking in social geography (including geographies of sexualities). One of the consequences has been to put some distance between the two approaches. Secondly, geographers of sexualities, in common with other critical social and political researchers within social geography, have been suspicious of theories of affect. This impasse has arisen due to the perception of a gap between the 'ethical' approaches that are offered by thinking on affect and the 'political' approaches that are adopted by most social geographers. An attention to affect is seen as heralding a neglect of questions of politics. Thien (2005), for example, voices a common concern that the impersonality of affect (in the sense that it takes place *between* not just human bodies, but material bodies of various kinds, human collectives, discourses, memories, neurons, sensory surfaces

---

4   It might also prompt the question: why have queer theories not had wider influence in shaping the development of geographers' engagement with affect? Unfortunately, I do not have the space to address this question in this chapter.

etc.) amounts to a lack of concern for the subject and its relation to power – in other words, how various kinds of subjectivity are constituted socially and geographically. Tolia-Kelly (2006) worries about the neglect of historical memories that she asserts must constitute bodies' capacities to affect and be affected. She writes that social differences such as race, ethnicity, class and wealth must make a difference to how a body is able to feel pain or suffering. These social differences in the capacity for affecting and being affected in turn must be historically constituted. For both Thien and Tolia-Kelly, these are political questions. Resonating somewhat with Hemmings's (2005) concerns about the autonomy of affect, both voice a concern about the relative lack of organisation of the virtual field. Without the familiar coordinates of recognisable subjects, objects, emotions, socially marked bodies, and social institutions, these considerations of the virtuality of the body seem to some authors to be rather abstract (see Binnie, Chapter 2, this volume). The question they raise is how when thinking through affect does one understand social and political differences between different individuals, groups and collectives?

These questions and concerns raise some important issues, not all of which I have space to deal with here. What I would like to do, with reference to geographies of sexualities, is to sketch out three sets of arguments in response to this disjuncture between these two different theoretical approaches. These three arguments are:

- Theories of affect can be used to think politically;
- There is some worth in thinking about social and political change as arising 'immanently', but such a view has a number of implications for thinking about subjectivity and emotions; and
- Queer theories often exemplify ways in which a politically attuned body of thinking can be alive to the ethical potentials of affect.

These arguments will be tackled in next three sections.

Affect and Politics

Deleuzian theories of affect can be used to think politically, and, I would argue, offer new ways of thinking about power. The virtuality of affect does not mean that social difference and social contexts have no bearing on the multiplicity of things that might be done. Brian Massumi (2002) thinks of affect as a 'surface' between the virtual and the actual. One implication of this is that not only does affect offer up myriad potential actions and relations (some of which become actualised), but that some of these actual relations, once expressed, *feedback* into the virtual field. This process of feedback – not only of actions, but of social contexts and relations too – creates an incipient kind of organisation of the virtual. Conserved within the virtual field are various *refrains* of ways of affecting other bodies and being affected by them (regarding the refrain, see Deleuze and Guattari 1988, 310–350). One might think of the movements and expressions of being affected and affecting another with whom one is flirting – a flash of the eyes to make eye contact, or a response to a caress. In such encounters there may be certain complexes of movement and expression that those involved 'remember' or will have 'learnt' and are ready to actualise. Of course, the selection of a particular of affect is dependent on the context, on the qualities and actions of the bodies that are involved in the encounter. So, while what a body can do must still be thought of as indeterminate, there are also 'tendencies' (Massumi 2002, 30) to repeat certain actions, expressions and ways of relating to other bodies, although each repetition is enacted differently because it responds to the circumstances at hand.

William Connolly's (2002) conception of 'virtual memory' also helps in thinking about the relationship between affect and the social and political. Virtual memory encompasses the idea of how the effects of previous events are conserved within the body (as habit, trauma, pleasures, unconscious antagonisms etc.). Virtual memory is distinct from both the explicit mental images and stories that we recall to consciousness and the explicit narratives that circulate and are 'remembered' socially. Rather, it is a way of handling or processing the multiplicity of events: there is always too much to consciously remember or think about at once. Connolly uses the example of watching a soap opera: we do not have the time to explicitly recall previous scenes or episodes in order to interpret what we are watching right now. In the midst of action, we rely on affect – on our bodies' memories of ways of proceeding in different circumstances – rather than on active, explicit memories (2002, 33).

For human bodies (understood in a loose sense as shifting composites of matter and movement – biological, technological, and social), virtual memory subsists below explicit memory as a repository of past cultural life (Connolly 2002, 25). Because we cannot always stop to ponder over our actions consciously, virtual memory is crucial for negotiating everyday activities. It is part of a rapid process for subtracting from the multiplicity of things that can be sensed or done in any encounter. This process of subtraction enables just one action out of the multiplicity that can be done to be selected for actualisation. Thinking about affect in terms of virtual memory again emphasizes how repeating actions done before is not a matter of representing the past, but of acting it out in new circumstances. The events that are conserved within the body – that the body remembers – are always events in which the body affects and is affected by other bodies. It is for this reason that virtual memories are not images of the past, but are the embodied remembering of ways of acting and relating.

So, the idea of virtual memory allows us to think about the conservation of events within the body as 'tendencies' to act out particular refrains of movement and expression when encountering circumstances that resonate with past events. This matters because it points to how the virtual offers potential functions amongst the many things that may be done in any encounter. Sexuality and desire, then, might be thought of as various series of refrains of affect. These refrains might throw bodies together in diverse combinations of affect (affectionate, dominating, anxious, disgusting, disgusted, tender, caring, caressing, playful, grabbing, holding, thrusting, stroking…). Yet, such affects cannot be collapsed to an inevitable story, whether Oedipal or heteronormative, even if they may resonate with such stories.

I think that these conceptualisations of affect suggest that there is a political field to be explored. On such a field our attention might be directed towards the emergence of affect and its conversion into various kinds of sexualised relations, and towards the feedback of these relations into the field of potential, an incipient organisation of bodies' affects and desires. Here, the role of events in modifying the field of potential for perception and action is important. An event of homophobic harassment on a city street, for example, feeds into the field of potential affects. The visceral dimension of trauma gathers in the body to be encountered again in new circumstances as a trigger to feeling and action. The event of harassment comes to resonate with certain situations that the person who suffers such harassment might find themselves in at a later date. Although impossible to predict in advance, there are many aspects of the latter encounter that might resonate with the former: markers of place, the tone of voice of somebody shouting, an ordering of events, a small gesture, glare or facial expression. Such resonation enters the process of selection of how to affect and be affected by other bodies. There are many kinds of potential affects – defensive, watchful, aggressive, walking with purposefulness, bringing one's friends closer around you – and these affects are neither mutually exclusive, nor an exhaustive list. Rather, they are among the multitude of combinations of things that might be done in many different styles. Moreover, such an event of harassment, if witnessed by others or reported upon in circulated media, might merge into an "implicit urban memory bank" (Connolly 2002, 31). Indeed, multisensory media such as TV and film are particularly adept at invoking, circulating and modifying such virtual memories (Connolly 2002, 36; see also Massumi 2002; Thrift 2004). The event merges with other affective virtual memories and enters into the perceptive responses to events yet to come, and to stories yet to be retold.

It becomes important to think, then, about how virtual memories might be shaped – a politics of the intervention into their formation – and also how the social and material contexts of their actualisation are shaped, in turn conditioning the body's ability to act and relate. Such a politics might be termed a 'politics of the imperceptible' (Grosz 2005b) or a micropolitics (Connolly 2002, 20–21) that acts to connect and organise bodies' attachments, memories, practices, routines, affects and emotions. Micropolitics are the collective deployment of the techniques of the self that help define the sensibility of one's perceptions, beliefs, and judgements. It is clear that such micropolitics play a part in defining different bodies' capacities for action. Not only do these micropolitics take place through everyday institutionalised settings, but they also take place through the mass media. As Massumi (2002, 42–43) comments, politicians (as well as movie-makers, television programme makers, and a host of different kinds of web-based content producers) are particularly adept at harnessing the power of 'transduction', the transmission of affect so that it can be actualised differently in different places, for different bodies. Transduction maximises the indeterminacy of the images and sounds of movements, timings, intonations and expressions that are being transmitted. By this maximising of their virtuality, the reception of these images and sounds in many different places is able to actualise their affects so that they resonate with a wide variety of narratives, memories, emotions, practices and institutionalised relationships. These processes raise questions about the relationship between affect and political beliefs and investments. It becomes important to explore the flashpoints of tension between affects and explicit beliefs and aspirations (Connolly 2002, 18–21). For example, bodies, especially when plugged into modern circuits of media (TV, film, music, cyberspace) or when caught up in the joys of alcohol-fuelled bar and club nights, might not know their own desiring limits; yet the same bodies might also have to negotiate the inheritances of living within conservative and/or religious families and communities.

Immanence and Imperceptibility

One very important question regarding what a queer politics of affect might be like concerns how change is envisaged. Judith Butler's (see i.a. 1990; 1993; 1997; 1999; 2004) ideas on performativity have placed the problematisation of recognition and identity at the forefront of queer interrogations of sexuality. Within such understandings, subjects seek to constitute themselves by striving to have their identity affirmed through being recognised and desired by other subjects. This is a politics that has among its determinate objectives the production and affirmation of new identities and new subject positions. These are worthy and important goals, but there is a need to be able to do other things too. As Braidotti (2002) and Grosz (2005a) note, Butler's understanding of the performative is primarily discursive. The changes that performative acts might bring are conceptual changes – changes in identity – that involve a negative difference from another identity, even in the pursuit of becoming 'an other of the Other' (for example, becoming something other than 'homosexual' if this is understood only as the defining Other of 'heterosexual'). By attempting to create new identities, such performative acts act as if these concepts can account for desires and practices, or as if desire arises from identified bodies, so all we need are new identities. Moreover, by seeking social validation for these performatively produced identities and subject positions, such a political process re-invests in the status and values of those who are able to validate – the socially dominant (Grosz 2005, 167; Haritaworn, Chapter 8, this volume).

What might also be needed, then, is to be able to attend to what a body can do and how bodies can have a wider range of affects than can be conceived of within a primarily discursive proliferation of new identities. Desire is messy and does not obey the obligations of identity – think, for example, of men who identify as heterosexual but who have sex with other men. Elizabeth Grosz (2005b) attempts to outline such a vision when she sketches out the possibilities for what she calls a 'politics of the imperceptible'. For Grosz, change is not first and foremost about changing one's identity. Neither does she put much store by visions of change where one sets out the concept of the desired end state (a new identity, a new mode of social organisation etc.) and only then implements it. Rather, change is 'immanent' – arising from within situations, from events, from within the movements, actions and desires of bodies. A politics of the imperceptible, then, attends to affect and the potential to do new things that arise immanently within the event. It deals with forces that are affective and forces that are materialistic. The forces that assemble bodies and that enable bodies to do things are not necessarily available to perception. They include forces above and below the thresholds of human perception. Some of these forces are too quick to be perceived, others are subliminal, and others are too multiplicitous for us to deal with in consciousness. Yet other forces are super-human social, material, technological and economic forces. Clearly, these forces are more than subjective: change does not necessarily originate from individual subjects wishing it to happen. Rather, Grosz suggests that if we seek social and political change, then we might learn to harness these sub- and super- human forces. By doing so, we might increase the power of action of (our) bodies – we might increase (our) affects.

This kind of politics does not aim for specific goals, but rather attempts to embrace an unpredictable future. Grosz's thinking with respect to the indeterminacy that affect affords us is exemplified in her discussion of sexuality (2005a). She considers female, especially lesbian, desires and practices to be open to experimentation and, therefore, immanent kinds of change. Even today, there is a relative shortage of language to describe female and especially lesbian sexual practices in medical, sexological and even popular discourse. However, for Grosz the task is not to make lesbian sexual experiences more recognisable. This would define the sexual pleasures in question, fixing them within (often male and heterosexual dominated) forms of knowledge, making them functional. Such precisely defined pleasures would be pinned down so that pleasures, practices and experiences that varied from them would become abnormal, a deviation from the norm, the definition. So, rather than more precisely define and name lesbian sexuality, Grosz suggest it is perhaps better to affirm its vagueness – the way it is fleeting, intense, plastic (2005a, 226). By affirming this openness, Grosz is gesturing towards how sexuality offers a provisional multiplicity of affect. She is attempting to open up a space where sexuality can be explored – where we can find our sexuality in our next encounter (2005b, 213). In this sense, rather than being trapped by social norms, discursive definitions, and our sense of where our past sexual experiences locate us on some kind of pre-existing map of sexual identities, we might instead explore an autonomous sexuality that embraces the potentials offered by events to come. It is through this process of exploration that the conditions and territories for subjectivity, desire and action might be recreated. While thinking about sexuality in terms of its futurity has some profound implications for how we might think about sexual difference and its relation to gender[5] (Grosz 2005b), it also presents us with the potential to recompose what our bodies can do.

In the light of feminist-informed critiques that Deleuzian theories of affect abandon both the subject (Thien 2005) and emotions and feelings (see Jacobs and Nash 2003), Grosz's ideas offer a way of thinking about affect while also attending to both the subject and to the 'feelings' of pleasure, desire, fear, anxiety, guilt, happiness etc associated with sexuality. If we follow Grosz, however, the subject can no longer be the main source of (sexual and political) knowledge and action. Neither can subjectivity and its recognition be the only thing worth fighting for in order that one's sexuality is recognisable and reciprocated by others. Rather, subjectivity is something that is still at stake. Subjectivity – being able to desire, take pleasures, know and speak – is an effect of the processes by which bodies and their socialities are continually recomposed. Subjectivity also feeds back into these processes, but it is important to remember that there are many other such sub- and super-human

---

5   These implications have only recently started to be explored. Grosz (2005a, 212–214) asks whether the idea of 'gender' does the same conceptual work as 'sex' if we acknowledge that there is no biological essence to sex.

dimensions to the emergence of affect. One of the problems of heteronormative understandings of sexuality is that desire is presumed to spring forth from already sexed bodies. Given that theories of affect problematise the body that desires, this also throws open the question of the subject that desires, for it is difficult to disentangle the subject from the body. The subject, like the body, becomes indeterminate: thinking about subjectivity demands a questioning of the affective forces that compose and comprise the potentials for subjectivity.

This is also true of feelings and emotion. Affect, in the Deleuzian sense, differs from emotion in that it is not yet actualised, and has not yet taken its place in our narrative lines of action and reaction – the expectations and normativity of how to feel (Massumi 2002, 28). Thinking about affect asks us to keep open the potentials for different powers of feeling. It allows for being able to feel different emotions than we might be expected to in response to a given situation or an interaction with other bodies; for example, the potential to feel something other than jealousy in polyamorous relationships might help such relationships to flourish. Thinking about affect, then, does not abandon emotion, but asks us not to prejudge what emotions we might be capable of. It also asks us to attend to the processes of how affect becomes actualised as emotion, for this might also suggest how affect/emotion is regulated. In this sense, theories of affect do not compete with conceptions of emotion, but complement them and deepen them.

Queer Politics, Ethics and Critique

This chapter has sought to attend to the potential intersections between queer theories and theories of affect, as well as thinking through some of the tensions involved in bringing these theories together. Concepts such as 'virtual memory', 'tendencies', and the feedback between the actual and the virtual can aid our understandings of politicised differences in capabilities for action. Indeed, such ideas can deepen our understandings of how events resonate over time, and of how sexuality might be thought of in terms of refrains of affect, repetitive but not inevitably the same at each repetition. Ideas about affect are also suggestive of different ways of thinking about social and political change. Rather than focusing on the recognition of specific queer identities, we might heed Elizabeth Grosz's (2005a; 2005b) call to cultivate the indeterminate future of our sexualities, a future that arises from within the eventfulness of our social and sexual practices. I would argue, however, that while Grosz discusses the exploratory embracing of the plasticity of pleasures and desires with respect to queer (predominantly lesbian) sexuality, such an openness to new affects is perhaps just as necessary among heterosexuals. This is especially so given the proliferation of pornographic images, ideas and pleasures, whose circulation proffers a rather limited set of affects that often, but not always, tend to become actualised as shame, humiliation, objectification, and male domination. Heterosexuals, too, need to cultivate an openness to the multiplicity of affect in order to explore new pleasures, but also to cultivate reparative relations with queer bodies.

Such a call suggests that there might be a need to explore a more reconciliatory relationship between critical politics and ethical and reparative stances. Some proponents of theories of affect question whether critical stances can add anything new to the world, or whether a critical stance allows an openness to the affective potentials afforded by events. Critique is thought to unintentionally validate, through the attention it pays to them, the oppressions it seeks to overturn. Massumi (2002, 12) warns that there is a danger in organising political action around questions of critical representation: such political action takes place in relation only to what is taken as a given political order, what is already recognised. It cannot sketch out other ways of living. Sedgwick (2003) describes critique as dominated by a paranoid tendency that sees specific forms of oppression everywhere in all relations. A paranoid obsession with opposition to and exposure of oppression sees this oppression as inevitable and implacable, which has the paradoxical effect of allowing no scope for things to be done about it.

In addition to taking inspiration from Melanie Klein, Sedgwick (2003) brings to bear the theories of affect developed by Silvan Tomkins, and it is interesting that these theories of affect connect so well with the questions raised by Deleuzian ones. Sedgwick advocates taking a 'reparative'[6] stance rather than only a straightforwardly critical one. A reparative stance, while recognising oppressions, does not spread the paranoid idea that they are inevitable. Rather, it seeks to attend to aggressive or oppressive bodies in an attempt to assemble or repair their relations into something coherent, although not necessarily like a pre-existing body or assemblage (2003, 128). Taking a reparative position entails seeking nourishment, comfort and pleasure from the relations that one fosters with other bodies (2003, 137). A reparative position adds new and various affects, and allows for the surprise that events bring (2003, 146).

While such a call implores us to move away from the paranoid constitution of queer thought (Sedgwick 2003, 146), it does not follow that we must separate ethical and reparative projects from critical ones. To do so would risk relegating sexualised, racialised and other underprivileged Others solely to the realm of critical investigation through narrative and discursive means in a narrow sense. Without critique, how would we perceive power relations? How would we know whether some bodies are excluded from a full participation in events? Immanent ethical judgements might not be sufficient to this task if the judgements immanent to a situation are already prejudicial to some bodies. An ethical openness to the potentials of the event needs to be coupled with an awareness that violence subsists amongst the potentials arising from the event and that this violence need not yet be indexed to any pre-given narrative. Such violence might be described as *senseless* violence.

Although we have seen in this chapter the value of an openness to affect, then, it is important that critique is not jettisoned, but rather that some ways of bringing together ethical and political sensibilities are found. Hemmings (2005, 558) suggests that there are many modes of critical thinking that already have a tradition of fostering ambivalent and indeterminate ideas, affects and significations that are not simply oppositional and that are not simply precoded into dominant representational structures. Queer considerations of shame, for example, do not advocate a passing over into its opposite – pride. Rather, they enable queer bodies to resonate with the

---

6   Such reparative stances should not be confused with the 'reparative therapies' that are promoted as being able to change 'sexual orientation' in order to 'heal homosexuality'.

effects of heterosexual dominance *and* to create community through connection, empathy and shared experience (2005, 549–550). This concern with how shame can open up to new, un-narrated affects – affects that cannot be determined in advance – suggests that a full range of affects must be attended to, not just positive ones. Critical thinking is capable of thinking otherwise, of thinking anew, but it is affect that enlivens it, enabling it to add new desires to the world.

Acknowledgements

I would like to thank Róisín Ryan-Flood and Ben Anderson for their inspiration with respect to many of the ideas in this chapter.

Chapter 5

Developmental Desire and/ or Transnational Jouissance: Re-formulating Sexual Subjectivities in Transcultural Contact Zones[1]

Hanna Hacker

One of many initial considerations for this chapter is rooted in my interest in the subjective experience of geographical and cultural rupture. How is such rupture able to shake the individual sexual/erotic sense of self (and, also, do so productively), so that 'afterward' it seems different, perhaps less universalised than it had before?

Living somewhere for a time, experiencing life in transnational or diasporic encounter zones certainly forces us to come to terms with the experience of quite different local settings and specifics in our everyday practice. What new theories of the self do these encounters generate in the field of the reformulation of the sexualised 'I'? What structural conditions enable, and also demand, such redefinitions and their representation?

To quote a straightforward passage from a pilot interview I conducted on this theme (the speaker, travelling on an Austrian passport, is located at the intersection of Western lesbian feminism, Africa-based straight development work, and international academic bonds with queer and cultural studies):

Question: What connection do you see personally between geography and sexuality?

Answer: Well, I've thought about that for so long that now I know pretty much how I want to answer. First, I would say that everyone, including myself, relies an amazing amount on their surroundings in performing their sexual or erotic subjectivity. This is, of course, often, or even always geopolitically marked. But you don't really see that until you live 'somewhere else' for a longer time. I am lesbian differently in Berlin than I am in Paris or Prague. Being 'out' is valued differently in each context; codes are weighted in a completely different way, and I can very well imagine that one city

---

1   An earlier version of this chapter was presented as a paper at the conference *Queer Matters* at King's College London, May 2004. I thank Lisa Rosenblatt for her patience in constantly reworking the translation.

makes me more femme and the other makes me more butch and in still another city I can't even remember what these labels might have once meant for me. Or Ouaga... what sexual performing is possible there anyway? In Vienna I can't be anything other than lesbian, and that's it. In any case, as a whole, these are experiences that I bring together under the term 'queerness'. The term 'queer' is at my disposal to use as an intelligible category. That's very important, and I can use it just as much for explaining myself as for representing myself, and I can also use it so that my self doesn't somehow totally fall apart.[2]

Using 'subjectivity' and 'space' as two key reference points, this chapter explores and reviews different dimensions of queer transnational and transcultural desire. First, I ask how these desires relate to practices in international development discourse, and vice versa. In what ways is 'development' tied to structures of desire? Can it be said that desire functions in development, and if so, how and why does it function? Subsequently, I raise the question of whether there is transnational (queer) desire that goes beyond developmental thinking altogether. What are the (textual) genealogies about that articulate pleasure and enjoyment in transcultural contact zones, and how pleasurable are they?

One leitmotif arises from the question of how desire is (or can be) made discursive when dealing with relationships of global inequity and relationships of transnationality, translocality and transculturality.[3] I work with theoretical categories from the field of postcolonial studies; with feminist developments that have furthered debates on hybridity, métissage, nomadology; and with approaches from queer theories, which hold a central position in this chapter. Just like my interviewee, I use the term *queer* as an articulate notion comprising, for the most part, a radically deconstructive attitude towards sexuality, genders and desire; sometimes, though, I do refer to texts that employ *queer* as an umbrella term for any gay and lesbian identity politics.

Particularly interesting for the purpose of my argument is Cindy Patton and Benigno Sánchez-Eppler's challenge to understand 'place' itself as a 'form of desire' and to interrogate the relationship of identities and spaces, "the intersection, the collision, the slippage between body-places" (Sánchez-Eppler and Patton 2000, 4). In a slight modification of this challenge, in the following I consider the production of transcultural sites as a form of desire: I put forth two structural patterns, which, in my opinion, contribute significantly to the production, the narrating of transnationality and sexual selves. In brief, I call these two figures *developmental desire* and *transnational jouissance*. My aim is to show why these patterns are highly relevant for both political practice as well as theory.

---

2   Unpublished interview, author and N.N., Feb. 2004.

3   These three concepts are not simply interchangeable with one another. Nonetheless, I will not go any further into differentiating them here, as they often have a similar content for the course of my argument.

Thus, some concluding reflections will lead us back to the question of how to construct spatial notions for dissident or subversive transnational desires. In what ways can we conceive passion, alliances and bodies in constant transition within this field of *-zones* and *-scapes?*

Development: Framing the Issue

The idea of 'development' functions, as it were, as an important orientation point in post-World War II history, and in postcolonialism in general. For its critics, 'development' is considered a largely un-deconstructed certainty found in the social imagination, one of the most powerfully effective master narratives in the era of globalisation and corresponding power/knowledge systems. In my opinion, the conclusions of development critics of the 1980s and 1990s are still valid: national (and multinational) actors unfolded control of discourse as an 'anti-politics' in cooperation with the machinery of development. In other words, development is the eradication of the possibility to think and speak in political terms (Sachs 1992; Ferguson 1994; Escobar 1995).

"Scarcely twenty years were enough to make two billion people define themselves as underdeveloped," says the voice-over in Trinh T. Minh-ha's film *Reassemblage,* an interrogation of media representations of 'underdeveloped' regions – more concretely, the Casamance in Senegal.[4] *Reassemblage* is now more than twenty years old; nonetheless, in my opinion it is still possible to say that the 'global North', or the 'global West', has a monopoly on representation in terms of the depiction and perception of the 'non-West', the 'global South', and that an important concentration of this monopoly lies in the trope *development versus underdevelopment.* The social and economical 'backwardness' of the 'underdeveloped' cultures, bodies and persons is thereby depicted and perceived not as an effect of political systems, but instead as something that can be bureaucratically drawn up, improved and overcome. This is not so much about the classical image of poverty and those 'affected' by it, but much more about those marked by 'underdevelopment'; the concern is less with exoticised desire and subordination, and more with efficient 'development' and efficient organisation of the sexual.

I would like to describe the specific ways of functioning of desire/demand – ways of functioning that combine with these development processes and their depiction – as *globalizing development desires.*[5] My theses related to this, in brief, are as follows. Development is tied with structures of desire/demand on all sides of the development encounter. These desires/demands have an economic and a social dimension that has to do with the 'demand' side, with demanding and being forced. They also have, perhaps primarily, a sexual dimension and a gendered dimension. There is no reason to assume that we are always dealing here with heteronormative demands or are always speaking of heteronormalised desires. On the contrary, I would postulate that it is quite definitely also about 'perverse' desire, 'queer' desire.

---

4 *Reassemblage*, Dir: Trinh T. Minh-Ha, 1982. The quote can originally be found in Illich 1980.

5 I formulated my first thoughts on this theme in Hacker 2003.

A developmental approach to desires/demands/wishes is classically one which demands, as it were, that wishes must orient towards possibilities and must follow a rational logic. Or at least that is the pretence … François de Negroni points out that perhaps the entire process of developmental aid is complexly permeated with desires and the wish for enjoyment, both of which the main actors want simultaneously to maintain and to leave unfulfilled:

> Il y aurait finalement une connivence tacite entre les 'opérateurs' du Nord et les 'décideurs' du Sud pour refuser le développement. Le développeur, tout comme l'amoureux selon Lacan, proposant quelque chose qu'il n'a pas à quelqu'un qui n'en veut pas.
>
> (Negroni 1992, 232)

> [There would be a silent agreement between the 'operators' of the North and the 'decision-makers' of the South to reject development. A situation in which the developer, just like the lover for Lacan, is offering something that he does not have to someone who does not want it.]

One counterpart to this assessment at the unconscious level is the depiction of reality by Gustavo Esteva. He addresses development aid from an indigenous position and claims: "In der Welt, in der ich lebe, ist die enge Beziehung zwischen Möglichkeiten und Bedürfnissen Voraussetzung für das Überleben" ["In the world in which I live, the close relationship between possibilities and needs is the condition for survival"] (Esteva 1992, 77).

It seems to me that, at least at a manifest level, the following rule applies: for those who are hungry – that is, those actors in development who are defined as 'hungry' – their wish should be for bread and not, for example, sex shops. Those defined as 'poor' in the third world (also in the third worlds of the first world), with all good sense, do not have any sexual subjectivity, not to mention queer demands. In the more progressive version, when development finally deals with sexual and gender models that 'less developed' groups of people design for themselves, then this development encounter is also about the intervention of rational planning and the formation of 'sensitised' and 'empowered' subjects who only desire what is feasible and not something else.[6]

"I do not intend to speak about – just speak nearby": one of the most famous quotes on postmodern speaking, originating from Hélène Cixous, likewise repeatedly orchestrates *Reassemblage,* Trinh T. Minh-Ha's anti-development film. But *dev-speak* (as development critics call the normalising language of development) is not really a speaking 'about'; *dev-speak* is a speaking 'in order to'. It becomes important to ask, then, when the relationships between sexual subjectivity and geopolitics are the issue, how does queer critique encounter this *dev-speak*? To what extent do queer text forms repeat it, and to what extent do they deconstruct it?

---

6   For a critical discussion of the concept of the subject in 'development', see also Triantifillou and Nielson 2001 and Lazreg 2002. Examples of the quite daring thematic approaches to sexual (queer) desire in development discourse, which, however, ultimately lead back to thinking in terms of planning, are Jolly 2000 and Jolly 2002.

Desire, Demand: Form(ulat)ing the subject

Towards the end of my field studies in the Russian Federation I had a vivid dream, writes the US-American Laurie Essig in 1999, a vision as it were: I had to cross a border and eventually passed as a man and a Russian, whereas at the same time I remained a woman and an American (Essig 1999, 161). In *Queer in Russia* – one of the most fascinating and controversial recent texts on the theme of sexual formations and cultural transitions[7] – the movements of the 'insider from abroad' are motivated by dense political and theoretical claims. In a space saturated by national and sexual transformations, she holds steadfast to her conviction that it is possible and desirable to define sex as queering the public realm. The desire of the woman who writes is represented in both a transgressive and reflective way; development, on the contrary, belongs to the Others' desire. In an era of postsocialist upheaval, she has no choice but to act in solidarity on the side of the Russian lesbian and gay research subjects and speak out critically against Western infringements, even if they come from her 'own' country and her own lesbian community.

What can be generalised about this example is that clearly not only capitalist consumption, imperialism and media globalisation are at issue, but also that the effectiveness of the development machine becomes manifest in LGBT (and feminist) realms of encounters: lesbian and gay NGOs practice institutional strategies for fundraising and learn to juggle with the jargon of lobbying for human rights. Feminist foundations and homosexual associations act as transnational donors in the contact zone between these 'developed' donors and 'not yet so well developed' lesbian, gay and feminist recipients. This leads to the creation of subject positions that had not existed previously.

A central passage from *Queer in Russia* crassly brings to light the globalisation aspects of *money makes identity*: at a start of all of the development relationships, practicing homosexuals allegedly did not give a name to their own sexual identities. They transformed along the way, abandoning the then collectively cultivated term 'sexual minorities', in the willingness for an, in part, tactical self-identification as 'lesbians' and 'gays'. With joyous coming-out workshops and characteristically Western cultural festivals, the US and German community developers imported into Russia material aspects of Westernising, closely related to sexual self-understanding and political style. A leading figure in the Russian Federation's queer culture said in 1990 in a conversation on this problem of dealing with local self-naming, alliances and identities: "What do I care what I call myself? If the Germans and Americans will give us faxes and computers if I'm a lesbian, then I'm a lesbian" (Essig 1999, 127).

Though not nearly as reflective as Essig in terms of attempting a writing movement situated 'nearby' or 'beyond development,' there are various field studies relating to 'North/South' encounters that read quite similarly in terms of content and that come to comparable results. They all analyze the re-formation of desire/demand structures and sexual identities through the intervention of US and Western European

---

7   For a controversial response, which charges Essig with harboring her own Westernising tendencies, see Braer 2002, particularly pp. 509–512. Many thanks to Adi Kuntsman for the information about this essay.

developers in 'developmentally needy' and 'developmentally capable' regions, for example, of the African continent.

Canadian anthropologist Robert Lorway in his studies of (homo-)sexual cultures in Namibia describes workshops run by the (Northern) NGO *The Rainbow Project* in the late 1990s, in which the participants were "*taught* about their sexualities: how to recognize and distinguish between gay, lesbian, bisexual and transgender identities" (Lorway, forthcoming, emphasis in original; see also Lorway 2003).[8] Of course, as he emphasises, at issue is not a simple one-way import of Western imperialism. In this process, identities will always be reshaped, parodied and actively performed. Still, desire and economic power are closely intertwined here. Poor black youths adopt new opportunities to forge solidarities, and benefit economically from networks based on their 'newly' named sexual identity. It is Western power that produces a tight association of emotional and economic desire within the newly created 'Namibian queer subject'. Lorway (forthcoming) concludes, "Relationships with foreigners have become both the imagined and real grounds for not only escaping poverty and oppression but also for fully uncovering and valuing the self. With so much at stake for local actors, unsafe sex with foreigners often becomes worth the risk…"

In the field of 'modernised' sexualities and HIV/AIDS policy, anthropologist Vinh Kim Nguyen shows, based on an example of 'local' actors from Abidjan and Ouagadougou, the ways in which and on what grounds Western confessional technologies produce uniform biographical narratives (Nguyen 2002; see also Nguyen 2005). Nguyen compares discourses of European development NGOs in the field of HIV prevention with (autobiographical) narratives of people affected by HIV in the same region. He maintains that, in his study, mostly French run AIDS prevention measures produce unified, 'global' biographical narrations among all those who, for example, when speaking in a self-help group or in a media context, increasingly adjust their 'positive self' to a Western model of public narratives. This Western model sets out how to speak one's own illness and the struggle against it, and how to conform to the maxim of *silence=death* (Nguyen 2002; see also Nguyen 2005).[9] Clearly obvious here is that these specific forms of speech about infected bodies have the greatest tactical relevance for access to medical care and other developmental means.

The 'queer' question here would be: what does that mean? In my reading, it means, among other things, that development functions on the basis of economic and sexualised desire (on all sides of the development encounter). Perhaps desire is the only thing about development that actually works, since postdevelopment theorists do quite convincingly show that failure represents a constitutive characteristic of development acts. One of the reasons that desire functions in development might be that any development encounter is equally about 'functional' and 'excessive' desires.

---

8   I thank the author for his permission to quote from the unpublished papers.

9   However, it seems that Nguyen did not sufficiently reflect on his own depiction of the affected persons and their environments. For example, he always introduces the 'locals' with their first names, and describes their surroundings, the apartments, the kitchens, the bars in West African metropolises as local colour, as it were, for a readership whom the author apparently assumes is 'not local'.

Thus it is a question of desire/demand based on efficiency, or based on a rational correspondence between the desire in the imaginary and the possibility in the material; but also (often) about desire in the sense of an excess, a surplus, a transgression of the appropriateness of wishes. Thus it does not seem very surprising that queer theoretical, critical, political reflection has great difficulty in escaping or countering this development/desire logic. The problem is not only a general one, namely that dev-think marks our globalised minds, which possibly all tend toward the previously described idea that socially disadvantaged people should orient their wishes within the borders of the possible. It is a particular difficulty for 'queer' because queer is based on 'desire' and *dev-speak* simultaneously appropriates 'desire.'

Pleasure, Danger: Transnational Jouissance

At this point, I would like to return to the queer 'I' from the interview quotation at the beginning of this chapter. This self certainly doesn't want to 'develop'. Instead, it is talking of transcultural confusions, of various forms of pleasure, of ethnic food and gender fuck, and implicitly of a radical chic cosmopolitan sense of being on the go.

Question: What connection do you see personally between geography and sexuality?

Answer: … Well, I think that in 'my' cultural rupture, in the confusion of returning after several years in a country of the South, I came up with the following saying: 'since I went away, all my friends have become transgender, they fuck with dildos and eat with chopsticks.' The accurate observation behind this, actually, vaguely confused me at first, and then this vagueness solidified as a bon mot. Eventually I, too, became bored by it. Not until I let it out for the last time, consciously resolute – 'that's it, the last time' – did I also listen to myself more clearly than ever before and comprehend all of the transgressions and appropriations that are in it. And I understood that all three are important and consistent, and actually no joke: the gender transgression, the slightly controversial sexual practice that is entirely true to consumer culture, and then, the way of dealing with consumption that is, at the least, ambivalent, with jouissance in the ethnic difference. Globalized lust, exploitation, and ethnic food fashions: that should be looked at a lot more closely, don't you think?[10]

What representational strategies speak of pleasure, of the fulfilment of wishes, of satisfaction in transcultural or transmigrant spaces, without being absorbed by developmental desire? Are there specific 'queer' figures of what I call here *transnational enjoyment* or *transnational jouissance*?

---

10 Unpublished interview, author and N.N., Feb. 2004.

In a popular lesbian-queer murder mystery recently published in Germany, the western European first-person narrator strolls through an eastern European country and abandons herself to the pleasure of the 'other' culture. She is young, super cool, lesbian (of course), naturally without commitments, *generation q(ueer)* and largely free of problems. The self-chosen, temporary life in a postsocialist society comprises pure enjoyment: her job, the master's thesis, casual sex, lounging around on the scene, eating, drinking, porno surfing and thermal baths.[11]

[G]anz im Vertrauen: In Deutschland hätte ich einen solchen Job nie bekommen. In Deutschland hätte ich Qualifikationen gebraucht. Diesen Job mache ich ohne, und nichts funktioniert. Da meine Vorgesetzten ebenfalls nicht qualifiziert sind, macht das weiter nichts.   (Kremmler 2001, 33)

[[C]ompletely confidential: In Germany I would have never gotten this kind of job. In Germany I would have needed qualifications. I do this job without any, and nothing works. Since my superiors are also not qualified, it doesn't really matter.]

Die Chefin ist atemberaubend [and, additionally, a Bulgarian academic and Kristeva expert] … Ich mag meinen Job … Markt ist hier toll … Die Konditoreien hier können dich vor Glück zum Weinen bringen … Wenn du nach zwei Stunden Thermalbad an einem frostklirrenden sonnigen Wintertag aus dem Király kommst, weißt du, was Glück ist.
                                                   (Kremmler 2001, 30, 41, 42 and 46)

[The boss is breathtaking … I like my job … The market here is great … The cake shops are so fantastic they can bring tears to your eyes … When you get out of Király after two hours in the thermal baths, on a crisp and frosty sunny winter day, you know what happiness is.]

In this text, as I read it, there is indeed no development. Westernising is not a problem for this hero; post/colonial desire is no big deal. She enjoys, she comes, she does. Foreignness doesn't get to her and, if perceived at all, it is a source of pleasant titillation. Transnational *pleasure-speak* simply seems possible. In the between spaces, however – and this is what I will argue in the following – as a prototypical strategy of perception and representation, it is about the experience of violence, gay-bashing and death threats against women. It is no coincidence that the corresponding literary model comes from the detective story genre. The tale of crime and the mode of 'detecting' deadly acts of violence orchestrate the entire *pleasure-speak*. It is the homophobic perpetrator who still defines the genre that legitimises the text.

The idea of pleasure that transgresses cultural and national borders is, itself, riddled with ruptures that grow wider the more closely they are looked at. This applies already to the term *transnational jouissance,* which I initially used heuristically. Renata Salecl and Slavoj Žižek have transferred the Lacanian concept of *jouissance* to issues of nationalism and national envy (see Žižek 1991; Salecl 1997). Enjoyment

---

11 Kremmler's second murder mystery, set in Sydney during the *Gay Games* 2001, operates with the thematic field of transnationality, migration and fleeing in an entirely different way once again, and in my opinion, in a much more differentiated and 'political' manner (Kremmler 2003). Without a doubt, both stories are thrilling to read!

is always the pleasure of 'more': the usufruct, the yield, the profit that others – in any case, supposedly – gain from enjoyment. According to Žižek and Salecl, militant nationalism and the hatred of members of 'foreign' ethnic groups is tightly woven with the phantasm that these strangers have access to (more) enjoyment, which is usually characterised as perverse, that is not available to 'us,' our nation, our culture. Additionally, the foreigners, the Others, will actively rob us of our own opportunities for enjoyment, for example, take away 'our' women, seduce 'our' children.

Historical analysis certainly does offer an important correspondence to this psychoanalytical interpretation with the concept of *colonial desire*. Robert Young understands *colonial desire* as "a covert but insistent obsession with transgressive, inter-racial sex, hybridity and miscegenation" (Young 1995, xii). The obsessive fantasy of blending forms the clandestine flipside to colonialist 'race' theories, which, on the surface, concentrate entirely on separating the 'races' and keeping them at a distance. Postcolonial theories as a whole do insist on the thesis that racism is in no way based solely on violence, but instead, is at the same time allied with desire for that which is constructed as 'racially other' and excluded or rejected. In spatial terms, Julia Kristeva's *abjection,* the process of producing a "rejected from which one does not part", is also something like a material geographical relation between historical subjects, on the one hand, and the excluded geographical sites that participate in constituting their subjectivity, on the other (Kristeva 1982, 4; see also McClintock 1995, 71–74).

The area that is perhaps very quickly associated with the idea of international amusement and its potential for sexual pleasure, namely that of tourism, can likewise hardly be identified as politically and historically neutral. Tourism, in essence, is tied with sexual norms, postulates Jon Binnie: "all tourism is sex tourism to the extent that tourist practices are sexualized and embody sexualized values" (Binnie 2004, 10). Lesbian and gay tourism has become a booming business: queer travel naturally includes the services rendered by the 'local' or immigrant population – jobs in catering, for example, as well as sex work.[12] In the entrepreneurial city, queer and ethnic venues serve equally as sites "for consumption at a cosmopolitan buffet" (Rushbrook 2002, 188), catering for gay and straight tourists alike, notes Dereka Rushbrook.

The search for sexual revelation in foreign parts is, just like development, a legacy and effect of colonialism. In an analysis of what he calls 'homosexual sexual tourism', Jarrod Hayes feels that it is important to learn how to navigate between the Scylla of the "colonialist implications of sexual tourism" (in the broadest sense) and the Charybdis of the "sexual conservatism of most critiques of it" (Hayes 2000, 26–7). This dilemma, whereby it is the transgressive drafts (in the sense of 'project', 'concept', 'outline') that become truly imprisoned in the contradictory reality of hierarchical relations, does not apply solely to queer tourism. It also applies to many further movements of the interim, transition, passing: the cosmopolitan flaneurs and flaneuses, a beloved figure of urban cultural history of (Western) modernity;

---

12 See the contributions in Puar 2002b. Incidentally, in my opinion, so-called 'female romance tourism' does not entirely lack queer aspects, as conventional gender attributions are shifted here; see i.a. Dahles 1998 and Phillips 2002.

the transmigrants, those who design social spaces between cultures of origin and migration culture; as well as Rosi Braidotti's feminist nomadic subjects, who reject the homeland and terminate their loyalty to the hegemonic culture; Gloria Anzaldúa's *mestiza* in the borderlands between geographical, ethnic, sexual affiliation; and others.[13] Each of these figurations speaks of transgressive pleasures and desires that cross borders and disrupt the functionality of development relationships, yet each is also implicated in violences, ruptures, and appropriations.

So, if drafts of transient passion that are outside of the 'development' logic outlined here tend, nonetheless, to be traversed or fractured by structures of racist, economic and symbolic violence, what does that mean in a queer perspective? What follows from such readings? To formulate a few speculative theses: in transnational contexts, queer *pleasure-speak* possibly serves as a strategy against *dev-speak*; yet, without an awareness of that which is threatening, of the annoyance, of the non-fun, such queer transnationality does not wholly function in the era of globalisation and development. *Developmental desire* means being able to move beyond the failure of 'development'; transnational desire includes, at least potentially, having to perceive and represent the simultaneous existence of fun and 'non-fun'. Donna Haraway's famous 'monsters' and their 'promises' – disturbances, destructions and consumption (see Haraway 1992) – orchestrate drafts, in theory and in practice, for transient spaces and the longing for them. Susan Pointon, in her consideration of Japanese hentai and its U.S. consumption, even suggests that the 'transcultural orgasm' of erotic pleasure both rests upon and is interrupted by violent 'apocalypse' (Pointon 1997). Such transgressive pleasures tap into deep anxiety and guilt, simultaneously troubling normative narratives of sexuality and perceiving the troubling nature of the 'imperialist' relationship between the US and Japan.

In the encounter with transcultural desire, several aspects of the transient, of the connecting self, appear endangered and threatened: the sexual dimension of the self, as well as the cultural, social, ethnic and other dimensions. Paradoxically, queer reflections, which actually do welcome the fragmentary and the fragmented, also contribute, in turn, to mending the ruptures. My previously quoted interview partner comments here on the function of the term *queer*: "The term 'queer' is at my disposal … as an intelligible category … and I can also use it so that my self doesn't somehow totally fall apart".[14]

Of –zones and –scapes: A Forward Look Beyond

How, then, would we come to speak of passion and alliances that undermine *dev-speak* as well as face the ambivalences of pleasure-speak and its threats?[15] A more

---

13 Mentioning only the 'classical' titles here, see Anzaldúa 1987; Braidotti 1994; Glick Schiller et al. 1995; Benjamin 1999. On the relationship between (gender) transgression and violence in the European modern age, see also Hacker 2002.

14 Unpublished interview, author and N.N., Feb. 2004.

15 This issue has witnessed a surprising update by the new trend in parts of development and social work to emphasise particularly the perspective of 'sexual pleasure' in the engagement for 'sexual rights' (Several cutting-edge initiatives were set in 2005 by members of the

complete consideration of the issues raised here would literally mean a change of the subject, so I will only touch upon a few suggestions for further inspiration. With respect to structural concepts, perhaps it would be productive to think of 'zones' less in terms of boundaries and foreignness, as is usually associated with border zones, encounter zones, contact zones; and instead to draft 'zones' more in terms of connections and axes. Katrien Jacobs (2003) attempts to do so in her analyses of 'queer voyeurism' and its subversions. Indeed, we have to take account of the fact that 'zones' likewise have their hegemonic aspect. Susan Pointon and Katrien Jacobs speak in this sense of transnational 'porn zones', of the connections between the porno industry's producers and consumers; but at the same time, an understanding of 'zone' as 'connection' includes a dispositif for multiple, also contrary practices (see Pointon 1997; Jacobs 2003). Here, *queerscapes* would be a similar concept – the perspective of queer transversal involvements.[16] Thus queerscapes might be thought of as transnational 'enabling networks' for elaborating queer understandings of space, gender, and sexuality. Representational strategies for transcultural encounters should focus on grasping the meetings between 'localised' and 'globalised' subjects in all their complexity. When we conceive -zones and -scapes subversively, it is difference that forms their most important element: collisions, mutual interventions, intermediary interfaces, passing and contested forms of encounter and their re/presentation (see Jacobs 2003). In a somewhat Deleuzian sense, the actors in such encounter zones would be 'minor', still-undefined subjects in the state of becoming: bodies/formations, which we will be or will not have been.[17]

Institute of Development Studies in Sussex, http://www.ids.ac.uk/ids/particip/workshops/rsexrights.html [accessed 20 March 2007]. It seems too early to place these strategies in the context of my argument. A few questions, however, suggest themselves: Is it a discourse of pleasure without (historical) threats? What about the legacy of colonial desire; how will the historical burden of attributing lust to a geopolitical Other be accounted for? And how about the Foucauldian critique of the illusion of understanding sex – and speaking sex – as a vehicle for liberation?

16 Thus not directly analogous to Appadurai's diagnosis of *global ethnoscapes,* see Appadurai 1996, 56–7.

17 Cf. Jacobs 2003 and Deleuze and Guattari 1988. Here, I was inspired by the essay's beautiful title: Körper, die ich nicht gewesen sein werde ['Bodies That I Will Not Have Been'] by Hanne Loreck (2002).

Chapter 6

Fucking Geography, Again

David Bell

Memories...

The Association of American Geographers (AAG) Annual Meeting in San Francisco, 1994 turned out, sometimes incongruously, to be a rather queer time and place. Though my recollections of it are patchy at best, I want to use this essay to reflect on that time and place, on the life of a paper from that meeting – my paper, which started life as 'Fucking Geography' – and the on-going trouble of talking about fucking in geography, a trouble that that paper sought to confront, but which it also fell foul of. Let me start with the story...

Getting to go to the AAG in San Francisco was an amazing opportunity, and lots of UK geographers made the trip. The AAG is one of those astounding academic mega-events: held in swanky hotels, attracting hundreds upon hundreds of people, most of them presenting and listening to short papers in countless parallel streams for days on end. It's overwhelming, giving a sense of the sheer bulk of the discipline and something of its diversity – or, at least, the version of diversity manifest in a US, mainstream academic context. It's a fascinating occasion to ponder what geography looks like, and – or so it seemed to me and my contemporaries, all postgrads or very junior faculty – a golden opportunity to push at the boundaries of the discipline.

Conference sessions on geographies of sexualities, which have in time come to be a more or less 'normal' part of events like this, were still virtually unheard of at the AAG in 1994. It had a gay group, but its function was purely social (and I guess also a little bit sexual). Nevertheless, someone convened a session and put out a call for papers, in response to which a number of us duly submitted our abstracts.[1] Something came over me: this was San Francisco, after all. I had had some fun the previous year at the equivalent UK conference, that of the Institute of British Geographers, talking about sadomasochism, public sex and sexual citizenship. These kinds of topics were still interesting; I was still trying to sneak them onto the geographical agenda. And yes, I was trying to *provoke*, to wake up the sleepyheads and make some people feel uncomfortable: this was a time when Geography as a discipline felt, to many of us, in need of some provocation in order to move forward,

---

1   Real apologies to the convenor and the other participants: I simply cannot remember everyone from this event. Jon Binnie and Julia Cream were there, I know, but who put together the sessions, who chaired and acted as discussants, I am unable to recall. I do remember the room being absolutely packed – at 8 o'clock in the morning – to hear us speak. And I remember drinking cocktails in the rooftop bar and lolling in the hot tub.

to be less 'safe'. This was the time of queer theory and politics, of OutRage! and ACT-UP, and provocation was very much in the air. So, inspired by an image by the famous gay artist Tom of Finland of a moustachioed gay 'clone' literally fucking the Earth, an image readable in some many ways, I had found the perfect title: 'Fucking Geography'. I formed a loose abstract around the title, paid my fees and sent it in. I was going to San Francisco!

Not long after, the first letter arrived. The AAG organising committee was troubled by my submission. Or rather, by a word. An f-word. They thought this word was inappropriate; that such an 'obscene' word probably did have its place, but that place was emphatically *not* the Association of American Geographers. I was told I could not use that word and that I was in danger of having my abstract rejected, scuppering my trip (I could only get funding if I was presenting a paper, moreover one that would lead to a published 'output'). I entered into dialogue, explaining how the title was a deliberate provocation, how it had obviously already worked, about academic freedom, blah blah. To no avail. Change the title or stay at home. Just goes to prove, as Martin Parker (2005, 47) very nicely puts it, that "you can not say fuck where the fuck you like."

Reader, I relented. But I did so in a way that I think made the paper and its title even more resonant. The paper became retitled '[*screw*]ing Geography: censor's version', a title meant to call up images of movies redubbed for TV audiences, where the swearing is moderated – 'fuck' becomes 'screw' – but in a totally heavy-handed, laughable way, with the actors mouthing 'fuck' even as the voice-track says 'screw'. The paper came to be all about academic censorship, all about geography's squeamishness – a lovely phrase borrowed from an earlier landmark commentary by Bob McNee (1984), still one of my favourite interventions (and, tellingly, about a previous AAG meeting in San Francisco). Prior to this, the paper was pretty loose – it formed around two images used as overheads (in those enjoyable pre-PowerPoint days): the aforementioned Tom of Finland drawing, and a photograph by Della Grace, *The Three Graces*, of naked, shaven-headed, pierced and scarred lesbians, plus a line or two from a Madonna song (this was the time of her flirtation with kinky sex). The lyric was used to suggest a desire to take geography from behind – I know, it was all a bit daft, a bit of fairly empty transgression. But *now* it had a dead serious point: why couldn't you say 'fuck' at the AAG? So the paper was about fucking geography as in fucking with the discipline, but it was also an exasperation: a rage at the discipline, fucking geography! And, of course, it was about the geography of fucking – about trying to move work on geographies of sexualities towards a consideration of sex, of what Jon Binnie (1997) perfectly called the 'messy materiality' of the sexed body, the fucking body. This triple meaning of 'fucking', and the slippage between meanings – in many ways like the slippery word play of 'queer' as both noun and verb – seemed to sum up the agenda I was trying to push with this paper. It was a common rhetorical device of the time, of course, used in sex-positive AIDS awareness campaigns and the slogans of queer activists (Crimp and Rolson 1990). So, in that way, too, this was a very queer time and place, a place where queer theory and politics rubbed shoulders (almost) with the massed ranks of the geographical profession. The paper was subsequently published as a 'guest editorial' in the journal *Society & Space*, minus the pics, where I guess it was supposed to make its provocation all over again (Bell 1995a).

A Queer Turn

The mid-1990s were an exciting time, even – or maybe especially – in geography. The porosity of the discipline, its magpie-like ability to pick things from elsewhere and put them to (usually) productive use, was in full flow. The so-called cultural turn in geography and the so-called spatial turn in the social sciences had both made geography 'sexy'. Or so it seemed from inside those turns; it transpires that much of the discipline was at best dimly aware of, and often quite irritated by, what was happening on these 'trendy' fringes. But for those of us who relished this fringiness, who made our intellectual home there, things were buzzing. Conferences, reading meetings and publications were blooming. Suddenly geography was a space where you could talk about almost anything. New trends were enthusiastically taken up, theorists and their theories were hoovered up, geography was everywhere.

The cultural turn, of course, was in reality a profoundly uneven development in geography. The so-called 'new' cultural geography didn't displace the old, and in vast continents of geographical knowledge, not much changed. But where it landed, the 'new' unsettled a few taken-for-granted things about geography and cleared the ground for new ideas, new theories, new methods. Journals like *Society & Space* were rammed with things turning cultural (see Crysler 2003 for a fascinating discussion of the journal).

Included in this category were insights from the still-forming transdisciplinary enterprise known as queer theory. The cultural turn made room for a queer turn in geography, or at least in those parts of geography most infected and affected by the cultural turn (a point I will return to later). Theorists like Judith Butler and Eve Kosofsky Sedgwick were being talked about by (some) geographers, and ideas like performativity were being toyed with, often to mixed reactions (see the fuss around Bell et al. 1994 as a bitter personal example[2]). There were attempts to 'spatialise' or 'geographise' queer theory, to varying degrees of success.

Of course, running ambivalently alongside queer theory was queer politics, which in some ways reversed that imperative, demanding instead the *queering* of geography. Jon Binnie (1997) made the most important intervention in this regard with his call for a queer epistemology in geography, a queering of geographical knowledge.

The confrontational tactics of queer politics, meanwhile, resonated through geography in attempts to expose the institutional and institutionalised homophobia of the discipline and profession, the homophobic uses of geographical knowledge, the sexual politics of geography. Queering geography was part of a mission that Mike Keith (1992) called 'angry writing' – a need to shake up the steady, dispassionate rationalism of academia and to expose the silences and silencings behind that facade.

---

2 The article 'All hyped up and no place to go' (Bell et al. 1994) was commissioned for the launch issue of the new journal *Gender, Place & Culture*. It represented four 'young' geographers wrestling with the idea of performativity in the context of sexuality and space. It was a tough paper to write. The journal immediately sought commentaries on the article, published alongside, which attacked it from various angles. This still feels like something of a betrayal, and prevented any of the authors from wanting to develop or revisit the themes of the paper. For me, this is another instance of geography's squeamishness.

Ty Geltmaker's (1992) discussion of ACT-UP and Queer Nation, two definitively queer political groups, represents clearly this trajectory, in its topic and its approach, opening with a vivid provocation:

> I would like to ask readers to imagine me speaking these words while stickering myself with the slogans and logos which have made AIDS activism as much a fashion statement as a political vision. I do this not only to remind everyone that 'Aids is Not Over', that 'Women Die Faster', that 'I am fag' and that 'Silence Equals Death' but also to appropriate my body as a politically trespassed public space.   (Geltmaker 1992, 609)

'Fucking Geography' was, to my mind at least, part of this agenda, as were other self-styled interventions made at conferences and in geography publications, including some by me, such as a paper hot-headedly called 'The Politics of Sex: Queer as Fuck?' that I presented to a day school of political geographers in search of 'new directions' (subsequently published as Bell 1994a). Then there are Jon Binnie's many landmark contributions, such as 'Trading Places: Consumption, Sexuality and the Production of Queer Space' (Binnie 1995) and the most important discussion to this day of the problems of queering geography – one that he revisits in Chapter 2 of this volume – his 'Coming Out of Geography: Towards a Queer Epistemology' (Binnie 1997; for a comment on the impact of this paper on the discipline, see Johnston and Sidaway 2004). Not to mention his paper on boy bands at *The Place of Music* conference, where Gill Valentine gave a paper on kd lang and lesbian listening (see Valentine 1995a).

This 'queer turn' was all about intervening in the discipline, all about testing, pushing, seeing what could be said where: where the fuck could you say fuck? Trying to expose the boundaries of 'proper' geographical discourse, and also to expose the policing of those boundaries, was a big part of this agenda: who is being silenced, what cannot be thought, and what's at stake? What was at stake wasn't just saying or not saying fuck, of course. No, this was about the inner workings of the discipline and the profession. The battle lines were established. And some battles were easily won, especially where the cultural turn had been most comfortably received, though even here there were clear limits and resistances that could be keenly felt. Other times, it was – and still is – a struggle to coax a productive encounter between queer and geography, sometimes a struggle that couldn't be 'won'. So the 'new' cultural geography was fairly open, urban geography and feminist geography too, though all of these only up to a point – it was and still is easier to write about sexual identities and politics than sexual practices, easier to get published on gay gentrification than fisting. Which meant, of course, that it seemed really important to talk about fisting, to make those interventions in unlikely places, even if now, with hindsight, it can look a bit naïve and foolish. And it was naïve and foolish; naïve for thinking it was worth taking geography on, foolish for the effects it can (did) have on careers and reputations in the mainstream of the discipline. The paper I did at the IBG was dismissed by a referee, when it was later submitted to a journal, as 'sophomoric'.[3] Yeah, a lot of it *was* sophomoric.

---

3   Which it was, to be honest; I will use the same defence Judith Butler used when questioned about an early piece of writing that she no longer identifies with: "I was young!"

Queer(ing) Geographies

Other parts of human geography have exhibited greater ambivalence or resistance to sexuality, sex and queering, though in most it has been possible to force open a small discursive space to talk about some sex and body stuff. Here, I do not propose any exhaustive roll-call or review of this work, rather a brief, highly selective namedrop of interventions in some parts of geography that I think have been significant, and potentials in other places. So, cultural geography has, in many ways, been at the centre of these discussions; sexuality is now more or less routinely and more or less unproblematically part of cultural geography, at least in its 'new'-ish variant, and appears in text books (e.g. Crang 1998; Mitchell 2000) as well as in journals and research publications. Urban geography has been home to research on, among other things, gay gentrification and gay urban politics, most notably in the work of Larry Knopp (too many papers to cite, but see Knopp 1992 as an example). Feminist geography has been sometime home to work on queer politics and theory, though I think it is fair to say there are on-going tensions between queer and feminism that are also playing out in geography (see Nast 2002 for one manifestation of this, and Sothern 2004 for comment on Nast). Political geography has accommodated talk of queer politics and of the spaces of sexual citizenship, though this hasn't really reshaped the main agenda or curriculum of the subdiscipline; arguably, the impact has been the other way round, in using the tools of political geography to think sexual politics with (Bell and Binnie 2006). A famous call to consider 'neglected rural others' (Philo 1992) has led to space opening up in rural geography to consider rural sexual cultures in what I think is one of the most interesting and lively areas of work at present (see, for example, Valentine 1997 on rural lesbian communes, and Little 2003 on rural heterosexualities). Historical geography has noted that sex has histories, too, as do sexed spaces (e.g. Ogborn 1997). Even parts of geography seemingly immune to the cultural turn seem (to me at least) ripe for queering: transport geography could learn from Puar's (2002c) paper on lesbian cruise ships, or Retzloff's (1997) work on the role of cars in producing gay culture in Flint, Michigan, or Bech's (1997) astonishing discussion of railways stations as sites of sexual exchange, for example.[4] Or, environmental geography might benefit from an encounter with Gaard's (1997) work on queer ecofeminism and Sandiland's (2002) on queer ecology as different, queered ways of thinking 'naturecultures', or Byrne's (2003) study of the landscape and amenities management issues resulting from public sex in urban greenspaces. And so the list goes on – even something as unsexy as 'diffusion studies' has been queered (Knopp and Brown 2003). While such activity might in some ways be considered marginal to the centre of these subdisciplines, it is nonetheless published in their representative journals and is coming to be folded into the canon and the curriculum in a lot of cases. Or, at least, it seems less unthinkable now to talk about

---

(Butler 1996). In fact, I was more affronted on first reading the referee's comment because I thought it said soporific.

4   On the lack of a cultural turn in transport geography, and possible ways forward, see Divall and Revell 2005.

sexuality, and even occasionally about sex, in some of these places. But I am starting to sound optimistic, upbeat. Time to remind ourselves that...

Sometimes it's Hard to be a Geographer

Has geography been fucked? Or at least fucked with, or fucked about? On the brief run through of evidence above, maybe it has: or, perhaps, we might say it has been queered a little bit.[5] But the institutional and institutionalised homophobia and erotophobia at the heart of geography is still there, arguably as strong as ever. Most people thinking and doing geography haven't been queered or fucked by the things I've sketched out in this essay. Arguably the greatest bit of fucking in geography at the moment is the call to breach the divide between human and physical geography, or at least to bring these two 'halves' back into a closer and more productive relationship – a very different kind of fucking (see, for example, Harrison et al. 2004).

I started this paper talking about my own experience of censorship ten years ago at the hands of the AAG. This was in no way an isolated incident, and it is still a routine part of dealing with the discipline. As a test and a final provocation, then, I would like to list a few other words I have been unable to say in print. *Homo. Cock. Fuck. Cunt.* And, of course, one more time: *Fucking.*[6]

---

5   As my friend Martin Parker found, writing in a management studies context, 'queering' can be seen as a more palatable, less obscene substitute for 'fucking'. His trio of papers on management are an excellent parallel to my discussion (Parker 2001; 2002; 2005). These papers landed him in hot water, too.

6   Homo was changed to Homosexual in a chapter title, following lengthy correspondence with the editors and publishers who thought the term unsuitable for an academic publication. Cock and Fuck were also deemed unsuitable in the same chapter. Cunt, used in a punning subtitle – 'Cuntry living' – in a journal article, was changed by a copy editor to 'Country', thereby totally spoiling the pun and making a nonsense of an explanatory footnote. Sophomoric but with a point to make, then.

SECTION II
Practices

Chapter 7

Playing with Restraints: Space, Citizenship and BDSM

RDK Herman

In October 2003, the Washington DC-based organisation Black Rose, which serves the region's community of people interested in bondage, discipline and sadomasochistic (BDSM) sexual practices, was forced to cancel its 7th annual conference. The conference was slated to take place at the Princess Royale Hotel in Ocean City, Maryland. Though the organisers had wished to keep the event private, word had spread among Ocean City residents, rousing protests from business owners, religious leaders and city officials. The annual conference, billed as 'the world's largest pansexual BDSM event of the year', was expected to attract over 1,200 attendees for three days of workshops, lectures, presentations and BDSM play. The hotel had already sold out its 300-plus rooms, with the overflow filling up other local inns and hotels. But public outcry led the liquor board to threaten revocation of the hotel's license, effectively shutting down the event (Smith 2003).

As Hubbard (2001) suggests, moral panics of the type caused by the Black Rose event are needed by the state in order to reassert its right to power. Thus the management of sexual morality is integral to the manifestation of citizenship. The case of the Black Rose event, and others like it, points out the special niche that BDSM occupies among 'alternative' sexualities, and speaks to the contested nature of BDSM in both sexual and civil arenas. Unlike categories such as gay, lesbian and transgender, BDSM is portrayed as a 'lifestyle choice' rather than an identity – a distinction that reflects the problematic nature of how sexual identities are discursively constructed and reified within the social order. Cast as a practice rather than an identity, BDSM is more open to persecution – social, legal and psychiatric. At the same time, while BDSM is clearly identified as a sexual activity, it is not the sexuality of it that comes under fire. Rather, it is the perceived 'violence' (physical and/or psychological) that brings moral condemnation and persecution from legal and psychiatric institutions, driving participants underground.

This paper explores how BDSM as a sexual lifestyle intersects critically with issues of sexual identity, citizenship and spatiality. The first part of this essay explores the construction of BDSM in discourses both external and internal to the community of participants. This leads to a consideration of the social and legal constraints placed on BDSM players, either at play or in their everyday lives. The spatiality of BDSM that results from these socio-legal constraints is then considered in detail. Throughout, this chapter examines the slippages between BDSM as a sexual identity and BDSM as a sexual practice, exploring how BDSM, more generally, and these slippages, in particular, fit into the overall consideration of non-normative sexualities and their spaces.

Constructing Perversion

The genealogical origins of BDSM lie, as Foucault would predict, in the discourses of psychiatry. Sadism and Masochism were originally defined in Krafft-Ebing's 1886 treatise on sexual pathologies, after which Freud coined the term 'sado-masochism'. This newly defined sexual 'disorder' then took its place within the clinical taxonomy of psychological and psychoanalytical theory. Taylor and Ussher (2001, 293–4) note that sadomasochism is still defined as a psychiatric disorder within the *Diagnostic and Statistical Manual* (DSM-IV) of the American Psychiatric Association and in the World Health Organization's *International Classification of Diseases*. In the UK, recent legislation (*R v Brown* 1993) reaffirmed BDSM's status as a criminal offence. It continues to be discussed in psychiatric textbooks alongside such things as rape and child sexual abuse.

Because they involve the infliction of pain and/or overt domination, BDSM practices embrace behaviours that are positioned as antithetical to proper, moral interpersonal relations. The UK's 'Operation Spanner', in which this deviance from accepted norms was judged as warranting state intervention, is a widely known case on this point (Langdridge and Butt 2004; Hubbard 2001). The judge at the trial resulting from the arrests made during Operation Spanner – and who ultimately imprisoned adult gay men for engaging in consensual BDSM sex – stated that the defendants had to be 'protected' from themselves, arguing that it was the role of the law and courts to draw "the line between what is acceptable in a civilized society and what is not" (Langdridge and Butt 2004, 36). More generally, studies of BDSM were concerned with extreme and invariably non-consensual acts, the analyses of which were then applied to consensual sadomasochistic sexual acts (Taylor and Ussher 2001, 294). Thus, public discourse on sadomasochism sometimes fosters images of Ted Bundy figures that sexually torture and even murder unwitting victims.

BDSM as a social phenomenon draws simultaneously on the terminology produced in the psychiatric and legal loci of power, and modifies and proliferates these terms into a panoply of identifiers. While the term 'sadomasochism' is used by scientific and professional communities, participants use 'S/M' or 'BDSM', terms that for them carry no pejorative connotations. It is much like the distinction between the terms 'homosexual' and 'gay' (Moser and Madeson 2005, 24–5). The acronym 'BDSM' combines the two related pairs of terms: 'Bondage and Discipline' and 'Sadism/Masochism'.[1] The former is identified as concerning the exchange of power

---

1   In this paper, I use the term 'BDSM' as a catch-all for the range of practices associated with B/D, S/M, D/s, and other 'fetish'-related practices among (particularly) the heterosexual/ mixed community. This term is widely inclusive, and suggests a 'membership' amongst peoples (referred to herein as 'players') whose tastes may vary widely but who draw from a common menu of activities. BDSM play is often referred to as 'scenes', in which role-play scenarios with specific types of activities are negotiated in advance. I use this term even when referring to the works of others, though other authors may limit their terminology to 'S/M'.

between participants; the latter with the infliction and receipt of pain. 'Dominance and Submission' (often written 'D/s') refers to the psychological and sexual power transfer, which can exist without bondage, physical pain or humiliation. 'Bondage and Discipline' refers to two separate but often related practices: scenes that involve physical restraints to heighten the intensity of the experience (bondage), and scenes that involve disciplinary punishment-and-reward scenarios (such as teacher-student). Humiliation is a factor frequently brought into play, usually requiring the submissive partner to perform acts that may be seen as degrading. 'Leather' refers specifically to the range of such practices as performed and prescribed within gay and lesbian communities (see Joshi 2003; Kamel 1995), which has its own rules of definition. But in any case, the term 'vanilla' is applied by BDSM players to conventional lovemaking without any S/M component (Moser and Madeson 2005, 24–5).

In practice, BDSM can encompass or engage with the enormous range of sexual practices that constitute what is generally called 'kinky sex'. So that participants can find partners, this range of practices is encoded: infantilism, role-playing, cross-dressing, pony play – the list of categories is extensive. Indeed, to understand BDSM, one must be familiar with the lexicon. *Alt.com* lists a menu of approximately 100 different specific practices. But the more immediate concern is the role of potential partners, and here again a range of terms are employed, denoting varying shades of meaning. 'Top' and 'bottom' may connote more about who is in charge than who is physically in either position (since one can 'top from the bottom') and mix uneasily with the terms 'dominant' and 'submissive'. BDSM players recognise the complexity of relations involved in BDSM and the inadequacy of simple terms to connote these complexities.

The BDSM community[2] has nonetheless co-opted the discursive production of identities initiated by the psychiatric community, and has become defined if not empowered through the proliferation of indicators and identifiers. A language of codes largely specific to BDSM helps to link players into a sense of community, creating a discursive space apart from, and yet parallel with, mainstream sexuality.

Players and Other Kinky People

BDSM discourse thus strikes a path between the appropriation and the contravention of normative legal-psychiatric hegemonic representations. While in the outside view, BDSM has been distorted by its conflation with violent, non-consensual sex crimes, the inside view of BDSM focuses strongly on consensuality, care and mutual commitment between partners. Taylor and Ussher (2001, 301), using interviews with 24 self-identified BDSM players, define BDSM as "comprising those behaviours which are characterised by a contrived, often symbolic, unequable distribution of power involving the giving and/or receiving of physical and/or psychological stimulation." It often involves acts that would generally be considered as 'painful'

---

2 The term 'community' is used here in its ambiguous, postmodern sense to denote people who may never meet, physically or otherwise, but who may nonetheless be identified collectively through ascription to something shared. It is recognized that ascribing 'community' to these disparate people is a heuristic device and does not denote a united, monolithic body.

and/or humiliating or subjugating, but which are consensual and for the purpose of sexual arousal, and are understood by the participant to be (BD)SM. Taylor and Ussher note that BDSM was often seen as "deliberately, consciously antithetical to hegemonic, patriarchal hetero-sexuality" (2001, 305). Generally, it was positioned as oppositional to conventional non-BDSM sexual relations ('vanilla sex'), which were positioned as "conformist, uninteresting, unadventurous and unerotic." They noted "a celebration of 'perversity', and of difference, consistent with many of the ideas embraced within queer politics" (2001, 305).

Most participants in their study positioned BDSM as escapism from the ordinariness, drudgery, boredom or alienation of everyday life. Some described the 'high' of BDSM activities as "a heightened state of consciousness or as in some way making them more astute, more enlightened or more alive" (Taylor and Ussher 2001, 305). All interviewees invariably positioned BDSM as 'fun' and as 'play'.

Although by definition BDSM has sexual connotations, it is important to note that BDSM practices are not necessarily directly sexual and BDSM 'scenes' may involve little or nothing that would be defined as sexual activity (genital stimulation). Instead, the defining characteristic is the power exchange whereby one person willingly relinquishes some degree of control over his or her body to another person. And this can range from soft bondage or role-playing to formalised slavery complete with contracts stating the terms of submission. Because of the wide range of practices that fall within this rubric, an important question when studying BDSM is whether or not people who engage in such practices self-identify as doing BDSM. And here, BDSM players run up against the hegemonic power of sexual normativity and the condemnation of 'perversion'.

People can engage in BDSM activities without ever defining them as such (Madeson and Moser 2005, 59; Wiseman 1996, 13). Any erotic play involving force, restraint, domination or erotic pain (scratching, biting, spanking) falls under the broadest definition of BDSM. As Wiseman (1996, 23) notes, "a pervert is anybody kinkier than you are." It is easy to move the bar that separates the acceptable from the unacceptable. Madeson and Moser point to the social unacceptability of BDSM, as well as the stigmatisation and pathologising by the psychiatric and medical communities, as reasons people avoid self-identifying as being into BDSM. It is an identity that positions one outside the norm and into a realm popularly represented as 'perverse'.

However, this non-normative behaviour is more popular than its stigmatisation suggests. Moser and Levitt (1987, 95) cite Hunt's 1974 questionnaire survey of sexual behaviour involving 2026 respondents in 26 US cities, which found that approximately 12 percent of females and 18 percent of males indicated a willingness to try BDSM activities, with two to three percent having already tried some and liked it. While this suggests that potential players form a fairly significant portion of the overall population, the number of those who would actually identify as being 'into BDSM' is no doubt smaller. *Alt.com*, a major internet dating resource for BDSM players, has 3,532,504 personal listings from within the US,[3] or about 1.3 percent of the total US population (though experienced players suggest that perhaps 10 percent of those listings are duplicates). On the one hand, this certainly does not represent all those who have an interest in forms of BDSM play, but only those who have

---

3   Statistics gathered online on 13 November 2005.

recognised their interest, are searching for partners and have taken the step to reach out within this forum. On the other hand, a small percentage of these listings are from people who explicitly state that they have no specific interest in BDSM, but who see this website as just another venue to try to meet people and who apparently hope that the sexually explicit nature of *Alt.com* will enable them to find sexual partners more easily.

This struggle against the discourse of perversion can be acute and protracted. Accepting in oneself a strong interest in BDSM behaviours involves a struggle that has parallels to 'coming out' for gays and lesbians (Moser and Levitt 1987, 100; Madeson and Moser 2005, 55–6). Acceptance becomes much easier with the support offered by organised BDSM groups. These can play a vital role, forming a support network that helps members realise that they are not alone, and providing an environment where one does not feel 'freaky' or perverted. In so doing, BDSM groups provide a new norm for its members, a norm that fits BDSM players into a society with its own rules and expectations, aesthetics and protocols.

Citizenship and Restraints

As with other sexual identities that veer from the institution of 'traditional' heterosexual marriage, BDSM players are periodically demonised as 'bad citizens', thereby shoring up what is considered normal and desirable behaviour. Bell (1995) argues that sexual dissidents enjoy neither true publicity nor privacy and are thus denied full sexual citizenship. Indeed, self-identifying as a BDSM player can run high social risks as BDSM is not a protected category, and discrimination against BDSM players is often considered appropriate. The National Coalition for Sexual Freedom (NCSF), a US-based advocacy organisation serving alternative sexual communities,[4] states that such forms of sexual expression can result in discrimination, prosecution and even violence against practitioners (NCSF 2004):

- It can cause you to lose your children
- It can cause you to lose your job or your income
- It can lead you into a maze of antiquated laws and regulations you never even knew existed
- It is arbitrarily criminalised by state and local authorities
- It is used by the radical right to marginalise minority groups
- It can result in the invasion of your privacy by the government, both within your own home or in educational, social and group environments.

Madeson and Moser (2005, 61) report that:

> Some of us have lost jobs because of a party invitation inadvertently left on a desk or being spotted in the street in SM garb. Sometimes it's more direct: We confide in a co-worker, sensing a friend and wanting to share. The result is often the same: We find ourselves out of a job. Conventional offices do not want 'perverts' working for them.

---

4   The NCSF defines their purview as including BDSM, swinging and polyamorous practices.

They argue that the secrecy of the BDSM lifestyle is 'part of the package', pushing many participants into organised groups where there is no need for secrecy or subtleties (2005, 61). The social need for BDSM communities translates into the spatiality of BDSM: where to meet, where to play, and issues of being 'outed'.

Beyond these social constraints, BDSM practitioners may find themselves under scrutiny from either the mental health system or the judicial system. At the Federal level, the US government has ruled that individuals cannot legally give up their consent. This ruling stems from anti-slavery laws, but today is utilised against BDSM players. Beyond that, State laws can prohibit specific activities. In Massachusetts, it is illegal to strike someone, period – consensual or otherwise.

As Evans (1993) has discussed, the state exists in a tension between the quantitative benefits of commodification and the qualitative costs of 'immorality' and 'decadence'. The buying power of the BDSM community is increasing, as is the availability of commodities. Although state pressure forced the closure of the Black Rose event, hotels that host large BDSM events cite the enormous profits these events bring and usually resist public-outrage demands to turn these events away. The state must weigh whether the financial costs (to local businesses) incurred by closing down BDSM venues is outweighed by the social capital to be lost (through moral outrage) if they do not. BDSM players must negotiate this uncertain territory, one in which they are usually safe but potentially in danger of police raids. The spatiality of BDSM reflects this tension.

Spatialities: Where Kinky is OK

As Hubbard (2001, 55) asserts, civil society is conceptualised as "a heterosexual (as well as patriarchal ...) construction that serves to make entry into the public realm very difficult for those whose sexual lives are judged 'immoral'." Given the stigmatised, pathologised and legally troublesome status through which BDSM is widely viewed as immoral, if not outright sick, BDSM practices are, therefore, *spatially* marginalised. Public reactions to the Black Rose event attest to this positioning. But what is curious is that the Black Rose event was to be held in such a distinctly public venue. Indeed, large BDSM events are generally held in hotel complexes much like other conferences and conventions. It is this slippage between privacy and publicness that warrants exploration.

Hubbard (2001, 66) portrays two types of sexually dissident spaces: those that perpetuate the distance between mainstream and dissident sexual identities, such as gay villages and red-light districts; and those that operate within the interstices of society, producing what Hubbard calls 'ephemeral sites of freedom and control':

> liminal spaces that disrupt dominant geographies of heterosexuality by creating transitory sites for sexual freedom and pleasure where the immoral is moral and the perverse is normal ... [T]hese spaces would inevitably need to exclude those whose presence threatens the control and freedom which inhabitants exercise over their own sexual performances, allowing them to articulate their needs and desires in a safe and pleasurable environment.   (2001, 68)

BDSM, it would seem, falls in the latter category, but players have developed communities and networks for producing both temporary and fixed places for BDSM activities that are both private and public. Some of these replicate (or parallel) normative heterosexual structures, and all of them serve to normalise BDSM to the participants. These spaces include the home, public and semi-public events (in homes or as local gatherings, including at fixed BDSM venues), and large-scale BDSM events like the Black Rose conference.

Most BDSM takes place in private homes. Spatially, this can involve the organisation or modification of one or more rooms to accommodate BDSM play, which can range from putting eyebolts in the ceiling, to constructing entire 'dungeons' filled with BDSM equipment. An online survey conducted in 2005 by this author found that, of 1960 respondents, more than half said they had modified rooms in their homes for BDSM activities, with about 20 percent saying they had converted most or all of an entire room.[5]

The home is the 'normal' place for sexual activity as defined in the heterosexual model of domestic procreation, but, because BDSM involves practices easily deemed 'scandalous', the security of the home becomes a critical issue for BDSM players. Various authors have discussed the problematic nature of the home's alleged 'privacy' (Duncan 1995; Johnston and Valentine 1995; Robinson et al. 2004). For BDSM, there are several issues. Most important of these is legality – the question of whether private, consensual sex acts are subject to interference from a state that sees a need to 'protect' its subjects from themselves. This is a matter of legal battles between governments and BDSM advocacy groups. Given the potential for legal interference or social stigmatisation, players must consider how much to hide or display their BDSM lifestyle from potential visitors (including repairmen, landlords, friends and family). As with Johnston and Valentine's (1995) discussion of lesbian homes, BDSM players may feel the need to hide their books, toys and other gear so as to avoid being 'outed' by visitors. Furthermore, there are the neighbours who may see or hear BDSM activities going on and call the police, fearing that non-consensual acts of violence are taking place.

Privacy among members within the home is contingent on power relations (parents-children), spatial proximity, thin walls and other factors that dissolve boundaries within the home (see Duncan 1995). This poses problems for both parents and children (in households that contain both), with the degree to which parents are 'out' about their BDSM lifestyle becoming an issue regarding their fitness as parents, an issue that particularly comes to bear in divorce and child custody cases. Conversely, adolescents who may be exploring BDSM do not necessarily have the ability to hide their gear from their parents, although, fortunately, the crossover of BDSM gear with the accoutrements of Goth and Punk can render it invisible.

But once people enter the world of organised BDSM groups, there are a range of public and semi-public events in which members meet. These can be divided into play parties that involve overt BDSM play and 'vanilla' events in which no BDSM activity takes place. Vanilla events include the 'munch', an event where BDSM

---

5   Further findings and discussion are available from the author. A more complete discussion of this survey will be published in a future article.

players gather at an everyday restaurant for a meal and socialising. Participants do not wear any obvious fetish gear, and the group is often booked under a fictitious, innocuous name. No one else in the restaurant would know that there is a pack of kinksters at the next table. Some organisations also host educational workshops, also often in vanilla settings such as a conference room at a local hotel. Again, members are reminded that they need to keep a low profile. These events provide important cohesion for local BDSM communities. They also present ways in which the BDSM community is able to carry on its activities in the interstices of 'normal' places.

BDSM play parties can also take place in public and semi-public venues. In some cities, open BDSM clubs are legal. And for the gay BDSM scene, Leather bars provide a venue for interaction. But elsewhere, the operation of these venues is subject to persecution through zoning and other legal controls, and owners may have to resort to secretive means to conceal the nature of the real activities that take place there. In such cases, visiting such a venue can be compared to going to a Prohibition-era speakeasy.[6] In areas where BDSM activities face legal action, it is likely that private 'dungeons' play a more important role.[7]

Some BDSM venues may be permanent, while others are indeed fleeting, coming together just for the event, to be quickly dismantled and the equipment hauled away at the end. The latter is true of large-scale events, such as the Black Rose conference, that draw thousands of people, booking up to five hotels and running a shuttle service to ferry players throughout three days of non-stop activities including lectures, demonstrations and dungeons full of bondage equipment for open play.

The spatialisation of larger BDSM venues contains distinct elements. First, a major attraction of larger venues is the availability of equipment. BDSM scenes between a dominant and a submissive almost invariably involve some form of bondage, rendering the submissive immobile while the dominant exerts physical and psychological control. This often involves the infliction of pain through whipping, spanking or using a wide range of devices that prick, pinch, shock or otherwise cause pain. The restraint required for such activities is most easily, effectively and erotically accomplished using large and sturdy equipment designed for this purpose: racks, St Andrew's Crosses, bondage tables and chairs, suspension devices (such as slings), or cages for confinement.

A room full of BDSM equipment provides very distinct sense of place. To the outside viewer, it may give the impression of a sadistic theme park; for the BDSM player, it may be more like an amusement park with a lot of rides. Most people cannot afford to have a lot of large BDSM equipment in their homes. It is not just the cost or the spatial requirements, but the privacy concerns as well. You can't camouflage a St Andrew's Cross or spanking bench as something other than it is. Thus venues that offer a range of equipment are loci for larger gatherings of BDSM

---

6   Speakeasies were secretive establishments for selling and drinking alcohol during Prohibition (1920–33), a period in US history when selling or buying alcohol was illegal. Without signage, they were known through word of mouth, and doormen suspiciously checked patrons to keep out undercover law-enforcement officers.

7   In Baltimore, Maryland, where this research is based, BDSM venues are illegal. But, an hour away in Washington DC, they are not.

players – whether in the dungeon of a private homeowner, or in a BDSM club, or in a larger, temporary venue.

In addition to bondage equipment, elaborate BDSM venues – small and large – may include areas for specific types of fantasy role-playing, such as 'hospital' areas for doctor(or nurse)/patient scenes, cribs for infantilisation scenes, and so on. Usually these larger venues also have an area for socialising. Many who attend these events do not necessarily engage in play. Group events provide an important social milieu for BDSM players to congregate, learn about new techniques and equipment, and feel part of a community. Sometimes it's just an opportunity to dress up in fetish gear and go out.

Finally, players also emphasise the importance of having a space for 'aftercare' once a BDSM scene is over. Professional dominatrix Mistress Steel (no date) defines aftercare as "affectionate care and attention following any type of traumatic or mentally challenging [BDSM] event." She notes the psychological challenges and the intense degree of personal exposure that BDSM scenes can involve. It is therefore considered to be an essential component of BDSM that the dominant give support, nurturing, comfort and reassurance to the submissive after the scene. Thus BDSM venues should offer a space for aftercare, away from the play area where scenes take place.

In all group venues and in the BDSM scene generally, codes of conduct are clearly established and enforced. BDSM play can be dangerous, and mistakes can lead to serious injury or worse. Most group events have 'dungeon masters' who are experienced and often trained in identifying problems and interfering with scenes when deemed necessary. Rules of protocol are often posted, and sometimes participants are required to read them and sign a waiver. It is a facet of the BDSM scene, perhaps in part also due to the legalities that surround it, that people look after each other very carefully, rules are strictly followed, and troublemakers are shown the door.

Thus far, venues specific to BDSM have been considered, and it is readily seen how these venues infiltrate everyday spaces at the same time that they remain largely hidden. In addition, there are other marginal but mainstream spaces, like science fiction conventions, where BDSM players can 'go public' without drawing attention. Goth clubs are a particularly open venue for the BDSM community. A late-1970s offshoot of Punk, Goth retains a fashion dominated by black clothing, chains and spikes, fishnets and androgyny. At these clubs, Goth and BDSM communities interact freely and seamlessly. The venues may be 'Goth', but are those collars and chains real or just a fashion statement? Goth clubs present a venue where fetish wear – and limited fetish behaviour, as well as cross-dressing and same-sex dancing – are acceptable, since Goth shares with BDSM both a style and a marginal identity.

Also peripherally serving the BDSM community are 'Medieval' and 'Renaissance' events, the clothing for which includes Victorian-style corsets, chain mail, leather, and silver jewellery. Stores selling this type of 'Renaissance' clothing also often identify as selling Goth wear. At the events themselves, one can find retailers selling a range of both peripherally and explicitly BDSM gear. Wiccan or Pagan events can similarly be gathering places for BDSM players.

Like science fiction conventions, all of these events are gathering places for marginal activities and identities. They serve the BDSM community in a similar way to how Podmore (2001) describes ethnically mixed neighbourhoods as provided 'shared space' for lesbians in Montréal. Such shared space is "perceived as an accessible space for marginalised social and cultural groups." In this case, the shared spaces linked to BDSM are those that involve elements of dressing up in a certain style and/or that involve elements of fantasy role-playing.

Cyberspace provides shared space for all communities and is particularly suited to the communication needs of marginal groups. This is no less true for BDSM. Apart from the voluminous commercial websites selling pornographic images catering to all sexual tastes, the internet serves BDSM in three forms: dating, socialising and shopping.

Just like online dating services for other interest groups, a number of websites, such as *Alt.com*, help match kinky people searching for partners. Similarly to other forms of online dating, members create profiles and identify the types of activities in which they have interests – the menu is extensive. In addition, one identifies sexual orientation or gender identity (straight, gay/lesbian, bisexual, transgender) – an aspect that emphasises the ubiquity of BDSM across these categories and the acceptability of different sexual and gender orientations within BDSM as a whole. Finally, users can impose a range of the usual demographic caveats (e.g. age, ethnicity, location) on their searching.

Several BDSM groups have list serves for members that provide a medium for announcements of upcoming events and for general discussions. These online social venues contain both the ordinary chit-chat found in any community and serious discussions of issues pertaining to BDSM – such as the importance of aftercare, the nature of the D/s relationship, responsibilities in BDSM play, and so forth. It is here, as much as at the events themselves, that the forging of community identity can take place.

Also available online (and to a limited extent in actual retail stores) is the opportunity to purchase BDSM toys, equipment and even furniture. Discrete packing preserves privacy, and credit-card bills show only an innocuous retailer name. The products available range the full gamut from blindfolds and strap-ons to cages and large bondage furniture – even steel double beds with multiple points of attachment designed for intense BDSM play. Here the use of internet shopping and disguised shipping transcend the barriers of space and allow people to equip themselves to explore BDSM without leaving the home.

Conclusion

BDSM clearly shares similar spatial and social constraints to other non-normative sexualities. Yet, BDSM does not focus on the "incoherencies in the allegedly stable relations between chromosomal sex, gender and sexual desire" (Jagose 1996) or the "rebellion against the formation of identities around fixed poles of gender or sexuality" (Gunther 2005, 23) that are generally associated with the term 'queer'. It might be argued that BDSM does little to undermine essentialist identities; yet it might equally be argued that its disinterest in them is its own form of anti-essentialism. On the other hand, to the extent that queer encompasses the range of sexual practices that lie outside the mainstream (Sherry 2004), BDSM is certainly included within queer. *Queertheory.com* lists resources for BDSM among its many categories.

BDSM spans the range sexual and gender categories – heterosexual, homosexual, transsexual, group scenes, and swinging – and crosses class and ethnic boundaries. Those who also identify as gay, lesbian or queer in any other context no doubt do so within the realm of BDSM – though for some, like Califia (1979), the BDSM identity may outweigh the other. But it is unclear whether 'straight' BDSM players see their identity as 'kinksters', with all the openness to alternative sexualities and gender identities that that entails, as placing them in the same political camp with gays, lesbians, transgendered people and others who may identify as 'queer'. The historical usage and all-too-frequent academic conflation of 'queer' with gay and lesbian (Giffney 2004) creates a discursive gap between the queer and the kinky.

Certainly BDSM involves a celebration of the perverse, but to what extent is it a disruption of normative gender and power roles? Role-playing, role reversal, age play, cross-dressing, strap-on play, pony play – such practices involve the appropriation and reification of normative gender/power roles, albeit in altered form. Roles can be reversed and/or intensified, but the roles remain nonetheless. BDSM thus disrupts normative heterosexuality more by taking it to extremes – or even caricature – than by contradicting it. This paradox is suggested in the title of one book on BDSM: *Bound to Be Free* (Moser and Madeson 2005).

While public discourses in the US are ambivalent about BDSM, the public realm is bombarded by images of sexuality in increasingly various forms. BDSM is consequently infiltrating into our cultural consciousness. Taylor and Ussher (2003) point to various films, television programmes and popular music acts that contain BDSM themes. Add to these, firstly, the two decades of MTV images of Punk, Goth and heavy-metal rockers in various degrees of drag and fetish outfits; secondly, the recent interest in television programming focused on leather-clad 'medieval' figures such as *Xena: Warrior Princess* or kinky vampires and witches as in *Buffy the Vampire Slayer*; and, thirdly, dark super-hero themes that engage more or less directly with BDSM themes and fashions, and it is clear that BDSM is entering the living rooms of the American public as well as living rooms in many other Western countries. BDSM players even go on talk shows to discuss their lifestyles.

Weinberg and Magill (1995) argue that BDSM is no longer the 'last taboo', although admitting that this does not mean that Americans understand or accept it. But I suggest that there is a degree to which a taboo position serves the libidinal needs of BDSM as a practice. There is a significant parallel between the spatial restraints imposed by social condemnation and the restraints actually used in BDSM play. It remains to be explored whether the removal of social and legal restraints will significantly alter the manifestation of BDSM as a social and sexual phenomenon.

Chapter 8

Queer Mixed Race? Interrogating Homonormativity through Thai Interraciality

Jin Haritaworn

> It is not the nonheterosexist behaviour of ... black men and women that is under fire, but rather the perceived nonnormative sexual behaviour and family structures of these individuals, whom many queer activists – without regard to the impact of race, class, or gender – would designate as part of the heterosexist establishment or those mighty 'straights they hate'.
>
> (Cohen 2001, 218)

In 'Punks, Bulldaggers, and Welfare Queens', Cathy Cohen observes that queer politics has not lived up to its radical potential, but collapsed back into a simple identity politics whose sole lens of oppression is sexuality, or rather white middle-class gay sexuality. She challenges the narcissistic construction of the 'powerful heterosexual other' by reminding us of the historical prohibitions against heterosexual marriages between enslaved black men and women, and the continuing stigmatisation and control of poor and racialised families. Her essay coincides with the proliferation of queer writings on 'race'. Yet, many of these discussions remain within a 'homonormative' frame that re-centres the position of the most privileged gays (Puar 2006). As Malini Schueller (2006) shows with such prominent queer theorists as Judith Butler, Donna Haraway and Gail Rubin, this often happens by analogising race and sexuality. By isolating and comparing the experiences of 'gays' (white) and 'blacks' (heterosexual), these writers obliterate racialised queer subjectivities and the multiple allegiances that they potentially give rise to.

This chapter takes this critique further by arguing that homonormative theorising can also often take the form of collapsing the non-queer into the queer, an especially pernicious problem when racialised non-queer subjectivities and experiences are treated as if they are white queer subjectivities and experiences. Such moves, I argue, elide the particular forms of violence against those who do not identify as queer (an argument that has similarities with Prosser's (1998) critique of the inclusion of transsexuals into the category of queer). Moreover, the implicitly homonormative conflation of white 'gays' and heterosexual blacks that takes place within much queer theorising ignores the particular violence that queer people of colour experience, including that at the hands of white queers.

I develop my argument with the example of Thai interraciality, which at first sight appears to be dominated by figures – from the 'prostitute' to the 'lady boy' – that could be classed as queer. Queer theorists themselves encourage such a reading. Judith Butler's 'heterosexual matrix' (1990; 1993b), which has formed the basis of our contemporary theories of heteronormativity, treats interraciality and other cross-category sexual practices as potential transgressions not only of heterosexuality, but also of the boundaries between, and ultimately of, these categories themselves. This has influenced some interraciality theorists to claim that differentially racialised intimacy and lineage transgress the binaries not only of race (e.g. Asian/white), but also of gender (male/female) and sexuality (gay/straight).

The chapter interrogates this claim through the critical lenses of people of Thai descent who were 'multiracialised' – in other words who had non-Thai social or biological parentage.[1] The interviewees' critique contradicts the heterosexual matrix in several respects. Heterosexuality was indeed the main lens applied to their families, which were often read by others as the products of unions between Thai prostitutes and their white clients.

Central to such readings is a construction of Thai female heterosexuality that contrasts with the privileged, normalised heterosexuality of Butler's 'matrix'. It is in its heterosexuality that this femininity is seen as most threatening. The interview accounts show that far from being vulnerable, innocent and warranting chivalry and protection, this heterosexual femininity is constructed as dangerous, immoral and inviting unfettered sexual advances by white men.

However, heterosexuality was not the only default reading of Thainess that the interviewees critiqued. In the dominant discourse, men of Thai descent, along with other East-Asian men, are assumed to be gay. Both women and men of Thai descent are further subject to feminising discourses, which dominate not only heterosexual but also gay contexts. This further problematises a certain homonormative discourse that constructs 'queer' space as an intrinsically safer one.

Apart from female sex work and male homosexuality, discourses of transgender and transsexuality are central to Orientalist constructions of Thainess. Unfortunately, there was an absence in my study of accounts that were assertively trans. This absence must be contextualised with my (new) shifting position as a researcher whose gender identity has changed dramatically since carrying out the research. At the time of designing the research and interviewing participants, I identified as 'a woman'. Given the silences in my own life thus far about gendered possibilities, it is no coincidence that my initial research questions, sampling criteria and list of topics were designed for a world without trans people. At the stage of writing, I identified as a 'non-trans ally', and attempted to produce 'traiterous knowledges' from that perspective (Harding 1991). I am now, in spring 2007, rewriting parts of this chapter again, as a newly gender-questioning person. The lack of trans biographies in my discussion produces a gaping hole that I can mark but no longer fill. I have, nevertheless, found it necessary to include within this chapter a critique of transphobia in my discussion of Butler's heterosexual matrix since her concept so centrally draws on trans experiences.

---

1   The chapter is based on interviews with 51 people from part-Thai families in Britain and Germany that were conducted for my PhD between 2001 and 2003. My study focuses on commonalities between Britain and Germany, which, despite historical differences in citizenship systems, have given rise to similar Orientalisms. In particular, both have equated Thai women with stereotypes of 'trafficked' 'mail-order brides'. The focus of my study is on the 22 interviews with people in the 'children's' generation, most of who were 'multiracialised'. I treat multiracialisation as a process by which people become constituted as 'mixed race', a concept that I open up to contestation in my thesis (Haritaworn 2005).

My chapter problematises this matrix in several respects. First, it complicates the hierarchies it posits between heterosexual and homosexual, male and female, masculine and feminine. In the place of simple hierarchies, the interviews point to multiple differentiation, including between gay people and between women. There also emerge solidarities across multiple axes of gender, generation, sexuality and class – such as between a gay man and his formerly sex-working mother, or between a butch lesbian and her heterosexual feminine friends.

My polemical question 'Heterosexual or interracial matrix?' does not foreclose white queer alliances with queer and non-queer people of colour. Rather, it attempts to intervene in a sexuality studies that ignores all but the most privileged sexualities (Haritaworn et al. forthcoming). Minoritised feminists in particular (e.g. Collins 1990; Morris 1998; Skeggs 1997; Wilchins 1997) have described the exclusion of racialised, disabled, transsexual and working-class people, including heterosexuals, from many of the privileges and ideals described as 'heterosexual' by sexualities scholars. This points to the necessity of replacing a single-issue politics of heteronormativity with a theory and practice of positionality that invites multiple perspectives.

Heterosexual or Interracial Matrix? Racialised 'Transgression' in Judith Butler's 'Gender is Burning'

Judith Butler's early work (1990), which was foundational of the new discipline of queer theory, privileged interraciality and other forms of cross-category sex as 'transgressions' of the 'heterosexual matrix'.[2] Central to this claim were two male-to-female transsexual women of colour, Octavia St. Laurent and Venus Xtravaganza. They had been among the subjects of *Paris is Burning*, Jenny Livingston's widely received film about the Harlem house/ball scene. In 'Gender is burning', Butler (1993b) defended the film from black feminist bell hooks, who had accused Livingston of being an unreflected white ethnographer (hooks 1992). She also defended her own concept of transgression, which had been attacked as power evasive.

In the following quote, Butler asked whether Livingston – who, like her, was white, non-trans and a lesbian – desired Octavia St. Laurent, a black transsexual woman who participated in her film. This desire across differences, Butler proposed, would render Livingston no longer racist but, rather, queer and transgressive:

> What would it mean to say that Octavia [sic] is Jennie Livingston's [sic] kind of girl? Is the category or, indeed, 'the position' of white lesbian disrupted by such a claim? If this is the production of the black transsexual for an exoticizing white gaze, is it not also the transsexualization of lesbian desire?
>
> (Butler 1990, 135)

---

2   Butler defined her 'heterosexual matrix' as 'a hegemonic discursive/epistemic model of gender intelligibility that assumes that for bodies to cohere and make sense there must be a stable sex expressed through a stable gender (masculine expresses male, feminine expresses female) that is oppositionally and hierarchically defined through the compulsory practice of heterosexuality' (Butler 1990: 151, footnote 6). For a critique of transphobia in 'Gender is Burning' see Prosser (1998) and Haritaworn (2005).

There are a number of problems with this representation, beginning with Octavia St. Laurent's infantilisation as a 'girl' who is addressed by her first name only. What strikes me most, however, is its un-self-conscious abolition of the idea of positionality. Butler made the desire for the 'Other' 'transgressive' not only of the boundaries between whites and people of colour, trans and non-trans people, but of these entire categories themselves. Sexualising St. Laurent, in other words, made Livingston not only less racist and (though this was hardly politicised here) transphobic; it also made her desire 'transsexual' and her, in turn, less of a 'white lesbian'.[3]

St. Laurent's status as the raw material of transgression at first sight contradicted the tenor of the article, which treated (especially racialised) transsexuals as transgression's main 'Other'.[4] This becomes apparent in Butler's representation of the second transsexual woman of colour, Venus Xtravaganza, who was murdered by a transphobic client during the making of the film. Jay Prosser (1998) has highlighted the contrast between Butler's indifferent reaction to Xtravaganza's death and her passionate condemnation of Xtravaganza's supposedly 'heteronormative' wish to modify her body. If St. Laurent had become 'transgressive' in her sexual objectifiability, Xtravaganza's desire for subjectivity and self-determination rendered her *anti*-transgressive. In Butler's theory of the heterosexual matrix, then, trans people of colour were only transgressive if they satisfied the desires of the dominant subject. As Prosser argued persuasively, Butler's queer project included minoritised 'Others' only if they were willing and able to remain its objects. It included them by excluding them from real agency.

Nevertheless, Butler's thoughts on interraciality and transgression have influenced newer theories of multiraciality (e.g. Streeter 1996; Allman 1996; Spickard 2001). Theorists and activists have taken up the invitation to present multiraciality as a particularly good example of the postmodern category crisis. Multiracialised people are argued to tear down the binaries not only between whites and people of colour, but also between men and women and straights and gays. This often relies on a conflation of multiraciality, bisexuality (e.g. Kich 1996), transgender and transsexuality (Shrage 1997) as similarly 'ambiguous' positionings. For example, Lani Kaahumanu, a non-trans bisexual activist of colour, proclaimed in her speech at the Gay and Lesbian March on Washington in April 1993:

---

3   There are certainly contexts in which the relationship between a white non-trans woman and a black trans woman bears transgressive possibilities precisely because their mutual positioning transgresses dominant ideologies of reproduction and desire/ability. What I take issue with here are the terms of transgression in Butler's account, which locate transgressive agency exclusively with the dominant person and reduce the minoritised person to an object of her narcissistic fantasies.

4   See Jay Prosser (1998) for a critique of how this text served to divide transgendered, especially drag, identities, which Butler treated as queer *par excellence*, from transsexuals, whom she viewed as dupes of the 'heterosexual matrix'. The chapter under discussion (Butler 1993), which responded to her critics by stating the 'limits of drag', was central to this division. This description of transsexuality as 'drag' is extremely transphobic. At the same time, as Namaste (1996b) argued, Butler's representations of drag identities, too, were ignorant of actual drag realities. This illustrates a general problem with queer discourse, namely the predominance of writing by non-trans dominant lesbians and gays.

Like multiculturalism, mixed race heritage and bi-racial relationships, both the bisexual and transgender movements expose and politicise the middle ground. Each shows there is no separation: that each and every one of us is part of a fluid social, sexual, and gender dynamic. Each signals a change, a fundamental change in the way our society is organised. (quoted in Williams-León 2001, 150)

Laurie Shrage's (1997) 'Passing beyond the other race or sex' illustrates some of the problems with this approach. From her white and presumably non-trans perspective, she argues:

Do people who live in multiracial families have obvious racial identities? Should the white parent of black children be designated 'white'? ... Would the designations 'mulatta' or 'transracial' help [to describe the white mother of black children], in the way that the categories 'transgender' and 'queer' help to contest other pernicious systems of classification? (Shrage 1997, 187–8)

In the same passage, Shrage compares a transsexual man's claim to maleness to a white person's claim to blackness. This comparison is deeply depoliticising of the opposing extremes that these two identities occupy in each system of oppression. It also collapses the two positions into each other, as if trans people could never be of colour and multiracialised people never trans, and as if both had exited the social as a result of their supposed 'ambiguity'.

As with Butler's 'transsexualised lesbian', the result is a complete depositioning of privilege. Not only does Shrage question the boundaries between black and white, she *turns* whites into blacks. Just like Butler, she sexualises this process of de-positioning dominant identities. Thus, "we should [not] be prudish about behaviours that 'crossover' or 'pass' into conventionally forbidden race or sex territories" (Shrage 1997, 189).

The sexual excitement that Butler and Shrage attach to multiraciality, transsexuality and sexual practices involving racialised and trans people[5] is reflective of wider trends, old and new. Thus, the fascination with particularised bodies and sexualities is nothing new (Ahmed 1997; Streeter 1996; Wilchins 1997). What is new, however, is the celebration of these supposed 'ambiguities' in mainstream theoretical and popular discourses (e.g. Haraway 1997) as the progressive, and presumably young, identities and practices of the future. I have argued here that this celebration depends on an abolition of positionality. This abolition problematically coincides with the historical moment at which the oppressed enter into representation (e.g. hooks 1990; Ng 1997; Moya 1997). The notion of 'ambiguity' as a pre-social property of certain bodies and relationships that Butler and her followers – ironically – subscribe to becomes further problematic when we examine the symbolic and material violence in the lives of those who appear to embody this 'ambiguity'. In the following, I will do this empirically with regard to the interview accounts I generated with multiracialised people of Thai descent. These accounts illustrate how notions of 'ambiguity' and 'transgression' are socially constructed in multiple overlapping power relations and,

---

5   In this, some types of cross-category sex are 'sexier' than others. Thus, heterosexuality is exempt from this celebration and considered the very opposite of the transgressive.

thus, complicate one-dimensional and de-contextualised hierarchies such as Butler's 'heterosexual matrix'.

Positioning the 'Heterosexual Matrix'

> If I'm out with my parents, they [people in the street] will see the mix, like husband, wife, child and boy-friend. And, you know, the child's boy-friend, and they ... it's very hard to understand. And they do look at you. And, you do notice it more when there's been a documentary on TV ... when there's been something about the sex industry in Thailand, or even something about Thailand that's controversial.

In this extract, 'Phil Taylor',[6] a gay man of English and Thai descent in his mid twenties, raised similar questions to those raised by Butler. How are this older white man, older Asian woman, younger multiracialised man and younger white man related? What sexual and moral 'truths' can we read off their combined phenotypes? Unlike in Butler, however, the analysis co-produced by Phil and me was dominated by heterosexuality rather than queer sexuality.

Phil did not employ a reductionist homonormative framework to make sense of the stares that attempted to dissect his family and find its moral truth.[7] Rather, the multiple discourses on race, gender, sexuality and sex work that were repeated in the act of dissection required multi-focal lenses. Elsewhere in the interview, Phil identified sex-work phobia as a major basis of these reading practices. Thus, people's first reactions when he revealed his part-Thai descent was often: "Oh, your father's English, oh is your mother, you know, working in a brothel?"

Female interviewees in particular, both lesbian and heterosexual, were forced to struggle with the sexualisation of their racialised femininities. Many were aware of exclusionary practices on the levels of both the state and civil society. Bee Sornrabiab, a Berliner in her late teens, critiqued intrusive immigration interviews: "Oh, Thai-woman. So what are you planning to do here? Where do you live? What are your intentions?" It was not in their homosexuality but in their heterosexuality and interraciality that Thai women were immoralised. Unlike dominant heterosexual femininities, however, that are constructed as the objects of protection by patriarchal institutions (Walby 1989), this type of heterosexual femininity was constructed as a moral danger from which these institutions had to protect the nation (see hooks 1981; Collins 1990; Skeggs 1997).

Bee described the street as a further place where Thai women were excluded from the privileges associated with dominant femininity:

---

6   To protect the anonymity of the interviewees, I have chosen pseudonyms for them.

7   I have adopted Fanon's (1986) concept of dissection since it denaturalises the intrusive stares and questions that many of my interviewees and their families had to negotiate. Rather than accept them as 'normal reactions to abnormal bodies', it questions the entitlement that some feel to evaluate others and the historical relations through which some embodiments, subjectivities and genealogies become constructed as 'knowable' and recognisable through this dominant gaze.

Bee:	When I go out with my women friends, many men go like 'Yeah, and how much? How much do you charge? *[In English:] How many?'* and stuff.
Jin:	They talk to you in English?
Bee:	Yes. Yeah, as if they'd met them on holiday, like 'Yes, and *[in English:] how many? I will knock* with you'.
Jin:	Like, more to your friends or to you?
Bee:	Nah, to my friends because they dress like, they run around, for me they run around quite normally. Some say [laughing] they run around with too much make up. Or really a bit like such women, but still, it's typical. Or one of them shouts: 'So, how about a massage?' Thai massage. And gestures like this: 'Hahaha.' Yes, my friends get a lot of sexual harassment. As if Thai women were only good for one thing.

Bee's account contradicts Butler's account of feminine heterosexual women as privileged. It also contradicts accounts of heterosexual femininity as entitled to chivalry (Walby 1989). Bee's feminine heterosexual friends were not given privileges or protection by men, but were, on the contrary, seen to invite their unfettered sexual harassment. Bee herself, who was a butch lesbian, did not in this part of her interview describe herself as the object of such attention. Her gender presentation was indeed unintelligible in the heterosexual gaze. However, this unintelligibility was a result of the whiteness of this gaze as much as of its heterosexuality. In a dominant imagination where Thai women could only be hyper-feminine, Thai butches simply did not exist.

Bee's racialised gender identity was not simply unintelligible as butch, but also illegible as female. Again, the matrix that rendered her gender unrecognisable was racialised as much as gendered. This was a strong theme in her account – her being forced to negotiate the legibility of her gender not only with white men but also with white women. Her following description of a situation in a club illustrates this:

As soon as I walk into the loo [toilets], the girls start screaming: 'Hey, you're wrong here! Boys' toilet is next door.' 'No, I'm in the right place, you know.' 'You wish, I bet.' [laughs] I go like, 'No, I am right.' [pause] And they just look at me, walk past me and get the bouncer ... [describes planning her outings by avoiding the use of public toilets] A girl once grabbed me by the collar, properly. This big ... girl, this German, in a club. Pulls me by the collar and goes: 'Hey. Shortie. Ha ha. You want me to show you where the boys' toilet is?' I'm like: 'Please take your hands off my shirt, yeah? I'm in the right place. At least you could be nicer about it.' And she goes: 'I'm sorry, I thought you were a boy. I just want to defend us, yeah. Boys aren't allowed into our toilet.' I go, 'That's ok but ...' I don't mind that they don't recognise me. But I do mind when they touch me and stuff. And I don't really like it when the bouncers come. It gets so embarrassing then. All the girls come and watch what's going on in the girls' toilet.

Bee's account can be read within a growing body of theory and activism that opens up toilets as spaces of contestation between feminine women, androgynous or masculine women, and transgendered people; as well as between male-to-female transsexual women and non-trans women; and between working-class and middle-class women (Flintoft 2004; Skeggs 2001). This body of work qualifies a simple queer theory of the heterosexual matrix by pointing to the centrality of transphobia, gender phobia[8] and competing gender identities in the policing of gendered spaces. At the same time, Bee's account highlights the limitations of these writings, which have so far failed to interrogate how these spaces are also centrally racialised. Thus, the women's toilet from which Bee was excluded was not simply gendered but gendered according to white norms of femininity. In this white space, 'Thai women' were only recognisable and, thus also, permissible as hyper-feminine.

While Bee did not frequent the lesbian scene, she might have had similar experiences there. JeeYeun Lee's (1996) pioneering work pointed to the invisibility of Asian femininities in white lesbian spaces. Her interviewees had been treated as heterosexual or at least as bisexual, femme, or newly 'out'. They were not legible as 'lesbians' who had a right to enter 'lesbian' spaces. The tropes through which white lesbians 'recognised' Asian lesbians – as 'hyperfeminine' and 'heterosexual' – were the same as in wider Orientalist discourses of Asian femininity. This also rendered Asian butches invisible. Since 'Asian' and 'masculine' are a contradiction in terms, Asian girls could not be butch. They were either ridiculed as inauthentic or misread as men. This mirrors my own observations of white lesbian and bisexual women's spaces in Britain and Germany. One of my masculine friends, for example, was regularly told by bouncers in front of dyke bars that this was a women only night.

In what I have quoted from Bee's account so far, Bee identified herself as a target of racism and gender phobia at the hands of white women. She did not identify herself as a target of sexual harassment at the hands of white men, but as a witness of this sexualised type of racist violence against feminine Thai women. In the following description of why she had left her old school, however, she talked about how she herself was sexually harassed by white boys. Her gender expression and her racialised background emerged as the central targets of this harassment:

> I was thirteen or fourteen, and the boys were so stupid. I was already a tomboy after all. And the boys bullied me, saying dumb things like 'What's your mother's profession?' and stuff, and that's then how it started ... [Describes how she turned to teachers who withheld their support] I'm like 'Thanks a lot, you can't help me either. I can't just sit there when one of them whispers from behind "Hey. When are you gonna give me that massage?"' That's what was happening – it was really quite bad. [Describes leaving the school and going to the benefits office] She said: 'I think th e problem is not the schools. You're the problem.' I go, 'Why don't you go to a school like that for a month? As a foreigner, as a Thai, and as a woman. See for yourself what's going on there.' 'So, what's going on there?' 'Um... Have you ever been sexually harassed?'

The white boys referred Bee back into a heterosexual womanhood that was neither 'normal' nor 'privileged' and that was disentitled to chivalry or protection from the authorities in charge. Bee deviated from a heterosexual hyperfemininity to which white men claimed free and unfettered sexual access. In contrast to a simple heterosexual matrix, Bee was not simply referred back into a 'respectable' heterosexuality that reminded her of her 'natural' role as a 'good' moral woman. Rather, her 'natural' role

---

8   Transphobia is the oppression of transgendered and transsexual people. 'Gender phobia' describes discourses and practices that enforce adherence to dominant norms of masculinity and femininity.

into which she was referred was that of an interracial sex object. This explains the brutality with which Bee's gender and sexuality were punished. Bee deviated from a subject positioning that was already intrinsically and irreparably 'disrespectable'.

Lesbian gender identities are an increasingly popular topic for queer theorists (e.g. Halberstam 1998). Bee's account, however, exceeds and contradicts 'queer' notions of heteronormativity. Her analysis of her own situation and that of feminine Thai women bore more similarity to a politics of solidarity than one of deconstruction and transgression. It opened up interracialised sexuality, not as a queer practice that put the essentialist category Thai femininity into crisis, but as a racist practice that constructed Thai femininity as inferior and available to white people.

Heterosexuality in these accounts, then, was not the privileged, powerful 'Other' that queer theorists have described. Bee's powerful 'Others' were not the heterosexual Thai woman who was harassed in the street, but rather the white girl who policed 'our toilets', the white bully who sexually harassed her, and the white bouncers, teachers and social workers who collaborated in these acts.

The simultaneity of sexism, heterosexism and racism also emerged from Watcharin Ekchai's account, a gay man of white German and Thai parentage:

Jin: What kind of stereotypes have you come across in German people about Thainess or Thai people and Thailand itself?

Watcharin: Well, that Thais are especially nice and friendly or laugh a lot. And that they are really good, like, at servicing, serving in restaurants or [laughs], whatever, stewardesses and stuff – that is always much appreciated. Um, and then stereotypes or certain ideas that Asian women are somewhat loose or at least you think that an Asian woman could also be a prostitute, yes, this idea exists very, very strongly in the German frame of mind. Um, and that Asian men aren't real men anyway – there is also always this idea you see. Like 'They're all little wusses' and you have to prove it to them. [Watcharin and Jin laugh]

Jin: Prove it to them?

Watcharin: Yeah, yeah. That you're a real man.

Jin: ... Does that also have effects on relationships between white men and Asian men, you think? Like this idea that Asian men aren't real men?

Watcharin: I keep wondering how the roles are distributed between these two. Between Asian men and European men. Whether that's always influenced by a certain role distribution. Whether that's always defined the same way, Asian man equals passive and European man equals active, you know? Which is often the case but needn't always be like that. Yeah, and somehow I find this quite stupid. But this is the idea that people have, I think. Because they judge according to physique, according to looks. Like someone who is petite is automatically passive. You know? That basically every [laughs] second Thai man is somehow gay. I think this also exists in people's heads.

Watcharin linked the heterosexism that constructed Thai femininities with the *homo*sexism that constructed Thai masculinities. The common link between these constructions was not the heterosexual, but rather an interracial 'matrix', which positioned Asians sexually and otherwise at the service of – straight or gay – whites.

Like Bee's, Watcharin's analysis departed from a homonormative 'heteronormativity' concept that views interracialised sexuality as an engine of boundary deconstruction. Rather, sexual encounters between whites and Asians were negotiated against the backdrop of a colonial division of labour where Asians were ascribed servicing positions not only on airplanes but also in bed. They were negotiated in the context of a gay scene that, despite its claims to equality, treated Asianness and femininity as one and the same and inferior. Watcharin's account highlights the need to depart from the dominant queer practice of celebrating racist and sexist gender ascriptions as raw materials for dominant acts of transgression, and to search for a language that is able to name and critique such acts as violent.

Conclusion: How Do We Tell the Difference?

Resisting the queer call to 'transsexualise the white lesbian' also means resisting the collapse of all Thai multiracialities into one omniscient standpoint. One of the central findings of my study (Haritaworn 2005) was that Thai multiracialities are internally differentiated and complicit in hierarchical discourses of race, class, disability and gender. This also extended to the interview encounter. I became aware of this when reflecting on something that Phil told me about being the son of a former sex worker:

> I don't deny it – I laugh about it, I joke about it. I just say 'Oh god!' You know, like my friends will say 'Oh, yeah, Phil and his podium dancing.' I say: [camp voice] 'Yeah, I got it from my mother, she was a whore.' [Phil and Jin laugh] And it's okay because ... and they're quite shocked the fact that I've just said that. But yet because I, I've accepted it, and I haven't got any qualms about ... that side of it, I'm not embarrassed any more, I'm old enough to be able to choose my friends, to be able to accommodate that background.

Phil chose to describe his relationship to his gay friends to me by drawing on the discursive repertoire of coming out and the style of camp, a repertoire that originated in our shared imagined queer community. As a 'hasbian',[9] I grammatically mastered this language, yet my understanding of the lives that it described was limited by my privileges vis-à-vis sex workers and their children. While borrowing from queer discourse, Phil's analysis departed from a naïve queer collapse between homosexuality, Thainess, and sex work. I read his, instead, as a multi-issue project of alliance, which acknowledged his differences from his white gay friends; his non-queer, ex-sex working, heterosexual mother; and me. Such alliance was not the automatic outcome of any shared 'queerness' but of attempts to position oneself both socially and politically. It also involved coming to terms with pathologisation and the shaming injuries of the past, and it involved becoming an ally to those one

---

9   My concept 'hasbian' borrows from Arlene Stein's (1997) interviews with women who had claimed lesbian identities during the hegemony of radical feminism in the 1970s and the 1980s and who had later returned to heterosexual lives. I extend it to other female assigned identities in various historical and biographical contexts, including my own currently shifting identity.

had previously left behind in order to fit in with queer and other single-issue identity politics.

If Ken Plummer (2003) has described the liberating effect of sexual stories, the most important ones have been those that have broken the silences of minority discourse itself. These include lesbian narratives of disability (Shakespeare et al. 1996) and lesbian, transgendered and transsexual ones of child sexual abuse (Allison 1996; Cvetkovich 2003; Forge Forward 2004), all of which resist their sacrificial offering by those who claim to disprove the dominant aetiologies that represent these groups as dysfunctional. Similarly, overlapping narratives of sex work, racism and multiraciality are entering the public realm. In 'Femme-Inism: Lessons from my mother', Paula Austin (2002) honours her mother, a Latina sex worker, as the role model for her own lesbian femme gender identity. Jacqueline Ching Black's (1994) '24 Frames' is a collage of her white father (looking sternly ahead in his military uniform), her Asian mother (dressed in a silk blouse and smiling at the onlooker), and her own naked torso. In writing, too, Amerasians have made sense of their complicated American military genealogies (Williams 1992; Ward 1997). Asian women who migrated by using marriage agencies (Preecha 1989; Larsen 1989) have started to share their stories autobiographically, albeit so far mostly by distancing themselves from 'real prostitutes' (see also Hayslip and Wurts 1989). The 'typical true story' is claiming its space at the margins of the overlapping social movements of anti-racism, feminism, disability rights, sex workers' rights, and sexual and trans liberation. It speaks of complexity rather than ambiguity, and it opens up the possibility of contradictory positions that are not in themselves essentially transgressive but interact with others in ways that can be both painful and liberating.

The growing accessibility of these narratives helps me honour the courage with which my interviewees invented viable identities from the ashes of multiple pathologisation.[10] The interview accounts challenged a theorisation of Thai and other interracialities as queer transgressions of the 'heterosexual matrix'. First, the heterosexual femininity described here entailed few privileges. Rather than warrant patriarchal protection, it was constructed as dangerous, immoral and sexually available to white men. Second, the homosexual readings of Thai men extended to – indeed were most visible in – gay spaces.

These positionings cannot be collapsed into a single queer space. The narratives in this sexually mixed, non-trans and non-sex-working sample included alliances with racialised heterosexuals, differences from white women and gay men, and solidarity with or disidentifications from sex workers and transsexuals. They contradict a vulgar constructivism that can tell neither these differences nor these similarities and that collapses all of these positionings into one another. In its place, I have proposed a discourse of resistance and alliance, which entitles people to name and challenge the systems that subordinate and exclude them and the people with whom they share their lives.

---

10 See Skeggs (2004) on the evaluation of identities as valuable or pathological. It should become clear from this passage that I am challenging negative and disempowering constructions of survivors of intimate or collective violence as victims by pointing to all these instances of agency.

Chapter 9

Drag Queens and Drab Dykes: Deploying and Deploring Femininities

Kath Browne

Geographies of sexualities have traditionally focused on the heterosexual/homosexual divide in the control and policing of spaces and territories. Yet, categories of sexualities (heterosexual, homosexual, lesbian, gay and bisexual) are all dependant on dualistic understandings of man/woman, usually as opposite sexes. Consequently, terms such as 'same-sex desire' only make sense within dichotomies that view man and woman as separate sexes. In this chapter, I wish to highlight the complex ways in which specific 'women' and 'men' can deploy and deplore[1] gender/sex (terms I will explain below). Here, femininity is the focus with respect to male and female embodiments. This chapter will work across two spatialities – that of the sexed body (as sites of femininities) and that of the Dublin Pride festival, which describes itself as a celebration of the "diversity and richness of the Lesbian, Gay, Bisexual, Transgender and Queer community" (Dublin Pride (2006); see also Browne forthcoming). I will draw upon an incident that occurred at Dublin Pride 2003 and the subsequent 'heated debate' between drag queen Annie Balls and Pride organiser Izzy K in the Irish 'gay' (sic) magazine *Free!* to explore the oppositional narratives of femininities that (re)created the 'drab dyke' and the 'drag queen'. This chapter will not examine how bodies come to be sexed within the man/woman dualism, but will instead explore the performances and hierarchies that feminise bodies *irrespective* of man/woman boundaries.

In this chapter, I seek to explore the moments in which gender is brought into being, an exploration I undertake through a complex argument that juxtaposes drag queens and dykes and understands them as diversely (re)forming femininity[2] both in and beyond the presence or absence of vaginas, breasts and other embodied femininities. I will suggest that, rather than solely examining individuals who exist between the categories of man/woman, we also need to attend to those who

---

1   I use these terms throughout the paper to juxtapose the use and reuse of sex in intentional and specific ways with the animosity that can accompany 'other' potentially contrasting and conflicting deployments and practices of sex. These contrasts and conflicts are often part of the deployment of sex and serve to illustrate the fluidity, hybridity and performativities of sex.

2   I use the terms 'femininity' and 'femininities' loosely here to refer to those performativities that (re)make 'woman', 'lesbian' and 'drag queen'. The point is to argue for a heterogeneous understanding of how femininities can be diversely deployed and deplored across (being constructed by and reproducing) male and female embodiments.

(re)affirm these categories. Thus, whilst those who exist between these categories will be examined, the empirical section of this chapter will focus upon those who seek to reaffirm these categories. Prior to this, I will explore why gender fluidities need to be addressed in discussions of geographies of sexualities.

Gender/Sex Fluidities and Geographies of Sexualities

In deconstructing heteronormativity, it is important not only to render sexualities fluid, but also to contest the assumptions of sex and sexed bodies that often structure definitions of 'heterosexual', 'lesbian' and 'gay'. This is significant because where gender/sex are fluid, basing one's identity on sexuality becomes like "building your house on a foundation of pudding" (Scott 1997, 67). Challenging the primacy of sexual orientation and identity in queer studies, Stryker argues, "All to often queer remains a code word for 'gay' or 'lesbian', and all too often transgender phenomena are misapprehended through a lens that privileges sexual orientation and sexual identity as the primary means of differing from heteronormativity" (Stryker 2004, 214).

This chapter does not specifically address trans issues, but it does take up Stryker's point that to contest heteronormativity, we need to explore geographies of sexed body spaces beyond sexual orientation and identities. I seek not to argue for new formations, categories and categorisations of gender/sex, but instead aim to understand gender/sex performances, deployments and readings as bringing sexed bodies, identities and spaces into being. Emphasising the ways in which sexed bodies, identities and spaces are always becoming – and, specifically, the need to conceptualise these as momentary, fleeting and needing to be re-performed – enables the instability of the dichotomies of gender and sex (male/female, man/woman) to come into view. This approach is particularly important given that sexed and gendered body identities are often naturalised through these processes such that they are invisibilised. Performative geographies (see, for example, Gregson and Rose 2000; Rose 1999) that recognise the fluidity of place and space can further contest the dichotomies of man/woman, recognising the contextual contingency of bodies and their (re)formation.

Femininity is clearly a contested concept. Here, it is not understood as a set of attributes that are traditionally associated with women, but rather the performances that constitute 'women' in a variety of forms – in this case as both drag queens and lesbians. The term femininity is usually associated with gender, and both femininity and gender are used here not to suggest a biologically sexed body onto which gender is inscribed, but rather for the purposes of focusing on gendered enactments that (re)form sexed bodies (see Butler 1990). In understanding sex as a complex set of reiterated performances and materialities (see Butler 1990; 1993; Browne 2004; 2005), this chapter seeks to explore how embodied sites can be sexed through enactments and readings. I argue that performances solidify (around)[3] categories of

---

3   I use '(around)' to indicate the contingency of gender, race and sexuality as categories that are never solidified but that are often used to understand our everyday lives – hence, we (re)create our lives *around* such categories as 'lesbian' and 'woman'. These are, however,

sexuality and gender/sex in complex ways. Thus, in relation to recent discussions of performativity that see performativity as an ontological concept whereby we are constantly becoming through the enactment of events (see Latham with Conradson 2003; Thrift 2003), this chapter seeks to understand how the world is continuously (re)composed yet retains an emphasis on power relations as hierarchising and bringing social differences into being. In other words, sex is not merely performed, it is policed. Examining the re-placements of hierarchies highlights the fissures and fluidities of supposedly coherent gender-sex links. These fractures are often evidenced in the policing of gendered embodiments, actions, spaces and politics. This policing takes a variety of forms. In this chapter, the ways in which femininities come about through mutual processes of deploring and deploying will be explored.

Deconstructing normative categories of sex can reveal how all sexed bodies and gender enactments are fluid, performed and policed. The role of heteronormativity in the (re)creation of sexed bodies, identities and spaces has been widely challenged through explorations of those who live between the sex and gender dichotomies. Here, however, I seek to show how such a challenge to heteronormativity can also proceed through an understanding of the practices and lives of those who reaffirm sex and gender dichotomies in non-normative and ambiguous ways within the categories of LGBTQ (lesbian, gay, bisexual, trans and queer). Thus, in contesting the unity of gendered categories and looking at the ways in which they both solidify and breakdown, I seek to further illustrate the momentary construction of sexed bodies and their spatial contingencies.

Contesting the Fixity of Gender/Sex

This section brings together geographies of gender that explore the fluidity and (re)construction of gendered space with queer theories that conceptualise the fluidity of gender/sex boundaries. Theories of performativity have been used extensively in queer examinations of the metaphoric, lived and bodily spaces between man/ woman, illustrating that the binary division into male/female, man/woman is not necessary, given or stable. Rather these divisions are constantly remade and reused to create sexed places, bodies and spaces (see Browne 2004; forthcoming). This section will explore how different gender transgressions inform us of the limits of the dichotomous gender/sex system that (re)produces man/male and woman/female and (re)makes the hegemony of heterosexuality. It will challenge the separation of sexed bodies as the surfaces upon which gender is written, examining how intersexed individuals, trans individuals and those that live as a 'third sex' contest and resist normative dichotomies of gender and sex.

Intersexuality contests the 'naturalness' of a dichotomous system of sex. Intersexed individuals have sexual organs that have aspects of "both biological

---

ideals that can never be attained (Butler 1990). Rather, we both adapt such ideals and 'work around' or negotiate these unattainable representations, imaginings and performances. This, then, creates the categories themselves, as this chapter will argue, in diverse ways, and, in turn, makes us never quite within, yet not entirely outside, categories such as man, woman, heterosexual and homosexual.

sexes to a greater or lesser degree" (Mackie 2001, 186). Intersexed individuals do not simply form a third category of sex, but instead suggest the dubiousness of biological definitions of a dichotomous system of sex. These individuals across the globe have become increasingly subject to medical surveillance and intervention as their bodies are considered wrong, abnormal and indeed dangerous (see Roen 2001; Whittle 2000). These interventions reconstitute the fantasy and 'naturalness' of two anatomical sexes at the site of the body (Barnard 2000; Cream 1995; Hird 2000). 'Corrective' (and painful) surgeries, where persons have their gender reassigned later in their lifespan, are related to gender stereotyping of an individual's personality. For example, individuals who enact traditional masculine characteristics are 'given' their 'right' male body (Nataf 2001). In this way, the sexed 'geography closest in' (Rich 1986) is (re)made in accordance with social and cultural norms (Cream 1995).

Trans individuals are usually defined as people who live as members of the opposite sex to which they were born or as people who live *between* the categories of male/female or man/woman. West (2004) argues that the term 'trans' should be used in place of 'transgender' and 'transsexual' (usually defined by operative status) to indicate multiple and complex embodiments and subjectivities. Therefore, trans is more than a one-dimensional model, and diversity within trans should not be reduced to the pre/post operation transition nor simply understood in terms of the body (Turner 2001). Some trans individuals may seek to reinforce gender/sex categorisations such that gendered understandings result in surgical procedures that 'correct' gender dysphoria (Wilton 2000). Others, for numerous reasons, exist between man/woman, for example taking hormones but not undergoing surgical procedures that (re)construct genitalia. Stryker argues that trans identities and bodies "cut across existing sexualities, revealing in often unexpected ways the means through which all identities achieve their specificities" (Stryker 2003, 214). Trans bodies and transgressions can also dispute the process whereby the dichotomy of gendered norms are read off from what is taken as a permanent and uniform sexed body. It is often gendered social attributes (such as a 'caring personality') that are used to argue for an identity and body 'opposite' to ones birth gender. In other words, whilst it is assumed that bodies dictate gendered characteristics, it may be that gender characteristics, at times, dictate embodiments. Perhaps a more nuanced explanation for the interactions between gendered characteristics and sexed bodies would recognise their mutual interdependence such that they would not be able to be disentangled (see Butler 1993; Halberstam 1998).

In challenging the dichotomous categories of sex, I also wish to contest the understanding that sexed bodies or gender identities exist along a continuum. Even the introduction of a 'third sex' is problematic as this can serve to fix the other two sexes, balance the binary system and homogenise a plethora of gender identities within the category of 'other' (Halberstam 1998, 26, 28). Moreover, I would agree with Hird when she contends that, "Replacing a two-sex model with a ten sex (or twenty or thirty) model does not in itself secure the abolition of gender discrimination, only perhaps the mental gymnastics to justify it" (Hird 2000, 358; see also Grosz 1994). A continuum or a ten or twenty or thirty sex model would both assume fixed points. Rather than look to make new taxonomies, categories or divisions, this discussion will move on to explore the ways in which gender and sex have been conceptualised as fluid, before exploring how these categories are diversified and (re)constructed by practices that deplore and deploy them.

Drag kings and queens enacting hyperbolic femininities and masculinities for the purposes of parody have been read as questioning the normative linking of genders and sexes, illustrating that gendered identities are not naturally associated with specific bodies. Butler (1990) sees drag as subverting and playing with gender norms and identities and, in this way, having the potential to expose them as performative contingencies. Drag, Butler (1990) contends, has the capacity to imitate – just as gender itself is an imitation – but with a difference, and herein lies its subversive potential. In terms of space and place, drag kings and queens – in performing these hyperbolic impersonations whilst on stage in bars and clubs – overtly contest the normative genderings of these places. They illustrate that just as bodies are performatively gendered, so too is place and space. In other words, gendered embodiments create the very spaces they occur in. Playing not only with gender but also with gendered spatialities, drag acts can illustrate that how we act produces gendered spaces and that this production is not inevitable. For example, drag kings have been read as hyperbolically impersonating men in order to illustrate that masculinities are performed rather than inevitable (Halberstam 1998; Volcano and Halberstam 1999). During their performances, drag kings can change their gender identities as they 'move' through scenes. Such performances illustrate that bodies are mutable and can 'pass' as male or female depending on clothing, actions, and also context and place (see Halberstam 1998; Browne 2005).

However, and importantly, the subversive potential of drag also lies in its reading, a point debated in Bell et al. (1994). If the subversive potential of drag is contentious and if it is not read as contesting the normative gendering of places, the question that might be prompted is to ask whether drag merely reinforces dominant understandings of what is 'natural'. Butler (1993) problematises drag in *Bodies that Matter*. She argues that drag can potentially reinstate the presumed normality of heterosexuality and normative genders by acting as the 'other' and the 'abnormal' to those observing the performance. Moreover, drag acts are often located in particular places and spaces such as gay bars or specific drag restaurants; in this way, they may suggest that gender is fixed except in such spaces – on 'unreal' performance stages – and is not (re)performed daily within regimes of power (Browne 2005).

Whereas drag can be recuperated as a performance that is confined to particular places, often those such as nightclubs, bars and restaurants as well as floats during LGBTQ Pride parades, there are those who on a daily basis choose to live as or are mistaken for members of the 'opposite' sex to which they were born (see Browne 2004). Women who are mistaken for men have been read as contesting sexed embodiments and their naturalised link to gender categories. The ways these women's bodies are (re)read renders their sexed bodies unstable (and often 'not human' – referred to as 'it'), illustrating the need for bodies to be both performed and read in order to be 'properly' sexed (Browne 2004). Yet, there is the excess involved in reading a body that exceeds categorisation, and the attempt to shore up that categorisation through the performance and policing of reading *and its context* illustrates the contingency and limits of sex. In addition, the policing of 'mistaken' bodies shows that sex-specific spaces, for example toilets, need to be continually re-practised (Browne 2004). These places do not simply exist as women or men only spaces. Rather, they are recreated as such through reiterative acts and the normative assumptions they enact. Gender transgressive bodies can challenge (often unintentionally) the assumptions that these sexed spaces rely on, so questioning the very stability of sexed space. In this way, rather than being confined to particular places, contestations of normatively gendered and sexed spaces can occur in everyday spaces. Of course, acts and bodies that draw attention to the tenuousness of such norms are also resisted, drawing attention to the attempt to expel that which destabilises self/other dichotomies. Such resistances illustrate the diverse ways in which particular spaces and bodies are brought into being as intelligible and human within sexed dichotomies.

The celebration of drag as transgressive, contesting the assumed permanency of gender, has emphasised the playfulness of this art. In exploring the interactions between a drag queen and a lesbian in Dublin, this chapter seeks to continue the project of rendering explicit the re-creations of particular genders. In examining the diversity of femininities, I will attend to the practices by which these femininities are hierarchised, drawing attention to the centrality of policing and performativity to the constitution of sexed bodies. This problematisation of sexed identification is not posed in order to challenge the necessity of collective action. When rightwing Christian groups protested at Belfast Pride in 2003, there was a reaction from people across Ireland who came to Belfast Pride, swelling the numbers at the parade. This illustrated that LGBTQ populations still need to unite in the face of discrimination, prejudice and violence. It is, however, to purport that relations of power do not solely correspond to axes of gender and sexuality that differentiate them (straight or heteronormative) from us (LGBTQ). Thus, adding painful and constitutive power relations into our understandings of the (re)formation of gender identities beyond dichotomous sex requires an examination of the processes of power that (re)make and discriminate between 'us'.

Drab Dykes and Drag Queens: Deploying and Deploring Diverse Femininities at Dublin Pride

Pride spaces, where the LGBTQ 'community' supposedly comes together to celebrate, can also be where the diversity between 'us' can lead to disagreements. As the political and commercial 'gay' scene changes, Prides can offer a stage where the contestations between commercialisation and communities, between politics and party, and between the margins and the mainstream of 'gay' can play out. Pride festivals often trace their histories back to 1969 when drag queens 'rioted' after police harassment in New York. The politics of visibility and calls for equality, it is often claimed, are now lost in particular cities where the emphasis is on the celebration and the 'party' (see Browne forthcoming for a full discussion of these tensions). Changes in the leadership and 'ownership' of Prides have led to significant battles regarding their purpose and form. In 2003, Dublin Pride was experiencing a change in leadership. Two women who had been heavily involved in the organisation of Pride since the mid 1990s were stepping down. In the run up to Dublin Pride, they came under fire for supporting diverse community groups and for blocking the commercialisation of the event in the form of large sponsorship deals. They were accused of putting 'too much effort' into two weeks of 'community festival', which contained poetry readings, ceili and other low key and diverse events, rather than focusing on the 'main event', the parade through Dublin and the post-parade party in the park that took place at the end of the two week festival.

This section will explore the climax of this debate after Dublin Pride 2003, during which the question of how one performs the party took an interesting turn and came to highlight the (gendered?) power relations that (re)make diverse femininities across male and female embodiments. In addressing gendered body spaces through the performance of Pride celebrations, I will not reduce these power relations to discourses of patriarchal structures, which often rest on specific sexed embodiments (see Jeffreys 1994; Devor 1989), but will instead seek to explore how femininities come into being through mutual processes of deployment, performance and recognition. This requires moving beyond assumptions of sexed subjectivities that are grounded only in their embodiments. As the last section of this chapter contended, 'man' and 'woman' are not fixed, but continually come into being. Thus, this chapter contests the notion of gender as built onto already sexed bodies. I will explore the site of the body that deploys, deplores, performs, reads and is read, (re)creating multiple versions of 'female', 'woman' and 'femininity'.

During the park entertainment in Dublin, one act (not a drag act) chose to use their 'Stand by your van' skit, which was a comedic reference to the travelling community in Ireland. This act regularly performs in Gubu, a gay bar in Dublin. On this occasion, however, the act offended members of travelling community[4] (along with others sympathetic to their cause) who were present, some of whom approached the organisers of Dublin Pride. As Izzy K., one of the main organisers of Dublin Pride, accepted the thanks for her six years service to Pride, in her leaving speech she offered an apology for 'any offence caused' during the acts (K. 2003). This was met by boos and jeers from the drag queen performers who were lined up behind the stage on the verge that overlooked the amphitheatre and stage (Balls 2003).

In the subsequent August publication of *Free!* magazine (a magazine for the 'Irish Gay scene', edited and produced at the time by three men), prominent drag queen Annie Balls, who had performed in the post-parade entertainment (but not the act in question), wrote an article about the incident. S/he made sure it was clear that "most of the performers ... were really men in frocks" and proceeded to defend all drag queen performers, arguing that "we take the piss out of others all the time and ourselves frequently" (Balls 2003, 6). Clearly, humour is being used here as a technique of power. By contending that mocking and laughing at people is appropriate in these performance spaces and by justifying such mockery on the grounds that it is not 'serious', Annie Balls renders those who are 'mocked' unable to challenge this performance. By challenging the joke as offensive, one makes it 'serious' and thus misses the 'joke'. However, in contrast to Bakhtin's (1965) affirmation of the

---

4   In Ireland, racism towards the travelling community has a long history. This is a particularly sensitive issue where gay and lesbian travellers feel they do not 'fit' into either the LGBTQ or the travelling 'communities'.

inversion of social norms and of the parody of elites or the powerful at carnival events, Annie Balls lists those at whose expense the jokes were made as including "fat people" and "women", who were "being sent up and imitated" by drag queens (2003, 6). These "afflictions" are "fair game" because as Annie Balls (2003, 6) argues, "It may seem bitchy, nasty and disrespectful, but when done without malice and with a sense of fun – it makes fucking great comedy." Here, it is the 'sense of fun' and the intentions of the actors that are important. These, according to Annie Balls, make insults 'comedy' and, in turn, make drag queens funny. Annie deploys, amongst other things, female 'bitchiness' to create her persona as a drag queen and argues that this type of comedy is "an important part of being a drag queen ... and greatly define[s] the gay sense of humour/attitude" (2003, 6). In this way, Annie moves from this type of comedy as constituting drag queens to it being a defining element of 'gay'. The deliberate use of the term 'gay' could exclude those who do not identify with this category but instead relate to the more woman-identified 'lesbian' – a point that Izzy addresses and to which I will return to below.

However, this comedy can also be exclusive and marginalising. Annie Balls's final few comments highlight the tensions that not only underpin her assertions of fun but also the complexities, contestations and negotiations within a diverse LGBTQ collective that comes together to 'have fun', in this case at the Dublin Pride festival. In doing this, she (re)creates other, lesser 'women': "Izzy ... you are the weakest link – take your minority basket-weaving, drab dyke poetry and out-of-date oversensitive activism and leave the stage ... Goodbye!" (Balls 2003, 6). Here, Annie Balls invokes the critiques of the Dublin Pride committee that was headed by Izzy and her partner Hayley. Both these organisers had put time and effort into organising the 'community' activities of Dublin Pride that took place in the two weeks prior to the parade and after the events in the park, rather than mainly focusing Pride's resources and attention on one day or weekend.

These frank (and abusive) comments illuminate gendered power relations that cannot be understood within simplistic man/woman divides. The presence of 'fun' as a defining characteristic of Pride justifies Annie's position, both as an entertainer and as a female-impersonating drag queen. The importance Annie Balls places on her comedy is contrasted with "drab dyke poetry" and "out-of-date oversensitive activism" (in the Irish context, 'out of date' probably refers to the early 1990s when campaigns for equality were the focus of Pride events because homosexuality remained illegal in Ireland until 1992 – see Crone 1995; Connolly and O'Toole 2005; Kamikaze 1995). In juxtaposing these different emphases, it is clear that 'activism' (or even political action) is no longer central to some of those who attend Pride festivals. In Annie Balls's view, this is 'out-of-date', and Izzy, as a 'drab dyke', becomes the deplored female. The exclusivity of 'pleasure' is delimited by 'trendiness' as well as gender, marginalising the 'drab dyke' who does not fit into the ideal of 'gay chic' and who is not sexually attractive: "those around me wet their knickers gazing at her, not due to sexual attraction of any sort – lesbian, gay, bisexual, transgendered or queer" (Balls 2003, 6). Izzy is not only a drab dyke, but an unattractive one. The celebration of 'young' 'beautiful' wo/men bears striking similarities to Whittle's (1994b) assertions that Manchester's gay scene is being (re)made so as to exclude older gay men and so as to be based solely on consumption for pleasure. What I wish to argue is that these spaces of pleasure also (re)create specific genders, including those that do not 'fit' the man/woman divide.

Whilst I have argued that Annie both deployed and deplored femininities in her article, when she replied in the next edition (September) of *Free!*, Izzy K. read Annie Balls's remarks as patriarchal: "There wasn't a dry seat in our house [laughter] when we read her [Annie Balls's] impassioned plea for Freedom of Speech – just so long as she decides who's free to speak. Annie won't tolerate any nasty lesbians thinking they have rights too!" (K. 2003, 18). Izzy uses the female pronoun throughout her discussion, respecting Annie Balls's persona as a woman. She also uses humour, discussing Annie's 'knotted knickers' ('knickers' is a term used in Ireland and Britain to refer usually to feminine under garments). However, Izzy sees Annie as attempting to control how women other to her can speak. In particular, Izzy refers to herself as a 'nasty lesbian', reading her womanhood through Annie's eyes. By (re)constituting herself in relation to Annie, she renders the diversity between women visible and, perhaps more importantly, points to the power hierarchies that constitute these 'nasty lesbians', 'drab dykes' and Annie's 'drag queens'. Throughout Izzy's response, she notes Annie's reputation for 'lesbophobia', and she also remarks, "I have always presumed, incidentally, that Annie's obnoxious views represented a clumsy attempt at an unpleasant character. But recent events confirm that these are Leo's views!" (K. 2003, 18).

Izzy sees Annie as an "unpleasant character" but one that she addresses in her response. This character can have 'obnoxious views' on the basis of her drag play, but these views only become offensive once Izzy attributes them to a 'real' person – in this case Leo (Annie Balls). She moves Annie/Leo between female/male, arguing that her/his act is "a dykebashing act delivered straight out of the cave of Queer Pre-History, ineptly delivered by Fred Flintstone-in-a-frock" (K. 2003, 18). In order to do this, Izzy deploys Leo's masculinity here to support the claim of misogyny (not that it is only men who are misogynists!). She also deploys the *difference between sex and gender itself* in the assertion that these are *Leo's* real views – an appeal to sex, to biological reality – which, by appealing to this non-performed 'reality', works to subvert Annie's appeal to humour and parody. In this way, Izzy implicitly refers to the misogyny directed towards lesbians and specifically towards dykes, a misogyny therefore directed at specific kinds of deployments of gender. This is a misogyny that is dressed up 'in a frock' and that is couched in fun, laughter and comedy, hence the invocation of the comical character of Fred Flintstone. The invocation of a hypermasculine caveman character and a 'pre-historic queer' is used to associate Annie's attitude with something that should have been located in the past and not in contemporary 'queer' sites. It also deploys humour in a gender specific way, relating Annie to her 'real' body in a comical way that refers to Annie's own assertion of 'men in frocks'. Annie and Izzy, therefore, both try to consign each other to the history books (c.f. Annie's 'out of date activism'), and both, in attempting to negate the other, also (re)construct contemporary understandings of LGBTQ Prides, scenes, groups and individuals. Both look to 'rights' – in Annie's case for a particular form of 'fun'. Izzy's rephrases Annie's invocation of 'rights', drawing attention to how it only seems to pertain to those who, Izzy avers, belong to the past because they still enact misogynistic lesbophobic femininities through their drag acts.

Both Izzy and Annie see Annie as a 'man [Leo] in a frock', yet both allow for Annie's persona. They both, therefore, move between gender and sex in relying on Annie/Leo's 'real' sex but continuing to (re)create Annie as female. While redeploying 'woman', Annie continues to associate herself with male genitalia, both in her chosen name, Annie Balls, and also in the title of her column 'Ask me Bollox' ('bollocks' being a slang term for testicles). In other columns, Annie discusses herself in terms of female embodiment, such as having periods, and deploys innuendos about 'pussies' and stories about visiting the gynaecologist. She, thus, plays with the boundaries between man and woman and exists in the interstices of male/female.

Izzy's piece in *Free!* suggests that she respects Annie's choice of femininity but that she also distances herself from it. While she defines Annie as something other than a 'nasty lesbian', she still refers to her using the female pronoun. From Annie's point of view, on the other hand, Izzy becomes, at least in part, assured a gender status that is linked to her sexuality. Izzy's 'drab dyke' becomes rendered inferior to the 'fucking good comedy' of Annie and her drag queen associates. Yet, it is Izzy who claims the 'nasty lesbian' label, perhaps using self-depreciating humour to claim the moral high ground. Thus, the 'drab dyke' is contrasted with 'men in frocks' in a movement similar to Annie's that simultaneously deploys and deplores for, perhaps, different (more sophisticated?) purposes. Izzy returns to her embodied sex status in the title of her response, 'Ask my box', which juxtapositions her female anatomy ('box' is a slang term for the female genitals) with Annie's male 'bollox'. Thus, it is through the act of naming ('dyke', 'box' and 'lesbian'), as well as through associations with female genitals, that Izzy becomes (re)made within sexed and, more importantly, sexualised (lesbian) structures. Izzy's lesbian, dyke status is referred to more often than her femininity. Annie draws attention to Izzy's dyke status in a move that deploys this status as being 'out of date' and 'drab'. Izzy, on the other hand, lays claim to her dyke status to emphasise her lesbianism in the face of Annie's 'lesbophobic' act.

This exchange unsettles the simplistic dichotomies of man/woman and male/female by placing both Annie and Izzy within 'woman' or 'female' and thus illustrating the diversity within this category. Rather than a return to an understanding of drag as merely the masculine reconstruction of femininity (Jefferys 1994), these readings illustrate the complexities of female identities and male and female embodiments. They implicate Annie, Leo and Izzy in the (re)making of multiple, conflicting females that mutually (re)create drab dykes, lesbians and drag queens. These readings might have been recuperated within the categories of 'man' (bollox, Fred Flintstone) and 'woman' (box, dyke), but, in moving between male and female, this argument illustrates the gendered acts of naming that bring 'man' and 'woman' into existence. The movements between 'man' and 'women' and the transgressions of these boundaries may be momentary, fleeting and recuperating, but this debate shows how appeals to sexed embodiments are simply that – appeals carried out for rhetorical and political purposes.

Readings of the fluidity of sex/gender and contestations of dichotomies do not always recognise the practices that can simultaneously deploy and deplore, (re)making not just embodied hierarchies but also specific gendered contexts. Let me be clear, I am not arguing that all drag queens are misogynistic or lesbophobic, and nor do I wish to argue that Annie's acts are not funny, unserious and potentially destabilising of gender. Rather, I wish to contend that these acts (and possibly the motivations behind them) are not only productive of femininities, but can also be power laden. I am contending that when thinking about 'playing with gender', it is important to explore the interactions that (re)make specific contexts, genders and sexualities. Izzy's response highlights the exchanges that constitute these acts, pointing to the offence they can cause not because they are 'impersonating women', but because they are degrading specific 'women'. In examining how both Izzy and Annie deploy and deplore certain femininities, the power *relations* that (re)make these multiple femininities have been the focus. It is therefore possible to explore not just how 'woman' comes into being, but also how the performances of femininity/ woman become deployed and portrayed as fashionable, desirable and 'fun' or as the other to this – drab, pre-historic and nasty. Thus, the 'spaces of betweeness' (Rose 1999) that bring bodies and identities into being through constant processes of becoming are (re)made through interactions that create 'others' in complex and nuanced ways.

There are two further points that I would like to draw out of this debate: firstly, the importance of 'fun'; secondly, the place of gender/sex in these arguments. Firstly, those who do not take offence at the 'joke' and who wish to party without 'oversensitive activism' and 'poetry' dominate commercial scenes across the United Kingdom and Ireland. These pleasure based gatherings are (re)made through relations of power – not only heteronormativity, heterosexism and homophobia from 'outside', but also contestations within LGBTQ groupings. These power relations have yet to be extensively addressed, not just in terms of territorialisation and commercialisation but also in terms of how some forms of fun and enjoyment are seen as more desirable and are catered for more than others. What I have argued here is that these power relations have brought diverse and contested versions of 'woman' into being.

In relation to my second point, however, it would be too simplistic to read this example as merely patriarchal because to do so would cast Izzy in a subordinate role as a victim, whereas the processes of power that were operationalised were far more complex than this. Izzy did not seek to be included in Annie's 'fun' or even to address the gendering and sexualising hierarchies that Annie (re)created, but rather she challenged the prejudices and the personal attack she read in Annie's article. It could be contended that this debate was not even about 'gender' in the man/woman sense. Would Annie Balls have had a problem with a dyke who had laughed at her 'humour' and 'enjoyed the show'? Conversely, a white, middle class, gay man who had wished to listen to 'drab dyke poetry' might have felt marginalised by Annie's discourses (see Elder 2002). Rather than argue that these individuals can be exhaustively specified by age, class and sexuality, and rather than provide over-generalised categories and explanations, it is perhaps more fruitful to explore the discourses that exclude and/or seek to render activities and individuals 'out-of-date' and 'out-of-place'. Using the debate between Izzy and Annie, it has been possible to explore the flows, confusions and contradictions that re-create women, dykes and drag queens, illustrating a variety of sexed embodiments, gendered enactments and the messy entanglements of power that create gendered and sexualised identities that do not always neatly align along axes of man/woman, male/female or even gay/straight. More broadly, there may be a need for discourses that seek to create different, various and competing versions of the queer contemporary and that centre on acknowledging the performativity of gender and sexuality. Perhaps the confused appeals to embodiment discussed in this chapter speak of a problem whereby only *performance* itself, rather than the *performative*, has become validated as a tactic of queer life, queer becoming and queer fun.

Conclusion

In deconstructing heteronormativity from the perspective of gender/sex rather than sexual orientation and identity, this chapter has argued that gender transgressions highlight the contingency of gendered identities and embodiments, such that we need to explore how the dichotomies of man/woman and male/female come into being. Such moves can be politically problematic. Valentine and Skelton (2003) and Elder (2002) note the dangers of exposing fractures and fissures in queer communities and, in particular, of publicly representing power relations and tensions between queer individuals and groups. Overtly pointing to power relations within LGBTQ groupings must be undertaken cautiously as it may fuel the heteronormative and homophobic aspects of the discipline of geography and beyond (Elder 2002, 988). In this chapter, I have chosen to investigate the argument that appeared in the gay newspapers of Dublin in August and September 2003, highlighting the tensions that (re)made drab dykes and drag queens. In doing this, I hope to have illustrated that power relations operate within the queer community and should not be subsumed for the 'greater cause' (Pritchard et al. 1999). While acknowledging that homophobia, heterosexism and heteronormativity continue to homogenise us, it must also be acknowledged that relations of power internal to LGBTQ communities can also hierarchise 'us' within this 'sameness'. For the purposes of this chapter, I have explored how these processes, in part, bring gendered categories into existence but do not do so uniformly or within the boundaries of simplistic man/woman dichotomies.

The chapter contends that alongside Nast's (2002) contestation of queer patriarchies, which is firmly rooted in embodied subjectivities (white, gay men), we also need to explore the hierarchisations internal to LGBTQ groupings that constitute diverse genders and sexed embodiments. These relations of power may solidify (around) other categories such as race, sexuality and dis/ability, but it is important to explore the eventful creations of these categories and the assumptions that underpin – as well as differentiate – them. A discussion of how bodies become gendered – as well as classed, aged, racialised, ablised and so forth – and consequently marginalised and rendered inferior moves us beyond rigid identity boundaries and contests sexualities that are built on gendered assumptions. In other words, rather than discarding the salience of gender, we need to consider how (unequal) gendered/ sexed bodies are brought into being in particular spaces and times and not necessarily within man/woman dichotomies.

Chapter 10

The Queer Unwanted and Their Undesirable 'Otherness'

Mark E. Casey

Lesbian and gay commercial districts existing in the major urban centres of North America, the UK, Australia and Northern Europe have been understood since the late 1960s (see Achilles 1967; Levine 1977; Castells 1983 etc.) as providing places of shelter from a hostile and violent heteronormalised society and as providing sites of consumption, community, and political action. However, since the mid 1990s, a number of social, cultural, economic and political developments within metropolitan centres, particularly in Western countries, have had significant consequences for the position of lesbians and gay men within society and for the lesbian and/or gay scene. Consequently, which lesbians and gay men are to be either desirable or wanted – and, conversely, which are no longer to be desired or wanted – within contemporary gay and/or lesbian urban scenes (which can be characterised through their intense commercialisation) is increasingly problematic. So, too, is the related question of who positions them as desirable or not, wanted or not. This chapter draws upon data from 23 interviews[1] undertaken with lesbians and gay men living in Newcastle-upon-Tyne, UK,[2] and will explore the experiences of those lesbians and gay men deemed to be 'undesirable'. The chapter will also discuss how claims to inclusion and visibility within contemporary gay and/or lesbian commercial sites are mediated by age, gender, the complexities of disability and by ideas concerning what is considered 'sexually shameful'. Underpinning the discussion will be the key themes of (new) *inclusions* and *exclusions*, with the chapter borrowing from Binnie's (2004) theorising of the 'queer unwanted'.

The chapter will open with a discussion of the current literature focused upon contemporary gay and lesbian urban spaces. In particular, it begins by outlining how the increasing visibility of heterosexual bodies and the intense commodification and presence of 'big business' in gay and lesbian districts can be theorised as (hetero)normalising such sites, while positioning some gay men's and lesbians' visibility within these sites as increasingly undesirable. The second half of the

---

1   The research that this chapter draws upon was undertaken during my PhD (2000–2004), which was funded by the Economic and Social Research Council (ESRC). The research sample consisted of 12 lesbians and 11 gay men, seven of whom were aged 18–25, eight 26–40 and eight 41 or over. Within the research sample, 12 self identified as working class, with the other 11 identifying as middle class.

2   Hereafter simply referred to as 'Newcastle'.

chapter focuses on the complexity of the multiple identities lesbians and gay men inhabit and on how these can be pivotal in claiming inclusion within or experiencing exclusion from commercial lesbian and gay spaces. The chapter will demonstrate how age, and in particular the aged body, is positioned as undesirable within the lesbian and gay scene of Newcastle as it increasingly targets the 'youth pound'. The chapter then problematises the exclusion of lesbians from a commercial scene that is dominated by both gay male venue owners and gay male clientele, a domination that places lesbian spaces and interests on the periphery. Finally, the chapter addresses how discriminatory practices within the lesbian and gay commercial scene – with its presumption of the able body – marginalise those lesbians (and gay men) with disabilities.

De-sexualised Commodities?

Since the late 1970s in North America and the early 1990s in the UK, lesbian and gay communities, gay and lesbian events and the material spaces they occupy have increasingly been theorised as yet further 'niche' markets for capitalism to target and exploit (Warner 1993; Whittle 1994a; Binnie 1995; Gabriel and Yang 1995; Miller 1998). For Diane Richardson (2005, 526), the movement of the market into non-heterosexual 'lifestyles' is particularly visible in the growing move from community based sponsorship of lesbian and gay events to an intensification of corporate sponsorship by 'mainstream' corporations such as Absolut Vodka, Red Bull, American Airlines, Virgin Atlantic and Qantas. Spectacles and events that celebrate lesbian and gay sexualities, such as the Sydney Mardi Gras, are, Valentine (2002a, 147) reminds us, no longer exclusively developed for local lesbian and gay communities, but are rather marketed for (global) lesbian, gay *and* heterosexual consumption. However, such mainstream sponsoring and global marketing of these events (once created to 'celebrate' the difference of non-heterosexual identities) towards heterosexual consumers can be understood as problematising which identities have the strongest claims to and visibility within such events. The visibility of heterosexual bodies and relationships at Pride marches, for example, may not be a reflection of a growing heterosexual tolerance of 'other sexualities'. Rather, it may be at the cost of lesbian and gay male visibility, reclaiming sites that might have been temporarily queered as dominantly and/or naturally heterosexual.[3]

The movement of the market into the previously criminalised and stigmatised lives of lesbians and gay men, although potentially problematic and exploitative, has undoubtedly contributed significantly to the creation and development of visible lesbian and gay districts. Borrowing from the success of gay identified districts in 'world cities' such as New York, Sydney and Paris (Connell 2000; Sant and Waitt 2000), secondary cities, or what Rushbrook (2002) has termed 'wannabe world cities',

---

3   The three year deal announced in 2004 for the Sydney Mardi Gras to be sponsored by the predominantly gay male website 'Gaydar' (where men, after paying a small membership fee, can log-on to meet friends, partners or other men for (easy) casual sex) from 2005 – 2007 may signal a gradual reclaiming of Mardi Gras as an event targeted at and created for non-heterosexuals, especially gay men and men who have sex with men.

have engaged in competitive strategies to attract capital. Such cities have, to varying degrees of success, re-created themselves as places of culture and consumption in attempts to lay claim to a certain cosmopolitanism that labels them participants in the global economy of the new millennium. In particular, Manchester and Newcastle in the UK and Melbourne in Australia have embraced the 'gay as now' model in their re-brandings – a present defined in global and cosmopolitan terms, mirrored in their respective high investment, stylised and heavily marketed 'gay districts'. However, Lisa Duggan, for one, is particularly critical of such urban spaces. For Duggan, sites such as the Castro district in San Francisco that were once central in the forming of gay political visibility, identities and feelings of community (see Castells 1983) are now complicit in reducing the 'gay public sphere' to sites of consumption and gentrified neighbourhoods. Districts that once existed on the periphery of both city centres and most heterosexuals' urban lives, through gentrification, a growing desire to engage with lesbian and gay night time spaces (see Chatterton and Hollands 2003) and a cosmopolitanisation of cityscapes and lifestyles, are witnessing the removal of unsafe and/or threatening sexualities. Concomitantly, visibility in these sites is increasingly limited to desirable and/or (hetero)normalised performances of a lesbian or gay identity, a trend that is part of what Duggan terms 'the new homonormativity' (2002).

David Bell and Jon Binnie (2004), in developing Duggan's concept of a 'new homonormativity', argue that it is now central in producing specific forms of commercialised, mainstreamed and non-threatening 'gay spaces' and identities. They suggest that as cities throughout the world weave commodified gay spaces into their promotional campaigns, a repertoire of themed 'gay villages' that are all too alarmingly similar is being produced. In turn, visibility and an ability to make claims to citizenship rights – once denied to lesbians and gay men, but now increasingly being extended outwards – are, nonetheless, limited to 'non-threatening' lesbians and gay men. Bell and Binnie remind us that "the availability of forms of public space is contingent, meaning that aspects of queer culture are rendered invisible and are denied access to the same public space being claimed as a *right* by the newly empowered sexual citizens" (2004, 1811, emphasis added).

As specific gay urban sites are assimilated into the wider urban framework, one door is opening and another is being closed as to who may enter and be visible within such sites and as to what spaces and sexual identities can emerge. For example, through current large-scale gentrification, which many city councils and city planners are embracing in the re-development and re-branding of the cityscape, gay sex zones are being sanitised or removed. Those who have previously partaken in these sites and their activities become rendered invisible, unwanted and/or criminal. For those community groups and researchers (e.g. Dangerous Bedfellows 1996; Delany 1999) who have focused on such gentrification and its consequences for sexualised zones, the loss of 'sex and sleaze' from Times Square, New York and the influx of mainstream, large scale corporations – its 'Disneyfication' – has, within the context of a wider de-sexualising of New York, been a primary concern. However, for both Rotello (1997) and Signorile (1997), gay men should adopt a non-threatening (homo)normative lifestyle through rejecting partaking in sexual encounters that can be deemed 'casual', rejecting multiple partners and supporting stricter controls of sexual activity, particularly within public sites (see Leap 1997).

The city of Newcastle can also be characterised as experiencing a growing sanitisation of public sex zones and, in turn, the removal from public sites of undesirable (visible) gay male sexual identities and practices. Such sanitisation has been celebrated within the city's press as the transformation of 'notorious gay haunts' into 'redevelopment showpieces'. This drive for respectability was particularly intensified during the city's bid for the title of European Capital of Culture 2008 (Young 2003)[4] when a heavily stylised, safe and desexualised gay district was used to epitomise an increasingly tolerant and cosmopolitan Newcastle. The city of Newcastle since the late 1990s can be theorised as adopting a post-industrial redevelopment trajectory similar to those adopted by other UK cities such as Manchester in the late 1980s (see Quilley 1997). As part of this strategy and in its attempts to move away from its image of a city in long-term economic decline, the city has embraced (very successfully) the new leisure, tourism and lifestyle focused industries of the new millennium. As the city council and its city planners have embraced the 'gay as now' model in the development of its own 'gay village' and in the redevelopment of the wider city, it has also actively pursued the development and gentrification of cruising areas within the city into 'respectable' sites. This move can be clearly witnessed in the fencing and gating of many of the city's parks to limit access after dark and in the loss of the city's most infamous cruising to the development of a number of large scale 'city-living' apartment complexes. Somewhat ironically, these apartments have been targeted at the 'pink pounds' of the respectable, mainstreamed and, most importantly, rich lesbians and gay men in the city.

Such a de-sexualising of gayness and the current rise of a *'politics of normalisation'* is behind earlier arguments developed by conservative gay writers such as Bruce Bawer (1993) who asserts that the lifestyle of the mainstream gay is indistinguishable from that of most heterosexual couples in similar professional and economic circumstances (cf. Sullivan 1995). Richardson (2005, 519) argues that these writers have articulated a gay (although somewhat predominantly male) agenda that seems to de-radicalise political perspectives on sexuality and that argues instead for an assimilation into 'mainstream' society with enduring centrality for marriage and 'family values'. As a consequence of the 're-drawing of the boundary of unwantedness' (Serlin 1996; Roofes 2001), aspects of 'bad' gay culture are increasingly threatened as the assimilation of more mainstreamed aspects of gay culture gathers apace. Cruising grounds that for decades have been used in search of anonymous sexual pleasure are increasingly policed as 'unhealthy' sites of unsafe

---

4   Newcastle and Gateshead's vision for 2008 was all about transformation and cultural regeneration. Its profile developed for the bid suggested: "With the collapse of its heavy industry, Tyneside is undergoing an amazing transformation as it re-invents itself as a modern city with a vibrant cosmopolitan culture ... Culture is at the heart of its regeneration plans, with £3 billion investment and £350 million cultural investment promised ... Take a walk along Newcastle and Gateshead's Quaysides and you'll see the air of excitement with the recently opened BALTIC Centre for Contemporary Art, and the new Sage Music Centre thrusting upwards into the skyline" (BBC Capital of Culture 2005). The city was unsuccessful in its bid, the award being given to Liverpool.

and deviant sex, in turn, designing the bad gay out the cityscape – "out of site, out of mind?" (Casey 2005, 38).

The 'normalisation' of the presence of certain types of sanitised and de-sexualised gay spaces within many urban landscapes is resulting in the large scale growth and visibility of gentrified 'gay districts'. As has been posited by theorists such as Serlin (1996), Duggan (2002), Bell and Binnie (2004), Casey (2004) and Skeggs et al. (2004), and by community groups (e.g. Dangerous Bedfellows 1996), commercial lesbian and gay sites are increasingly important if a city is to be able to position itself as 'cosmopolitan' within the global city discourse. Although these sites are often marketed by a city itself, as well as national tourist boards, as lesbian and gay 'districts', they are simultaneously commodified as asexual. A growing literature is emerging on the exclusion of the sexually shameful lesbian and/or gay from gay commercial sites, as discussed, but it is important to acknowledge that lesbians and gay men do not have to be 'sexually shameful' to be excluded or, more importantly, to experience feelings of exclusion from commercial lesbian and gay districts. Other identities that lesbians and gay men make claim to or are positioned as possessing – often thought of as more 'traditional' identities such as age, gender and disability – may be intersecting with Duggan's (2002) 'new homonormativity', resulting in a positioning of 'the old', lesbians and those with disabilities as the 'queer unwanted' alongside the sexually shameful. In the following section, the experiences of older lesbians and gay men will be drawn upon to discuss how (older) age intersects with sexuality and urban spaces in creating a growing number of unwanted and excluded 'others'.

Aren't You Really Just Too Old to Party?

In November 1977 the *Christopher Street* periodical was asking its readers "where do all the old gays go?" with a poster in mid-1980s Toronto echoing this question in asking "what happens to homosexuals over 50?" Pugh (2002, 162) argues that the sentiments expressed here are still as valid in the early twenty first century as they were in the 1970s and 1980s with sexuality studies and social gerontology largely ignoring older lesbians and gay men. Ageism, for Hunter et al (1998), interacts with heterosexism in constructing a particularly negative stereotype of older lesbians and gay men. The fear of becoming old within Western societies has over the course of the 1990s intensified significantly, with a proliferation of television and magazine adverts for cosmetics to avoid the onset of old age. For those who haven't avoided such an onset, television programmes such as the UK's 'What Not to Wear' (BBC 1) and 'Ten Years Younger' (Channel 4) and the US's 'Queer Eye for the Straight Guy' (Living TV) decree the benefits of botox, facial fillers and teeth whitening to reverse the appearance of old age. The domination of the youthful body as exclusively sexually desirable and attractive within (Western) popular cultural discourse has the consequence that those who can no longer at least *appear* to be youthful are deemed to be no longer sexually desirable. However, Pugh (2002, 164) suggests that the media presentation of Viagra has begun to challenge the assumption that older men do not have sex. Through the media's attention to Viagra, older men and, consequently, heterosexual women's desire for sex is increasingly entering everyday popular discourse. Nonetheless, the implication that older men need chemical enhancement to develop an erection and have sex has done little to challenge the widespread belief that only the young can and should have sex.

As many gay scenes have been increasingly stylised, branded and marketed towards the youth market, identities other than sexual identity are affecting access to such sites. The occupying of lesbian and gay commercial spaces, and the movement into and out of such spaces, is becoming increasingly restricted for those deemed to be 'undesirable' (Shakespeare 2003; Taylor et al. 2003; Chatterton and Hollands 2003; Weeks 2003). Age can mark out older lesbians and gay men as 'undesirable others' whose movement into and use of such spaces has become increasingly restricted. Chris, who is in his late fifties, in discussing his own experiences of venues located within the Newcastle gay scene, reflects upon how his recent claims to a gay male identity position had not allowed him to experience feelings of inclusion and/or comfort in a number of gay identified venues:

> Not my scene definitely ... there was an area for dancing that seemed far too small for the number of people who wanted to dance, and it was LOUD! ... I went because there was a group of us who went and they said "do you fancy coming for a drink?" And it seemed a pity to say no all the time ... but it is not an experience that I would necessarily want to repeat.   (Chris, 59)

Here, Chris's experience reflects how age and the aged body affect his ability to engage or connect with the venue he enters and with other among the clientele. His inability to appreciate the "loud" music and to find comfort on the "far too small" dance floor reflects how some venues only cater for a limited range of capabilities for action and connection from an equally limited range of patrons. When asked why he would not wish to repeat this visit, Chris stated:

> Well they [other clientele] were much younger than me ... And it became an issue after a while as the longer you were there you thought this isn't my age, and this isn't my age group ... it is almost like saying I don't belong there.   (Chris, 59)

Within venues that identify as lesbian and/or gay, just to identify as a gay man does not lead to feelings of inclusion or belonging for Chris. His discomfort and sense of 'I don't belong here' in the 'young venue' setting indicates that sexual identity is but one of a number of 'identities' or criteria that influence feelings of inclusion, belonging and/or comfort in a commercialised gay space. As some of Ruth Holliday's video diarists expressed within her research, discomfort and feelings of being evaluated, of being judged and/or of not belonging can be experienced when going out on the 'gay scene' (Holliday 1999, 482). The growing tendency for both heterosexual and gay identified night time venues to cater for younger audiences can result in problems for those no longer considered 'young enough'.

> [P]ubs are kind of segregating people. 'SIMS' has become the skinheads and the older crowd ... but downstairs it is very male orientated. 'The Cat' has become female orientated upstairs, 'Thirst' has become the younger crowd, the drag queens and the older guys in 'The House' and some of the students, it is kind of segregating slowly.   (Matt, 22)

> [I]t's very much geared towards younger people. A lot of my friends are older than me – you know, 35 plus. 'The House' has made a big difference I have to say … as 'The House' is one place that you can actually hear yourself think. (Lauren, 23)

> Because the others are *so* loud! I just feel so old saying that! 'Violins' – you go there and you just feel like you are 140 years old! … well 'Violins' is alright – it's me that feels 140… (Kath, 40)

The presence of 'people their age' and a lack of loud music in a bar such as 'The House' is discussed by Lauren as an indicator that the venue is catering for 'older' lesbians and gay men. But, this venue is talked about as unique within Newcastle's commercial scene – "it is the one place you can hear yourself think" – reflecting how older lesbians and gay men are being pushed into a limited number of commercial spaces in a youth obsessed scene. Sarah Thornton, in her study of British club and bar culture, reminds us that door staff have the power to inform potential club users if they are too old or too young, denying them access at the point of entry (Thornton 1995). For those who do gain entry, feelings of comfort and/or inclusion can be fulfilled or denied through the type of music played and the ability to get served by (younger) bar staff; even if an upper age limit is not enforced, Jon Binnie suggests that "if you are not young and pretty" you can still feel excluded (Binnie 1995, 198). An increasing segregation within venues, as commented upon by Matt, may not be such a serious issue within larger urban centres where there is a greater venue choice or the targeting of specific niche markets. But, in Newcastle and other small urban areas such segregation is problematic and may become increasingly intensified as more venues chase after the pink youth pound. Although the importance of youthfulness on the gay scene may be nothing new, many research participants discussed how the venues in the Newcastle scene were increasingly focused upon catering for a 'youth market'. This growing focus on 'youthfulness' was experienced as problematic for those no longer considered as being 'youthful' (or at least 'youthful' as defined by door staff, bar workers, venue owners, other clientele and/or as defined by themselves), creating previously unknown age restrictions, discomfort and exclusions for accessing venues in Newcastle's commercial gay district.

> I think there is a certain set, or certain group, that is very body conscious and I suppose that puts quite a bit of pressure on you and that goes with the desire for eternal youth. (Liz, 30)

> I think the pubs and clubs in the city are *very* loud and *very* much aimed at young people … and it just occurred to me recently that I am probably too old for that anyway. (Margaret, 46)

The views expressed here have been found within other studies and, as Stephen Whittle reflects, "ironically beautiful young people don't need safe and tolerant places – because sex is always going to be easy for them; they are, after all, beautiful and desired" (1994b, 38). As Mark Simpson argues, "despite the commercial hijacking of gayness, not everyone is invited to the party" (1999, 213). The gay male subculture, in particular, is youth orientated, although this is not some essentialist characteristic, but the result of intensive commercialisation, branding and the value

ST. PATRICK'S
COLLEGE
LIBRARY

given to youthfulness and the 'body beautiful'. Echoing this, within Padva's (2002, 290) research, one of her interviewees suggested that: "gay magazines[:] it's all young and slim or big and muscular". As tolerance and acceptance of lesbian and gay men has grown in Western cultures, the young, beautiful and urban (white) gay man and lesbian have been adopted by the media as representing the acceptable and desirable face of the sexual other: value and status increasingly equals being young, attractive and physically fit (Attitude Magazine 2003). With the aged body having limited value within contemporary Western cultures, for gay men, midlife may present a crisis of identity and social interaction, with anxiety developing about the loss of sexual attractiveness and/or sexual performance. Older lesbians can face sexual discrimination, with views of older women in contemporary Western culture generally more negative than those of older men (Hunter et al 1998, 156). For the older lesbian, lack of comfort in entering venues where younger lesbians and/or gay men are present may result in self-exclusion from certain sites. As Margaret reflects, "I am probably too old for that anyway," with Liz echoing that the body consciousness and desire for 'eternal youth' of others among the clientele "[put] quite a bit of pressure on you."

Where Have All Those Lesbians Gone?

To theorise 'gay urban spaces' as gay male and lesbian as the chapter has done so far is to forget that both men and women, regardless of their sexual identity, continue to have unequal access to economic and cultural resources and power within public and semi-public spheres (see Castells 1983; McDowell 1995; Binnie and Valentine 1999; Valentine 2000a). Urban spaces are gendered, with women's traditional association with the spaces of the private sphere (although the prevalence of this association is declining) and the placing of men in the spaces of work and other public spheres continuing to affect how both men and women gain access to and experience the contemporary cityscape.

Gay men have been documented as more visible than lesbians within the urban landscape, with streets, bars, parks and events characterised as male space(s), particularly at night (D'Emilio 1993; Higgs 1998; Puar 2002a; Skeggs et al. 2004). However, several factors – including dependent children, fear of violence and the continued limited appeal of the 'lesbian pound' to capital – have impinged not only on lesbians' ability to create lesbian-identified space, but also on their ability to make such spaces as clearly defined as gay male sites (Lo and Healey 2000). The increasing cost of entry into venues and expensive drink prices, tied together with gay male spatial domination, may act as barriers for lesbians wishing to enter and consume commercial lesbian and gay spaces, reflecting the complex interplay of gender and money when claiming a spatial visibility. As Skeggs et al. claim of Manchester's gay village, "there is little that marks the Village out as lesbian. It is predominantly represented ... as gay = gay men" (2004, 1843).

In Newcastle, at the time of the research no venues maintained exclusive women-only nights within the city's gay scene. However, during the 1990s, two bars had existed that had maintained either women-only nights or women-only policies –

'The Cat' and 'Moon'. The importance that these two bars had had to their female users was clearly communicated during several of the interviews. For Lisa, each bar's existence as a women-only site had been central to her inclusion and visibility within the gay commercial district and within the wider lesbian communities of Newcastle.

> [W]ell I used to go to 'The Cat' when it was a women's bar ... upstairs for quite a long time, and then another one opened called 'Moon', but that was really dire. But I used to go there as you could go on your own.
>
> (Lisa, 36)

> [W]e approached Emma who used to have 'The Cat' and set up upstairs as women only, and been there since the first night ... it was Friday, Sunday and Wednesday. So, seeing the history of it and the evolution of it and the success of it – really nice. And the ethos if someone new came in and they were sitting at the bar, we would be like 'would you like to join us?' You know that kind of friendly atmosphere.
>
> (Pauline, 35)

The creation of women-only spaces within a dominantly gay male orientated scene had been generated by women desiring to have a venue in which they could both socialise and be safe. The establishment of the upstairs space within 'The Cat' as a women-only one reflected the desire and the need for the creation of an exclusive women-only space and of lesbian visibility within a scene catering primarily for gay men – in this case, with gay males acting as the 'push factor' in their claims upon and domination of the upstairs space above 'The House'. The establishment of 'The Cat' and 'Moon' as women-only venues were not discussed by those lesbians interviewed in economic or business-like terms. Although lesbian identity has undergone some commercialisation and branding (e.g. lesbian chic, 'boi culture'), this has yet to be on a scale similar to the commercialisation of both heterosexual and gay male sexual identities (Binnie 1995; Kates 1998; Richardson et al. 2006). The social and inclusive character of both 'The Cat' and 'Moon' had been a key factor in the experience and use of these venues for their female customers. The development of women-only spaces had allowed their clientele to enter and use the traditionally male preserve of the pub safely, something that was not a common phenomena in the then 'macho culture' of early 1990s Newcastle (see Lewis 1994).

The inclusive character attributed to both of the women-only venues within some of the interviews was not echoed in the experiences of all lesbians who had entered these premises. Access to such spaces does not necessarily reflect equal entitlement and power over access to lesbian territories (Retter 1997). The 'would you like to join us?' conviviality discussed by Pauline had not been a reality for all of the women. According to some of the women, a lack of a true inclusiveness at 'The Cat' was central in its demise as a women-only venue and in the development of the second women-only space, 'Moon'.

> [T]here was an issue about deaf people ... [T]here was a movement built around the pub that evening where it all got a bit heated. Where half the pub was telling them to piss off and go somewhere else and the other half of us were saying turn the lights up and the music down – it is a reasonable request and they would not. And there was like a stand off, and they would not do it, and in support of the deaf women in Newcastle, there was

a number of women who boycotted 'The Cat' for a long time and that is why 'Moon' was created. Like, if we are going to boycott 'The Cat' we need another space, and the upstairs of 'Moon' came about.   (Kath, 40)

The venue's unacceptable discriminatory practices in its refusal to turn up the lighting and turn down the music so that the deaf women could communicate with one another had marginalised these women and marked them out to be an undesired clientele. Through their initial presence, these women had claimed an inclusion – both through their identities as women and lesbians as well as through possessing the necessary financial resources to enter a commercial venue. However, their disabilities had then been used to position them as undesirable and unwanted within 'The Cat', reflecting the complex interplay between gender and disability when entering and consuming spaces. Inclusion within the space of 'The Cat', then, had been dependent not only on having a female body – so as to identify as a lesbian – but a body free from certain disabilities so that customers had been expected to be able to see within dark interiors and hear over any loud music played.

Lesbians and gay men with disabilities are likely to face numerous barriers in their leisure and social lives, which can be major obstacles to accessing everyday environments where non-disabled people often make contacts that lead to friendships, sexual encounters or romantic relationships. Where venues do not prevent barriers, people who have a visual impairment, for example, can be excluded from activities, such as cruising for a sexual partner, that are based heavily upon eye contact (Shakespeare 1996, 200). As Tom Shakespeare has observed, because disabled people are denied access to work, they are, in turn, denied access to expensive leisure settings (Shakespeare 2003, 147). This can then deny them access to lesbian and gay commercial venues that continue to reinforce the invisibility of the unwanted disabled lesbian and/or gay man.

The loss of 'The Cat' and 'Moon' as women-only spaces reflects not only the significant structural and cultural changes that Newcastle's gay districts continue to undergo, but also the weak position of those 'queer unwanted' who fail in their appeal to capital or to the latest trends. Such changes might merely indicate the declining need for gender exclusive spaces. However, I would argue that this is not the case; rather, developments such as those outlined in this chapter reflect a combination of growing branding and commercialisation, increased exclusions, and the domination of a specific gay male, youth orientated and wealthy culture within the Newcastle gay scene, all at a significant cost to some lesbian users. As Colly argued:

Oh, I think lesbians have to work much harder to create their own space and I don't think they are as welcome in general, no.   (Colly, 39)

Conclusion

As Newcastle has embraced the 'gay as now' model in its re-branding, previous and new inclusions and exclusions for its lesbian and gay population have become increasingly intensified, while the boundary keeping out those deemed the queer unwanted has been re-drawn. The gentrification, sanitisation, de-sexualisation and commercialisation that has come to characterise many 'gay urban scenes' since the mid 1990s is having the consequence that not only are those considered 'sexually shameful' to be excluded, but other identities, when intersecting with sexual identity, are affecting lesbians and gay men's claims to visibility and inclusion within commercial gay sites. For older lesbians and gay men – those no longer considered 'young enough' – the growing importance and value placed on 'youthfulness' is creating issues of discomfort, exclusion and isolation. As gay scenes and commercial ventures within these scenes have become increasingly stylised, gentrified and expensive, lesbians and gay men on limited incomes are experiencing deepening limits on their ability to claim an inclusion within these sites. Where 'having the look' has become essential in a youth and beauty-obsessed society, being poor or old just isn't sexy. The increased sanitisation and redevelopment of public sex areas and the decline of 'cost free' community spaces are reducing access to non-commercial sites, reflecting the very real and lived exclusions being placed upon those undesired by the commercial scene. The documented growth in gay districts is not necessarily an indicator of an increase of choice and tolerance or of the targeting of specific niche markets. Processes of exclusion continue to operate, particularly towards those deemed undesirable to capital. Through the limited commercialisation of lesbian identities and the perception of their having limited incomes relative to gay men, lesbians are easily deemed to be the 'unwanted', with their marginalisation intensified as gay urban districts chase the young, wealthy gay male pound. As the 'boundary of unwantedness' moves beyond the 'sexually shameful' and as the boundary of marking out those who are desired shrinks, the complex interplay of the multiple identity positions lesbians and gay men occupy is having significant effects upon experiences of spatial inclusion and exclusion. It can be posited that older lesbians and gay men, lesbians more generally and those lesbians and gay men with disabilities are not only part of a growing unwanted, but can also be theorised as being among the sexually shameful. As Pugh (2002) has suggested, sex is marketed as something only the young, beautiful and able-bodied should engage in. Older and/or disabled lesbians and gay men and lesbians more generally who claim a sexualised identity and who are sexually active have become positioned not only as unwanted but as sexually shameful because they break the association linking the act of (gay) sex with young, white and able-bodied men. For those queer unwanted living in small and geographically isolated urban districts such as Newcastle, exclusions are intensifying. Such exclusions and lived unwantedness are now crucial in understanding the politics of commercialisation within contemporary gay spaces.

Chapter 11

Straights in a Gay Bar: Negotiating Boundaries through Time-Spaces

Tatiana Matejskova

Through the marking of ethnically, culturally and sexually othered spaces as cool and sexy in recent decades, cultural and sexual differences have become increasingly commodified for mainstream consumption. Scholars such as Cara Aitchison et al. (2000), David Bell (2001), Leslie Moran et al. (2003), Dereka Rushbrook (2002) and Stephen Whittle (1994b) note the popularisation of gay parades and gay bars amongst straights, often as part and parcel of the broader commodification of gay cultures. Most accounts frame the expansion of heterosexual presence in gay venues within the production and consumption of symbolic cultural economies in late capitalism. Scholars in gay and lesbian and/or queer geographies, sometimes quoting directly some of their informants (e.g. Moran et al. 2001), interpret the presence of straights in queered spaces as, at best, diluting the queerness of these spaces or, at worst, as a straight spatial re-appropriation tending towards the complete dissolution of such spaces and their users' identities as queered. I do not want to deny that increased straight presences may threaten gay bars' long-term viability in the face of powerful structural factors that contribute to the fragility of many gay or lesbian venues. Instead, I want to draw attention to the limitations of such interpretations since they overemphasise the commodification of spatialised sexual differences as mere temporary fads (see for example Whittle 1994b) at the expense of closer attention to the diversity of gay spaces and modes and contestations of straight presence in them. For example, Rushbrook (2002), in analyzing how entrepreneurial city governments now promote queered urban areas as tourist attractions, erases the differentiations between straight locals and tourists. She also downplays the diversity of motivations for straights to go to gay bars in favour of focusing only upon an unreflective visual consumption of queerscapes by tourists-voyeurs.

More recently several scholars have analyzed gendered dimensions of straight presence in gay and lesbian space. Both Mark Casey (2004) and Beverly Skeggs (1999) point out that the majority of straight customers in queered commercial leisure spaces are women. These women are often befriended or acquainted with bars' gay male users and base their inclusion in this space upon the dominance of gay male presence over the presence of lesbians and other straight people. This chapter similarly focuses on the gay-marked and gay male dominated bar/club 'The Groover'[1] that has been sometimes open and sometimes closed to straight, mostly

---

1 For reasons of confidentiality all the names used in this chapter are pseudonyms.

female clientele. Yet, unlike most research, I also pay attention to the ways in which negotiations of the boundaries of gay space are tied to customers' attachments to and motivations for using this space at different times. Also in contrast to most research, which focuses on large cities in the Anglo-Saxon world that have long been primary battlegrounds for and sources of strong gay and lesbian identity politics, I examined these issues in Bratislava, the capital of post-socialist Slovakia. In this paper, I suggest that the processes of opening up gay bars to heterosexual clientele might not follow the linear and coherent script of the dissolution of such spaces as effectively queered and the re-imposition of a heteronormative order upon these spaces. Straight presence in a gay bar/club, like commodified consumption of cultural and other differences, is substantiated only through a variety of meanings and bears diverse contingent consequences. I assert that, in many instances, gay clubs are complex spaces where gay customers and staff negotiate admission of and relations between differently sexed and sexualised subjects. As a result, I call for a serious exploration of mixed gay commercial spaces, especially in non-Western geographic contexts, and an examination of these spaces not only in terms of issues of privacy and safety, but also with respect to time-dependant use motivations and various use practices.

In the first part of the paper, I briefly discuss conceptualisations of the gay bar as a particular kind of (semi-)public space that are relevant for the subsequent analysis of the boundary negotiations in The Groover. The paper then turns to the temporal dimension of thinking and researching queered spaces with straight presence.

The Gay Bar and Its Boundaries

Gay bars have become some of the most commonly researched gay spaces, not least because in many cities they have become visible markers of queer presence in public space. Bars, clubs and other commercial venues are popularly understood as public spaces. Yet, access to them, as to many nominally public spaces such as parks, is often limited for certain people (and) during certain times. Such restrictions of access effectively enact boundaries around a place in order to control who uses the space and in what ways. Because a gay bar is a place established primarily as a communal – if commercial – venue, it might be thought of as a particular type of a semi-public place. In one of the earliest accounts of gay bars, Barbara Weightman (1980) understands them as private places. Yet, as feminists came to point out the problems of the public/private space binary, it became less productive simply to label spaces as either private or public. Drawing on the work of feminist geographers Gill Valentine and Lynda Johnston (1995), I prefer to think about nominally private and public spaces through the concept of privacy.[2] *Privacy* here can be conceptualised as the realm of subjects' personal sexual identity and the performance of that identity, that is, as *something that can be had* and lost. And while privacy is certainly one of the attributes of an ideal private space, for many gays and lesbians privacy is often lacking in nominally private places, such as their (parental) home (Valentine and

---

2   Similarly, Staeheli (1996) analytically distinguishes between spaces and the actions taken within them and asserts that sometimes private actions are taken in public spaces. Yet, because Staeheli relies on the unquestioned categorisation of private as opposed to public actions, I prefer Valentine and Johnston's focus on privacy.

Johnston 1995). In contrast, a gay bar, while not a private place, is often just such a *space of privacy* for gay customers, where their sexual identities, which are out of place in other (semi)-public spaces, can be enacted and performed (McDowell 1995, Brickell 2000). In heteronormative societies, gay bars then offer an opportunity for meeting gay partners as well as for inhibition-free socialising and organising (Stewart 1995, Whittle 1994b). As Moran et al. (2003) document, the privacy aspect of a gay bar – or the perception of lacking privacy due to a potential heterosexual gaze – is clearly one of the reasons why straight presence on the gay bar's premises is seen as problematic by some gay customers.

In a somewhat different take on privacy, Michael Brown (2000) conceptualises gay bars as a type of extended closet. At the scale of the subject, the closet designates the very space of one's homosexuality, unknown to the outsiders. As such, firmly established boundaries are at the very heart of Brown's notion of the closet. The closet-like character of a gay bar, its usage as a space of privacy, then becomes especially important for those customers who are not publicly 'out': they can (temporarily) leave their straight 'public selves' beyond the bar's threshold, which forms its boundary with the surrounding heteronormative space (Moran et al. 2001).

A gay club or bar's clearly established boundaries separating it from the larger heteronormative spatial regime have also been associated with safety and protection from heterosexual violence (e.g. Whittle 1994b). Some research shows that there is a higher probability for gays to become victims of physical and verbal abuse in and in immediate vicinity of areas identified by the heterosexual public as gay, such as gay neighbourhoods. Yet, more policed and bounded gay and lesbian spaces, such as bars and clubs, are often simultaneously perceived by their gay users as safe (Myslik 1996).[3] Within this context, it has been noted that a substantial straight female presence might weaken such perceptions of safety and privacy amongst lesbian customers (Moran et al. 2001). Finally, notions of safety are often tied to what Johnston and Valentine (1995, 102) call gay and lesbian 'ontological security' or the notion that only through experiencing queered spaces can gay and lesbian identities be internalised as actually existing.

Case Study and Methodology

The information presented in this paper is based on participant observation, and numerous informal and several formal, semi-structured interviews conducted with The Groover's customers in June and July 2003. All of my informants were, like the great majority of the clientele, in their mid-to-late 20s or early 30s and middle to upper-middle class. My access to the bar and its customers was eased by my previous connections to several frequent gay male users that I had befriended during my university studies in the city and with whom I had visited the bar relatively regularly, once or twice a month over several years. In this respect my experience of gay commercial venues resembled that of many straight users who, as mentioned

---

3   For alternative findings on perceptions of gay spaces as dangerous see Moran et al. 2003, although they do not differentiate between gay areas, in general, and gay bars, whose boundaries are usually more guarded.

in previous literature, are often women acquainted with gay male customers (see Casey 2004). These initial contacts with several customers enabled me to recruit further informants whom I formally interviewed outside the bar's premises and on one occasion in the bar itself on a quiet weekday afternoon.

Initially, most informants wondered greatly why anyone, especially a straight female, would be interested in studying gay communal spaces in Slovakia. In the first place, many informants felt there was not a gay 'scene' in Bratislava, which lacks many communal establishments and a strong sense of gay community. It was suggested that I switch focus to the bars in Prague. But this sense of bemusement was also permeated for most with a slight insecurity about being a research subject. I attributed such unease to a still strong fear of privacy invasion, linked to the intolerance of the pre-1989 communist regime and to the persisting sensitivity surrounding gay and lesbian issues in the country. Discussions of confidentiality, a consent form and an assurance that this was not a research project commissioned by any Slovak official authority assuaged this concern amongst the patrons although I was not able to gain access to the management of the bar who, however, were aware and tolerant of me conducting research amongst the bar's users. Since the bar is a rather small place, within a few weeks I was identified by the regulars as the researcher, and on occasion patrons would seek me out and volunteer their take on various issues.

While I attempted to interview different bar patrons, it was difficult for me to recruit any fully 'closeted' gays, and I found this to be one of the limitations of my research. Similarly, since I was particularly interested in perceptions of straight presence in the bar, of which as a straight woman I was an example, I am aware that some of my informants might have been more cautious with me than they would have been with a gay researcher. As Kim England (1994a) argued in her reflective piece on an abandoned research project on lesbian spaces in Toronto, fieldwork is always a dialogical process in that researchers are a visible and integral part of the research setting and their positionality and biography directly affect what kind of data they will be able to gather. Fresh from a methodology seminar that had had plentiful debates about the advantages and disadvantages of being an insider versus an outsider in community research, I grew increasingly uneasy about my position as an outsider to this gay community. My dilemma as a straight researcher in and of queered spaces, triggered by my then rather incipient consciousness of the contentiousness of the spatial politics of sexuality, stands in contrast to my pre-research presence in The Groover in the late 1990s. With similarities to England's position, the actual experience of this research, including the ethical dilemmas and limitations I faced, ushered me to my eventual decision not to pursue this project further in an expanded and comparative version.

The place where I conducted this research, The Groover, was one of only two gay bars operating in the city of Bratislava in summer 2003. The Groover was opened in the early 1990s as an exclusively gay bar/club. Other gay (male) (semi)-public spaces in the city included a low-budget hotel, a bathhouse that only two of my informants knew of, as well as two more formal, nation-wide gay support and activist organisations with offices, online forums and websites.[4] Most of my informants were

---

4   In addition, a number of websites dedicated to various issues of concern to gay men and lesbians have been established in recent years, as well as magazines, including the first

ambivalent about these organisations because they questioned the actual leverage that these organisations had for promoting gay rights and decreasing homophobia. Since 1997, these organisations, in cooperation with some members of parliament, have worked, so far unsuccessfully, to introduce a law allowing civil unions for same-sex couples. However, as a part of a new, wider gay and lesbian coalition 'Initiative Otherness' (Iniciatíva Inakosť), they have succeeded in including sexual identification into a 2004 anti-discrimination law.

In addition to these gay-designated spaces, there exist other gay-friendly spaces in the capital, especially some city-centre cafes. These cafes often become communal places where gay customers meet informally. It is usually the presence of gay waiters or managers that at first attracts their gay friends and further acquaintances. However, these cafes distinguish themselves from more enclosed gay spaces because their clientele cannot be easily controlled. In other words, the boundedness of these cafes as either heterosexed or queered spaces is harder to watch over and maintain. Overall, gay communal places at the time of research were relatively few and far between.

The Groover as a Space of Negotiations over Boundary-Making

Scholars examining gay and lesbian (semi)-public spaces, such as Weightman (1980) or Valentine (1993b), have often stressed the difficulties for the long-term viability of many such commercial spaces. In contrast, The Groover has been functioning, with an exception of one year, in the same location for almost a decade. Almost from its very beginning in the early 1990s, the bar included some straight clientele although it was originally established as a gay-only private club.

As noted earlier, owners and managers of (semi)-public spaces such as gay bars have the power to 'privatise' them by defining who is privileged to use these spaces. How gay bar managers assess 'who belongs' differs somewhat from how access to other (semi)-public consumer spaces is managed (see e.g. Goss 1993 on malls). Moran et al. (2001) explore different strategies of boundary-making in gay and lesbian spaces. Strategies for variably strict control of clientele include bouncers eliciting confirmation of customers' sexual self-identification at the door; bouncers 'policing' unwanted, ostentatious heterosexual presence inside the club; or bouncers requesting visible markers of membership, such as VIP cards. In The Groover, the bouncers are the most visible enforcers of the bar's boundary as a gay space. Restricted access is facilitated also by the fact that the bar is located in a 'historic' vaulted cellar with only one public entrance. Overall, The Groover does not have a very high visibility, unlike many gay bars in contemporary popular gay neighbourhoods in larger cities in North America or the UK. Due to a persisting legacy of the spatial marginalisation of queer places and to the fact that it was originally established as a private club, The Groover is located on the outskirts of the city's main zone of night-life entertainment in the historic centre. Unlike most bars and clubs in the centre of Bratislava, The Groover has no large signs that make it highly visible from the street.

(non-pornographic) gay (and lesbian) magazine in the country, launched with one year's financial support from the Dutch embassy in the Slovak Republic.

Most first-time customers, especially straight ones, are introduced to the bar by more established users. While the bouncers will occasionally ask about a new customer's sexual identity, entry is mostly regulated by the regulars who bring most of the new customers to the bar. The 'exceptions' for straights accompanying gay customers often extend to other straight customers arriving alone, sometimes after 'proving' ties to gay users inside.

The level of presence of heterosexual customers in the bar has varied over the course of the more than a decade of the bar's existence, depending to a large extent on periodic attempts to admit only gay men (and lesbians). Such efforts at regulation, aimed at keeping out larger numbers of straights (including, especially in the early days, homophobic and violent ones), focused on introducing membership cards and allowing entry only to their holders. Each time such efforts have been attempted, they have been successful for only short periods. The last attempt at restricting entry in the winter of 2002 was justified by the managers' claims that straight customers were taking away sensitive information about gay customers and were spreading gossip about alleged promiscuity in the gay community in Bratislava. Some of my informants, including the gay ones, explicitly and independently from one another dismissed the managers' allegations as untrue. Yet, the issue of verity is less interesting than the very fact that the managers were so very concerned about the image of the gay community amongst the broader public, and the fact that the managers could use this claim as a significant and legitimate reason to exclude straight customers. This, in turn, points to a persisting stereotyping of gays and lesbians in broader dominantly-straight society, both reflected in and perpetuated by a lack of gay and lesbian political representation and by the reluctance of many public personalities to come out publicly. In recent years, the Slovak Christian Democratic Party, especially, has become more vocal in rejecting homosexuality and civil unions (along with abortion rights). Support for this party, however, fell to 8% of the national vote in the 2006 parliamentary elections. Public acceptance of homosexuality, including civil unions and same-sex marriage, has increased (to 39% and 24% respectively in 2005)[5] and is stronger than support for heterosexual cohabitation without the intent to marry.[6] In everyday life, however, most gay men and lesbians perceive Slovak society to be discriminatory and homophobic. About half of gay men and lesbians restrict the public display of their sexual identification, such as holding hands or kissing, at all times, and, importantly, 40% report experiences with various forms of verbal harassment and 15% with instances of physical violence.[7] Returning to the managers' claims about the need to restrict straight entry, such restrictions

---

5   However, these percentages, while on par or a little bit higher than in Poland or Hungary, are significantly lower than in the Czech Republic (62% and 42% respectively), which, after five attempts, legalised civil unions and which is seen by many gay and lesbian activists in Central Europe as having developed the most progressive social and legal environment for gay and lesbian communities (www.altera.sk).

6   While 93% of a representative sample in 2003 felt same-sex partnership was acceptable (64% always and 29% sometimes) only 84% felt the same way about heterosexual co-habitation without an intent to marry (58% and 26% respectively) (www.ivo.sk, Institute for Public Issues).

7   See www.altera.sk.

reproduce the closet function of the bar – that of unmediated privacy – that some of the customers and/or managers might want to strengthen or, at least, not erode. The possibility of renewed admission restrictions became a topic of discussion in the bar early on:

> The word spreads around here, everyone talked about it … [I]t's ridiculous, I'd come here with friends and they have to stay out 'cause they don't have the stupid cards? I told Martin [one of the bartenders], they're good *customers* … one of my [straight] friends buys 'Champagne' bottles *all* the time, gets drinks for the bartenders. They spend some money here. It's not like they're sucking on one stupid gin and tonic all night long.
>
> (Saša, 26)

As Saša's quote illustrates, some regulars continuously signalled dissent regarding gay only admission to the bartenders with whom they were friends as well as to the managers. Somewhat jokingly, one of my informants described the regular gay clientele's engagement in a vernacular politics of resistance to regulation attempts as an 'uprising', regardless of the fact that they were expressing their refusal primarily individually, without a collectively devised strategy. Within several months, the strengthened regulation in the bar and all the talk about it abated due to the persistent lack of support for such measures from a substantial portion of the regular clientele. From Saša and others' accounts, it became evident that the regular gay clientele influenced the decision of whether or not to regulate straight access more stringently than that of occasional gay customers. In addition, my informants highlighted the importance of the spending power of more affluent patrons, both gay and straight. Spending power was an important factor in this decentralised decision making since the bar made almost all its money from drink sales. In this respect, several gay customers openly reflected on the fact that Bratislava as a rather small-to-mid-size city does not offer enough of a stable clientele for a profitable investment in a gay-only bar. More generally, then, whether because of its location in a mid-size city or not, The Groover confirms that customer purchasing power is clearly implicated in negotiations over the desired boundedness of gay bars, and, as Casey (2004) and others (e.g. Whittle 1994b) note, it is increasingly becoming so. As Mark Casey elaborates in Chapter 10 of this volume, the commercialisation of the gay scene has more broadly been increasing the exclusion of many gay men and lesbians from queered spaces as they become 'the queer unwanted' whether on basis of age, 'hipness' or income.

The history of the ebb and flow in the admittance of straights to The Groover points to the limits of thinking about the opening-up of queered spaces as a gradual dismantling of such spaces as queered. It rather reveals a lack of a clear consensus about how gay-exclusive the bar should be amongst its gay patronage. While opening the bar to straights is mostly problematic for those gay customers valuing the safety and privacy that the bar offers to them, for openly gay patrons it is the socialising aspect that they appreciate more:

> When I was in college, sure, I was there a lot. I'm still friends with some people I met there. My first year [in college], I think, someone told me that there was a gay bar in town, The Groover. It was great … I didn't know anyone [gay] in town at first, and then I came,

and I had never seen so many gays together … but they [gay bars] just aren't so much fun for me anymore. When Adam [his partner] and I travel we'd go to some, but usually we go to [names of straight bars] with friends.   (Peťo, 27)

This quote, however, also suggests that some gay customers' attachment to The Groover, which as a gay bar provides a space for the exploration of one's sexuality, has weakened with time. At the time of our interview, on a weekly basis, this particular patron, Peťo, preferred other mostly trendy heterosexed bars and clubs in the city because of their other qualities as bars and because he felt less restricted being affectionate with his boyfriend in these places. His long-term relationship and his being out in public were two crucial factors influencing his lack of support for the gay-exclusivity of The Groover. Gill Valentine and Tracey Skelton (2003) argue that queered spaces are highly crucial in the process of coming out. Yet, as Peťo's case shows, with an established gay household and gay 'public persona,' the importance of gay bars for some patrons wanes. When this, along with a loss of gay clientele to other gay bars, motivated the opening of the bar to straights, it bore direct effects for those patrons for whom the bounded space of the bar had provided much needed privacy. Nonetheless, the contestation of the 2002 attempt to exclude straights from the bar can also be seen as an expression of some patrons' refusal to be strictly re-closeted. This also points to the varying attachments to gay social spaces that gay men with diverse personal politics of sexuality form and maintain (or not). Much research suggests that the "gay (*political*) self" (Whittle 1994b, 38, emphasis mine) has been implicated in attachments to queered spaces in the US and Western Europe. The recent and slow attempts at the de-stigmatisation of gay and lesbian sexualities after decades of a communist regime, and a relative lack of queer politicisation in post-socialist Slovakia, might have contributed to the lack of attachment to exclusively queer space that I observed at The Groover. I do not want to suggest that the gay commercial scene and the politics of sexuality in Slovakia will necessarily develop in the way they have in an Anglo-American context. Such an approach would only reinforce an already existing Anglo-American bias in much of the literature, which takes – whether implicitly or explicitly – as a universal blueprint what is a particular geo-historically embedded process of the configuration of queered spaces and politics. Yet, constrained by the lack of research on the spatial politics of sexuality in post-socialist Europe, as well as by the scope of this particular project, I still find it important to point out those crucial insights from the existing literature that need to be reckoned with in the future development of alternative conceptualisations and strategies for the politics of queered space in non-Anglo-American contexts.

Going back to The Groover, while many gay customers refuted the exclusivity of The Groover as a gay-only bar, a few also felt that the community aspect of the bar was stripping down their multiple sources of identification, leaving only their sexuality. While the idea of a gay community, inside and outside of the bar, can carry a positive connotation as a site of affiliation, it clearly also functions as a 'homogenising' device (e.g. Woodhead 1995; Valentine and Skelton 2003). From the standpoint of everyday life experiences and practices, strong regimentation only on the basis of gay sexuality is perceived by some customers as at times highly problematic especially as the gay community is small. As Jojo (26) puts it, "Everyone knows [of] everyone else,[8] like some gay family or something. You do something and within three days every damn gay in town knows about it, like my life ends here ... you can never know who knows whom." Gays and lesbians aware of both the fragmentations and internal-controls of gay and lesbian communities might still value them as 'buffer zones' against heteronormative society (Valentine 1995b). Still, perceptions of individual gay communities, their effects on gay and lesbian spatial practices and the complexity of their attachments to gay communal spaces deserve more scholarly attention.

In addition to perceptions of The Groover's community as confining, informants also cited other reasons for their weak investment in it:

> I hooked up with a guy last summer in the Canary islands – he's from London. When he came to visit me in March, he's like let's check out the bars. I didn't want to take him there [to The Groover]. We've been to the [gay] bars in *London* together ... it's just so embarrassing, they always try to rip you off on the drinks, you know, with foreigners ... it *could* be a cool place but we're just not in the same league.   (Tono, 23)

As Tono points out, dissatisfaction with the service and the bar's interior design[9] dampens some of its gay clientele's attachment to it. All my informants, however, expressed even greater dissatisfaction with the second (then) functioning gay bar in the city. The last part of Tono's quote hints also at the importance of some gay customers' experience of a much richer gay night-life scene, especially from their summer trips elsewhere in Europe. More generally, the geographic and increasingly socio-economic proximity of Bratislava to large cities such as Vienna, Prague and Budapest is crucial for the part of the clientele that can afford not only regular night-trips but also extended weekends there. Two professionals in their late 20s, in particular, elaborated on their strong and growing preference for venues outside of the country, citing a more established and diverse night-life environment, including gay (friendly) 'artsy' cafes and bars; more specialised gay S/M venues; a more diverse crowd – internationally, professionally and sexually; and, lastly, greater opportunities for socialising.

What remains relatively little attended to in current research are the motivations and modes of straight presence in gay bars. In Manchester's 'Village', the gay clientele have been concerned with instances of straights breaking the unspoken rules of the bar or the "particular expectations of civility" by, for example, dancing in an inappropriate part of the bar in a way that could constrain gay customers' ease in their own space (Moran et al. 2003, 182–3). During my observations in The Groover, straight customers tended to hang out on the outskirts of the dance floor in – from my point of view – rather small and relatively inconspicuous groups. At times, straight women would use the centre of the dance floor although never as part of a straight couple showing affection, but rather with their gay male friends. Somewhat

---

8   *"Každý každého pozná..."* There is no distinction in everyday Slovak between 'to know someone' and 'to know of someone'.

9   The interior space was changed a few years ago when the dance floor was enlarged and the booths, seen as more conducive to intimate conversations or affectionate exchanges, were removed from the adjacent rooms.

surprisingly, none of my informants mentioned any recent incident with a straight customer needing to be disciplined. In respect to their motivations, Skeggs (1999) and Rushbrook (2002) find that straight female patrons value the (relative) absence of straight men. One of my informants (Andrea, 29) confirmed this by describing a gay bar as a place to "party at without constant harassment by [straight] men." Other reasons that heterosexual informants cited for frequenting The Groover included: ending a night of bar/club-hopping in a gay bar, in part because The Groover is one of only a few venues in the city open until dawn; spending the night, or part of it, out with their gay friends in 'their place' (the primary reason for my own past attendance of this bar); the presence of favourite DJs; or theme parties on special nights. Similarly to Skeggs's (1999) and Casey's (2004) studies, some straight female users felt they shared a 'dance culture' with their gay male friends. But unlike in Casey's study (2004), only one straight customer used this shared interest, along with her claim of 'being a good customer', to claim a *right* per se to use this space.

For both straight and gay customers, the bar clearly performs a multitude of purposes whose relative importance always varies. Descriptions of various attachments or lack thereof that gay and straight customers have to this queered space exemplifies how there is a multiplicity of interests and positionalities that patrons have within and across different social categories and personal histories. These multiple interests and positionalities influence the kinds of expectations customers have when they choose to go to The Groover at particular times.

Time-Spaces of The Groover

As several scholars have noted, the temporal dimension often plays an important role in furthering our understanding of how gay and lesbian spaces work as actually queered. Moran et al. (2003) point out, in particular, their gay informants' nostalgia for the past, when Manchester's 'Village' was a gay-only place. Gill Valentine (1993b; 1995a), discussing strategies of queering otherwise heteronormative spaces, describes lesbian spatial appropriations of concert halls on the occasions of lesbian musicians' performances. Since these spaces are only effectively queered for relatively short periods of time, Valentine describes them as lesbian *time-spaces*. I suggest, however, that limiting the concept of time-space to territorially mobile gathering sites reveals a certain failure to conceive of all spaces in conjunction with time and of the specific ways in which time operates in and through places. As David Bell (2001) shows, people's conception of a singular place differs depending on time. Moreover, through time-dependant practices, people actively create a multiplicity of places.

Temporal practices in and usages of The Groover did indeed vary in summer 2002. Thursday, Friday and Saturday evenings and nights were the times when the DJs would play dance music and the space was used mostly as a dance club. It was during these nights that, notwithstanding the bouncer-regulated entry, straight customers were most numerous, although still as a minority. On 'dance nights', the composition of the clientele differed mainly between Friday and Saturday, two longest open nights of the week. On Saturdays, the bar attracted a larger crowd, and especially younger gay men who would come from the surrounding towns and villages for a night out. While the greater presence of straights in the Friday night crowd did not deter most gay informants from coming to the bar, for quite a few of them the larger concentration of other gay men from outside the city was a drawing factor on Saturday nights due to the greater opportunities for socialising and meeting potential sex partners. In addition to the changing nightly uses between different days, The Groover would also offer special events once or twice a month, such as striptease or transvestite shows, lesbian parties, 'thematic' commercial parties such as Absolut Vodka parties, all of which attempted to attract somewhat different crowds.[10] Until shortly before my research, gay movies would be screened on selected afternoons, mostly without any straight presence.

Discussing the 'straight invasion' of 'the Village', Moran et al. (2003) quoted one lesbian patron expressing little consternation about this straight presence since it, similarly to the straight presence in The Groover, would happen primarily on weekend nights, apparently not her prime use time. Although it is most highly visible to the outsider's eye, for many gay customers, the weekend night time is indeed not the most prominent use time of gay commercial venues. In the case of The Groover, there were several customers who used it mostly as a place for a late afternoon or early evening coffee or drink on their own or, more often, with a friend or partner. These afternoons and early evenings, especially on weekdays, have, over the years of the bar's existence, remained most effectively queered in terms of exclusively gay male presence. At these times, the bar actually functioned more as a cafe, even though it is located in a cellar, which is more typical of bars/clubs in Bratislava. Notwithstanding the lack of enforced boundaries at this time of the day and thanks to the usual absence of straight customers, gay customers could enjoy more privacy.

The diversity of temporal uses of The Groover exemplifies how the bar effectively works not as a single queered space, but rather as a series of somewhat differently queered spaces at different temporal moments – a series of time-spaces. While such a bar remains territorialised and marked as gay, at the same time it consists of a bundle of time-spaces created out of the singularities of congregations of bar/club/café customers and their uses. This also means that privacy – as one of the most important aspects of gay and lesbian communal places – is not necessarily attached or fixed in space, but may be more fluid and sensitive to the particular configuration of bodies in time-spaces.[11] In contrast to Valentine's focus on the tenuous borders of some lesbian spaces, and in line with David Bell's work and Doreen Massey's (1993) usage of space-time, I emphasise here how the boundaries of some gay spaces – strongly grounded materially – also vary in their permeability. Put differently, I emphasise that *all* socio-spatiality is by definition temporally conditioned. In respect to queered spaces, an exclusive focus on gay spaces as a temporally and spatially coherent whole can obscure the complexity of the production and consumption of the space as 'gay', including the negotiations over its boundaries in relation to straight presence within it.

---

10   During the two months of my research, however, no nights targeting lesbians in particular took place.

11   I would like to thank Kath Browne for this point.

Conclusion

As Mark Casey (2004) writes, it might be that both lesbians and *some* gay men feel less safe in the presence of straights in queer spaces. Yet, because the voices of those gay men refusing stricter exclusion of straights are mostly omitted from the accounts of straight presence in queered space, I have chosen to foreground them here. More broadly in this paper, I have read one particular 'mixed' gay space – which, however, still marks itself as gay – as a space where its (gay) customers regularly negotiate the different meanings that this space bears for them. Many of my gay male informants did not express a concern regarding the erosion of privacy and safety by regular straight presence in the bar. In their accounts, they, perhaps surprisingly, elaborated on their changing, often fading, attachment to the bar as a *gay* space and on their attachment to The Groover as a *bar* in the first place. Alongside this relatively weak emotional investment in the bar, the overwhelmingly dominant presence of gay men rather than lesbians as a clientele may certainly have contributed to weak perceptions of straight (mostly female) presence there as a threat (Casey 2004; Skeggs 1999). I share Casey's (2004, Chapter 10, this volume) and others' concern with both the exclusion of lesbians from many mixed gay spaces and the overall lack of lesbian social spaces. The gendered dynamics of the bar under analysis in this paper has, however, always been much less complex since The Groover has never fully marked itself as a both gay *and* lesbian space in the first place. In this chapter, I have further proposed thinking about the temporally varying permeabilities of mixed gay spaces and have argued for the need to examine queered spaces as a series of gay-marked, yet malleable time-spaces. Like most (semi-)public spaces, many gay bars invite a variety of time-conditioned uses and prompt different interpretations of such spaces by their customers, uses and interpretations that do not necessarily clearly map onto established identity category divides.

I would like to conclude with Michael Keith and Steve Pile's (1993, 222) assertion that "politics is invariably about closure." I agree that moments of political decision-making are indeed organised around closures, yet I also maintain that politics is just as much about how subjects *arrive* at such closures. As Larry Knopp (1997, 182, emphasis mine) puts it:

> Oppressive institutions and practices surely do exist – at all spatial scales – as do the injustices which they perpetrate. But attempts to *fix* these whether in the form of particular identities, political subjectivities, social movements, *or places*, ultimately deny the contradictory dimensions of power relations. The political consequences of this can be serious indeed.

In other words, politics is also about what happens in between the moments of closure. Thus closure, or rather the reoccurring attempts at closure of The Groover for gay use only might be interpreted, at a certain level, as the resistance of gay customers to heterosexual appropriation of this social space. Yet, a focus on the gay clientele's negotiation of such closures and the regulation of access to The Groover reveals a more open range of gay customers' takes on the straight presence in the bar. It might also reveal a variety of competing understandings of strategies and practices for challenging the dominant heteronormative socio-sexual order.

Acknowledgements

I would like to express sincere thanks to Kath Browne for her patience and guidance in bringing this chapter to completion, as well to Karen Till, Tiffany Muller and Sam Schueth for all their help and advice throughout the project. Any remaining mistakes are mine, however. Last but not least, I would like to thank all my informants for sharing their experiences and ideas with me and for enabling me to conduct this research in the first place.

Chapter 12

Between Transgression and Complicity (Or: Can the Straight Guy have a Queer Eye?)

Phil Hubbard

Far from posing a radical challenge to current modes of thought, queer theory is becoming a game the whole family can play.

(Halperin, 1996, 4)

Compared with the geography of, say, the 1980s, human geography now seems more than just a little queer. Indeed, one could follow Halperin's line of argument to suggest that far from being of marginal concern to human geographers, queer theory has attained the status of a normalised discourse within the discipline. Witness, for example, the inclusion of an entry on queer theory in the most recent *Dictionary of Human Geography*, the way that revised editions of established student texts often incorporate material on queer approaches (e.g. Knox and Pinch 2001; Cloke et al. 2005) and the widespread citation of the book that arguably did most to introduce geographers to the possibilities of queer (Bell and Valentine's (1995) *Mapping Desire*). The Sexuality and Space speciality group of the Association of American Geographers may not be particularly well subscribed, but it is certainly well-established and is a vibrant part of the AAG landscape. This is not to say that resistance to queer has not persisted in some quarters, or to deny the fact that many undergraduate students are not exposed to anything but the straightest of straight geographies; but, certainly, queer approaches have moved (or been moved) from the margins towards the centre in a quite remarkable series of ways (see Oswin 2005).

One significant consequence has been an interest in the languages, methods and attitudes of queer among those geographers who do not identify as LGBT or have never experienced homophobia first-hand, yet who are passionate and angry about sexual injustices of many kinds. Further, it is notable that there are a number of geographers, straight or otherwise, who are interested in using queer theory to expose the fractured and multiple geographies of heterosexuality. In this chapter, I thus want to reflect on my own tentative attempts to develop a critical stance on heteronormality and to consider how this project has benefited from an engagement with queer theory. I want to do this by way of raising some broader questions about the mainstream assimilation or co-option of theories that were, in the first instance, both politically – and practically – motivated to disrupt the heteronormative.

In this sense, I want to explore if it is possible – or even desirable – to utilise queer theory (or more correctly, *theories*) to explore the geographies of heterosex.

Related to this, I want to question whether the co-option of queer is lessening the disruptive potential of what was, for many, conceived as a much-needed interruption of heterogeographical praxis. In short, are queer theories being bent out of shape by their assimilation into the geographical mainstream?

Queering the Mainstream

Though 'queer theory' connotes a multiplicity of approaches and standpoints, one might follow Cocks and Houlbrook (2005) when they argue that queer theory takes its lead from Foucault's argument that resisting identity of all kinds is a virtue. Recognising the inadequacy of extant frameworks of representation and language for framing the sexual subject, not least when theorising homosexual desire, queer theory thus constitutes a body of work that challenges established views of sexuality, which regard it as either biologically-rooted or, conversely, socially constructed. Central here has been the desire to disrupt and deconstruct binary oppositions – primarily homosexual/heterosexual, but also sex/gender, mind/body, self/other and so on. Queer theory has hence problematised any neat correspondence between gender, sexuality and biology, exploring the polymorphous sexual identities and practices that transgress and exceed 'normal' expectations of masculine and feminine sexuality. Closely associated with postmodern and poststructural thinking, queer approaches first took shape in the interstices of gay and lesbian performance, literature and critical studies, becoming more widespread in the social sciences as the work of luminaries as diverse as Butler, Irigaray, Grosz, Sedgwick, Rubin, Rich, Chauncey, Weeks and (especially) Foucault became more and more influential in the 1980s. Indicating the impossibility of expunging questions of sex and sexuality from discussions of gender, identity and the body, the radical agenda of queer became increasingly influential in an academy where structuralist accounts had long foreclosed vital questions of sexual desire and disgust in preference for 'socially relevant' analyses of class.

The emergence of queer studies as a significant field of academic and popular enquiry over the last two decades has undoubtedly provided a much-needed interruption to 'mainstream' geographies and has placed issues of sexuality firmly on the geographical agenda. One obvious consequence has been an exponential growth in the number of geographical studies exposing forms of desire that often remain hidden in conventional geographies, charting how particular lesbian, gay and bisexual identities emerge and take shape in specific spaces, often against a backdrop of political censorship, police repression and moral approbation. From the 1980s onwards, the mapping of LGBT spaces and networks thus added an important dimension to urban (and latterly rural) social geography. Emerging from such studies was a consensus that particular sites acted as key spaces in the making of marginal sexual identities, with networks of sociality and solidarity serving to give dissident sexual identities both spatial expression and political form. Studies of the residential spaces of LGBT communities, the 'ghettoisation' of their meeting spaces and their interactions with non-LGBT groups in public space thus became the stock-in-trade of a nascent 'queer geography'.

In retrospect, it is clear that many such studies nonetheless remained rooted in the language and methods of more established geographical approaches and continued to perpetuate many extant understandings of straight and gay identities as distinctive and antithetical. Hence, many pioneering works in the geography of sexualities literature are most appropriately described as geographies about LGBT communities rather than queer geographies. Nonetheless, collectively these 'strategically essentialist' studies did much to reveal the aggressively heterosexual nature of everyday spaces, describing the forms of banal homophobia that impinge on the spatial mobility and visibility of LGBT communities. From this base, therefore, geographers began to explore the way that LGBT sexual identities are made and remade in both private and public spaces, focusing on the processes of identity construction and becomingness played out in the context of heteronormalising forces (Bell et al. 2004).

Yet, beyond these pioneering geographical studies of (homo)sexual performativity, human geography in the 1990s seemed remarkably disinterested in the normalisation of heterosexual performance and practice. Moreover, it had little to say about the way in which heteronormality itself rested on the continuing repression and silencing of heterosexual 'Others' – 'bad' sexual subjects such as sex workers, pornographers, errant fathers or lone mothers. For sure, there had been some acknowledgment of heterosexual subcultures and identity practices among those geographers for whom the Chicago School remained a key influence. Likewise, it was customary for ecological mappings of urban space to identify red-light areas or twilight zones in which sexual Otherness was made visible, commodified and surveyed. Yet few, if any, had thought through the issue of how the separation and/or recuperation of 'scary' sexual identities reinforced *particular* notions of heterosexuality. Simultaneously, there was an almost deafening silence surrounding the sexualised nature of 'family life', despite the burgeoning feminist literature on mothering, domestic reproduction and the gendered nature of welfare systems.

On this basis, in 1998, I organised a session at the Association of American Geographers on the 'Moral Geographies of Heterosexuality'. The key impetus for this was my concern that geography's new-found (or at least rejuvenated) interest in the relations between sexuality and space was providing only a partial understanding of the constitutive power of space in the making of sexualities and identities. Perhaps rather simplistically, I was concerned that the focus on gay, lesbian and bisexual geographies was implying the existence of a monolitihic and hegemonic heterosexual identity, ignoring the fact that this, too, is riven with contradiction and complexity. The fact that the session attracted only a handful of papers suggested to me that perhaps this argument was simply not worth making and that many geographers felt that questions of heterosexuality could be dealt with through the lens of already-widespread feminist theories (and perhaps even Marxist ones). Nonetheless, the encouraging attendance at the session and the fact that other geographers were coming to similar conclusions to myself (though from markedly different perspectives – see Bondi 1998; Nast 1998) reaffirmed my belief that prising open the geographies of heterosex was vital. This was an argument I elaborated in Hubbard (2000), drawing sustenance from those who were simultaneously arguing for critical scrutiny of whiteness as a category.

Writing more than half a decade later, it is interesting to note that some are still bemoaning the lack of attention devoted to heterosexual geographies:

> The potentials of queer geography for scrutinizing the functions of sexuality regimes and ideologies … and the fundamental spatiality of the production of heterosexual identities remain underdeveloped. Geographers have neglected the functions of heterosexual space, although it is often posited as a dominant space that queers must negotiate.
>
> (Thomas 2004, 375)

Yet, to the contrary, I would suggest there has been a remarkable (if disparate) body of post-millennial work that has (either explicitly or implicitly) answered my call for geographers to explore the geographies of heterosex. There is now, for instance, much work on the importance of heterosexual norms in the making of national identities and citizenships (Nast 2002; Rand 2003); on the normalisation of heterosex in spaces of education (Thomas 2004); the geographies of family space (Aitken 2001); the encounters of heterosexualised bodies in public spaces (Iveson 2003); and the importance of the urban/rural divide in the idealisation of heterosexualised gender relations (Little 2004). While there has always been geographical research on landscapes of prostitution, this literature has arguably been rejuvenated through the engagement with queer theory, by, for example, relating the forms of spatial governmentality used to regulate sex work to heterosexist performances that normalise male access to women's bodies (see Pitman 2002; Sanchez 2004). Interestingly, there is also some work emerging on the relation between 'queer' and 'straight' identities that problematises both categories by stressing that identifications are constructed relationally in specific sites (Moran et al. 2001; Casey 2004). In effect, the perspectives of queer theory are no longer invoked only by those writing on the geographies of LGBT. Straight geographies have gone queer too.

Progress and Pitfalls

The widespread, if selective, diffusion of queer perspectives across the geographical discipline is surely worthy of more comment than it has attracted. For those who pioneered the study of sexualities in the discipline, either to hostility or (at best) indifference, the institutionalisation of queer perspectives must provoke some ambivalence. After all, one of the key precepts of queer studies was that the margins could be used as a site from which conventional sexual geographies and histories could be re-written. Queer geography has thus promoted the use of situated knowledges via a queer subjectivity that acknowledges the diverse experiences of those with 'dissident' sexualities. Drawing inspiration from poststructural writing, this approach demands that the 'dispassionate' academic voice is replaced, introducing Other voices into writing. Likewise, it implies that the geographer adopts the role of activist, seeking to undermine heterosexist assumptions through research and praxis. Typically, nuanced ethnographic work with non-heterosexually identified individuals has undermined essentialist and heterosexist accounts of dissident sexuality, revealing a 'polymorphously perverse' sexual landscape in which the subject of desire is never 'biologically' pre-determined, but always becoming. Queer geography has thus rendered 'normal' sexuality strange and unsettled, challenging heterosexuality as a naturalised social-sexual norm and drawing attention to the diverse forms of homophobia affecting different sexual dissidents.

In this sense, the use of the term 'queer' signals a key connection between this (geographical) project and the radical politics of groups (such as OutRage, ACT UP or Queer Nation) who promote a non-assimilationist politics. While not all those geographers working with queer theory have openly embraced these radical politics (see Bell 1994b), it is certainly the case that most have positioned their work within a transgressive project. This is particularly evident in the scathing comments that some have directed at those who sought to 'explain' the spatial behaviours of LGBT populations using heterosexist assumptions about the shared values and lifestyles of such groups (Binnie 1997). An example of the latter, for example, is the stereotyped assumption that LGBT individuals are inherently creative and hence 'pulled' to city centres offering a cluster of creative and cultural attractions (an idea explicit in Richard Florida's work on the 'creative class' – see Markusen 2005). The implication here is that writing about LGBT groups from a straight perspective is a merely a case of the mainstream getting 'a bit of the Other', implying that only a genuinely queer epistemology would break down heterosexist assumptions. But is the same true when queer theory is used to destabilise ideas that imply heterosexuality itself is fixed and monolithic? Is it really possible to adopt queer epistemology when we are studying 'dominant' sexualities?

While many of those working on geographies of sexualities have not expressed their opinions on such matters (at least not publicly), within the wider academy it is possible to find many who feel that queer studies constitutes the academic wing of the lesbian, bisexual, gay and transgender movement. To find the language and attitudes of queer adopted by those who do not participate in these movements (or who participate in a different way) troubles some: for instance, Halperin (1996) suggests there is "much to worry about in the current hegemony of queer theory – not least that its institutionalisation, its consolidation into an academic discipline, constitutes a betrayal of its radical origins." Against this, other protagonists have pointed out the tensions and contradictions of queer, questioning whether queer is a term that should only be invoked in relation to 'authentic' studies of LGBT identity politics:

> Queer is a complex term replete with different connotations for different groups. To think about queer space is to rethink the terms queer and space. Is queer a kind of irreducible strangeness, the repressed condition of apparently stable entities, the uncanniness of everyday life? Or does queer refer to the term of gay and lesbian self-identification that emerged around 1990 to describe a new constellation of political identities?
>
> (Colomina et al. 1994, 83)

My own take on the matter is clearly informed by the former understanding and the idea that queer theory is about disturbing the taken-for-granted connections between particular practices of sex, gender and sexuality. Indeed, following the lead of Butler, I am convinced that much – and perhaps all – of that which is assumed to be heteronormal is far from normal, and demands to be queered. Exploring the repressions, silences, strangeness and pleasures of straight performances is surely as important as exposing the differential practices and spaces of diverse LGBT populations. Rather than heterosexuality, we need to talk of *heterosexualities*. After all, in an era when the media is replete with stories about the rise of the 'bi-curious metrosexual', it seems odd that many academics do not follow suit by exploring the variety of desires and disgusts that adhere to a multiplicity of heterosexual identity categories. Conversely, demonstrating that stereotypical heteronormal environments (like suburbia) are actually sites where a variety of (sometimes excessive and perverse) sexual identities take shape is vitally important work, not least given the persistence of commentaries that overstate the sexual conservatism and 'inherent heteronormativity' of suburbia in an attempt to celebrate occasions of LGBT transgression (e.g. Gorman-Murray 2006, 55).

My own take on queer is thus to argue that all 'truth regimes' of sexuality need to be disturbed and that this form of *interruption* is as necessary for dispelling the myth of a monolithic and dominant heteronormality as it is for dismissing the idea that LGBT populations are abnormal. Queering space involves exposing its performative power and drawing attention to the way specific sites institutionalise and reify identities of all kinds, straight and gay. A queer geography, in that sense, is not about identity politics at all, but is anti-identity. For me, it is about the fluidity of sexuality, the diverse forms of desire that are aroused in different sites, and the ways these are enabled or repressed through different institutional practices. From this perspective, the fact that increasing numbers of geographers are seeking to address the critical silence surrounding geographies of heterosexuality is to be welcomed. In arguing this, I might be accused of seeking to rupture the connection between the LGBT-informed politics of queer and the adoption of queer epistemology. Perhaps so. But we should not rush to dismiss the widespread adoption of a queer approach as a banal assimilation of the queer agenda. Just because queer approaches are no longer the sole provenance of those who identify as LGBT does not mean that geographical studies of heterosexuality are politically reactive or passively complicit. Far from it. As Butler implies, the more performative confusion that is cast over the assumption of a straight/queer binary, the less likely it is that the naturalness of that binary will be upheld. Queering sexual identities of all kinds is thus vital if we are to identify possibilities for more productive sexual formations, performances and spaces.

SECTION III
Politics

Chapter 13

Pussies Declawed: Unpacking the Politics of a Queer Women's Bathhouse Raid

Catherine Jean Nash
Alison L. Bain

Introduction: Raiding a Sexy Space

On 14 September 2000 the Toronto Women's Bathhouse Committee (TWBC) hosted '2000 Pussies', its fourth women's 'Pussy Palace' event at Club Toronto, a gay male bathhouse just south of the 'Gay Village'. From 6:00pm onwards, women lined up in the rain outside a converted Victorian redbrick house, some sheltering beneath a rainbow awning, to gain entrance to this queer women's bathhouse event. Volunteer security guards regulated the queue and handed out sheets of paper listing rules, etiquette and legal rights. They checked identification and bags before letting women through the door to purchase tickets at the entrance kiosk. After sliding their tickets under a metal grill, women were buzzed into the bathhouse and handed a clean white towel and safer sex supplies. Music played loudly. In the steamy red glow of a glassed changing room, some women removed their clothes and stepped into the hot tub, sauna or shower. Inside, signs directed women to erotic activities such as porn, souvenir polaroids, lap dancing, S&M and massage. Up a flight of worn, grey-carpeted stairs chandeliers lit the way to the locker room where women who had brought outfits to change into swapped street clothes for combinations of leather, rubber, lace, mesh and cotton. A narrow mirrored hallway off of the locker room was lined with small rooms that each provided a narrow bed with a turquoise plastic mattress and a locker. Some of these rooms were available for rent, but many were left open on a first-come-first-serve basis. With each new floor, the lighting became dimmer and the hallways narrower. But everywhere the atmosphere was the same: sexy and playful, charged with sexual anticipation.

At 12:45am, five male plainclothes police officers from the City of Toronto 52 Division entered the Pussy Palace. They spent an hour and a half searching all four floors of the bathhouse, apparently undertaking an inspection for possible liquor licence violations. They knocked on closed doors. They searched rooms. They interrogated patrons and took down identifying information. And then they left.

Several weeks later, six charges were laid against two volunteer organisers who signed a Special Occasion Permit (SOP) for contravening the Liquor Licensing Act. If convicted of these charges, the two volunteers would have faced fines of up to C$100,000 and one year in jail. Although it took over a year to resolve, the charges against the TWBC volunteers were ultimately dismissed when Mr. Justice Peter

Hryn determined that organisers and participants had a reasonable expectation of privacy vis-à-vis men and that the police, therefore, should have used female officers (*R v Hornick and Aitcheson* 2000). The court found the actions of the police, in the use of male offices, analogous to "visual rape" and the court viewed such conduct as "flagrant and outrageous" and as going "against common decency." The decision was touted as a victory for sexual freedom by the bathhouse organisers (see Bain and Nash forthcoming).

Throughout their ordeal, the TWBC received a considerable amount of support. Response from within the local gay and lesbian communities and without was swift and sharp; many mainstream city councilors, attendees, veteran civil rights activists and bathhouse organisers framed the raid as homophobic sexual discrimination and harassment – a deliberate crackdown on homosexual sex and nudity that was uncalled for given that relations between the local police and the gay community had been relatively incident-free for over fifteen years. Several large-scale demonstrations, a number of high profile fundraising events and several media campaigns to increase awareness helped to raise funds to defray legal and other expenses and provided positive coverage of the TWBC's legal battles.

In this chapter, we focus on the political and social strategies employed by gay and lesbian community activists in defending the TWBC volunteers from quasi-criminal charges.[1] We argue that in the process of garnering financial and political support for the TWBC and in defending against alleged liquor licensing violations, the 'queerness' of both the spaces and the practices fostered in the bathhouse events were subsumed under a discursive mantle of acceptable and normalised gay and lesbian identities – what some scholars have labelled a form of 'homonormativity' (e.g. Bell and Binnie 2000). The result, we suggest, is the reinforcement of a localised and historically grounded form of conservative homonormativity that reflects acceptance, both within the Toronto gay and lesbian communities and mainstream interests, of what Bell and Binnie (2000) call the 'good gay citizen' over the 'bad queer'. This reflects the spatialised relations of power operating within well-established gay and lesbian urban neighbourhoods such as Toronto; marginalised sexualities are often rendered invisible in the service of promoting a more widely acceptable understanding of homosexual identities.

With this discussion, we intervene in the ongoing debates about what is meant by the 'queering' of the geographies of sexuality and highlight the tensions, erasures and transformations seemingly inherent in such a politic. It is not our intention to in any way take away from the incredible work and dedication of the many individuals engaged in working for the dismissal of all charges against the two Pussy Palace volunteers. What we want to do here is examine, in hindsight, the broader implications of the political and social response to the Pussy Palace raid and to consider the wider political issues relating to projects considered 'queer'. We begin by outlining

---

1   This articles is based on a series of in depth and semi-structured interviews with a number of the members of the TWBC as well an analysis of the alternative and mainstream press on the raids and their aftermath, the court decision dated January 2002 and attendance at several bathhouse events. For a discussion of the methodological issues involved, see Bain and Nash (2006).

what the TWBC understood to be their mission in 'queering' particular spaces and practices, and we position this within the larger academic literature on 'queer' and on Canadian gay and lesbian politics. We discuss the specifics of the political and social responses by gay and lesbian activists in Toronto and consider the mainstream response to the raids. We conclude with a discussion of the complexities of the politics of queer at this current juncture in the history of the urban gay and lesbian social movements in North America.

Spatialising Queer Identity Politics

The Pussy Palace was the first temporary women's bathhouse in Canada. It is an irregular bi-annual event held in a gay male bathhouse. The first Pussy Palace was organised in the fall of 1998 by a small group of queer women activists to "address the invisibility of queer women's sexuality" by creating a space that supports "casual, kinky, and public sex" (Gallant and Gillis 2001, 154). "Aside from launching thousands of orgasms" the TWBC organisers believe that the Pussy Palace is an event that validates minority sexualities that "do not conform to the middle-class norms of 'private, marital/sanctified (hetero)sexuality'" (Gallant and Gillis 2001, 156).

While a women's bathhouse may resemble, at first glance, the operations of a gay male bathhouse, for the organisers, the Pussy Palace is a project of radical 'queer' organising because they regarded the establishment of a sexy space, such as the bathhouse, as encouraging new sexual practices and behaviours amongst lesbians in ways that would lead to the formation of 'queer' rather than merely lesbian identities. The Pussy Palace organisers see their project as transgressive and subversive both towards the common conceptions of lesbian sexuality operating *within* gay and lesbian communities and towards conceptualisation of both lesbians and women more widely (See Nash and Bain forthcoming).

The TWBC's desire to queer spaces and identities draws on more broadly circulating notions of 'queer' in both academic and activist circles. The word 'queer' is often used by activists as an umbrella term to address a lengthy list of sexual and political articulations: lesbians, gays, bisexuals, transgenderists, transsexuals, transvestites, cross dressers, and drag queens – the category of queer is neither finite nor fixed (Phelan 1997; Peters 2001). The counter-hegemonic dimensions of queer mean that the term is frequently used to challenge accepted orders and social practices by embracing multiple and ambiguous identities, subjectivities and performances (Corber and Valocchi 2003a).

As we have argued elsewhere (Nash and Bain forthcoming), the TWBC interprets 'queerness' at the bathhouse in terms of flexibility in bodily presentation, openness to different forms of sexual expression, playful overtness with respect to sexual intention, and casual sexual activity. The TWBC encourages people to explore, experiment or play with gender and to challenge the rigidity of gender categories, stereotypes, norms and expectations, particularly those the TWBC see as arising from the narrow and limiting tenets of lesbian feminism. The TWBC celebrates what it sees as the more libratory aspect of gay male bathhouse space practices and a more 'earthy' and purportedly 'authentic' working class woman's sexual expression.

When the term queer is applied to space, Tattelman (2000, 224) suggests that "'queer space' involves the construction of a parallel world, one filled with possibility and pleasure." Queer space functions as an intervention in the world of the dominant culture, replacing its fixed principles and binary modes of thinking with the mutability of our everyday queer action. It is important to remember that while the TWBC is striving to create a queer space with all of the possibilities and potential of the idealised spaces, practices and identities described by Tattelman, the politics of those representations take place within the larger gay and lesbian community with its own norms, social practices and politics.

In Canada, gay and lesbian political activism over the last twenty years has engaged directly with the state at all levels to render the lives of gays and lesbians visible and on an equal footing with heterosexuals in all realms of public and private life (Warner 2002; Grundy and Smith 2005). A substantial portion of activist energy has been directed towards obtaining the inclusion of 'sexual orientation' in various provincial and federal human rights legislation, efforts which have been largely successful (Lahey 1999; MacDougall 2000). Most significantly, Section 15 of the Charter of Rights and Freedoms in April, 1985 brought the so-called 'rights' revolution to Canada (Blomley and Pratt 2001). At the time of its passage, Section 15 did not expressly include sexual orientation as a prohibited ground of discrimination. However, sexual orientation has been 'read in' into the section "by the majority of the Supreme Court of Canada, not in ringing tones but somewhat grudgingly" (MacDougal 2000, 1).

The impact of a rights-based politic has had a profound effect on the positioning of gays and lesbians in Canadian society. Many argue that this rights-based strategy has resulted in a conservatisation of the gay and lesbian political movement and a narrowing of the social possibilities for sexual expression (Herman 1994; Kinsman 1996). Success in rights claims negotiations with the state by sexual identity groups most often comes with what many see as the 'burden of compromise' that results in an agreed to "'acceptable' mode of being a sexual citizen" (Bell and Binnie 2000, 3). State and gay and lesbian acceptance of certain forms of homosexuality – forms of 'homonormativity' – reflect forms of discipline and constraint that effectively close off spaces that support various forms of alternative 'erotic citizenship' (Bell and Binnie 2000, 19). Alternative sexual practices and identities are pushed into the private (and invisible) sphere, causing "a division between 'good gays' and (disreputable) 'bad queers'" (Stychin 2003, 28).

Despite concerns that this so-called mainstreaming of gay and lesbian identities has resulted in a narrowing of sexual expression and practices, others argue that it has, in fact, spawned a countermovement of sexual dissidence and transgression (Duggan and Hunter 1995; Leap 1999). Lawrence Knopp (2004, 122) points out that in recent years an "even more radical anti-identity politics of hybridity and fluidity, ... variously termed diasporic, post-colonial, post-feminist and queer," has appeared within gay and lesbian political and social contexts – a politics in need of both material and symbolic social space. As we argue here, these attempts at imagining and creating spaces that facilitate the practice of a 'queer counterpublic' often find themselves policed not only by heteronormative surveillance and discipline but also by homonormative surveillance as well (Duggan and Hunter 1995, xi).

Events organised by groups such as the TWBC reflect deliberate attempts to establish spaces that support alternative or queer sexualities in response to what bathhouse organisers regard as the normalising tendencies of both lesbian feminist politics and mainstream prohibitions on women's sexuality in general (Nash and Bain forthcoming). We argue that gay and lesbian activists' responses to the bathhouse raids reframed the bathhouse raids within a historically specific gay male bathhouse culture that effectively erased the queerness of the events and replaced it with a homonormative version acceptable to both gay and lesbian and mainstream interests.

Rewriting History: Gay Men and the Pussy Palace

With the September 2000 raid, the TWBC faced considerable legal and financial difficulties. As a non-profit, volunteer-based organisation, they were not equipped to deal with the time-consuming and expensive pressures of a very public criminal trial. In order to deal with these difficulties, the TWBC sought help from the local gay and lesbian communities and, in particular, sought the advice and guidance of local gay activists who had considerable experience of political organising in Toronto.

The contemporary gay and lesbian communities in Toronto have a long-history of altercations with the police (Nash 2005; 2006). Although there were a number of clashes with the police throughout the 1970s as Toronto's gay district developed, perhaps the most dramatic, and the event most often referred to as crucial in the history of the Canadian gay and lesbian political movement, was the 1981 bathhouse raids. Called the "biggest and most terrifying of all" raids by Tom Warner (2002, 110), Toronto police executed simultaneous raids on four male bathhouses and charged 304 men as 'found-ins' and 20 others as 'bawdy house keepers'. The next day, gay activists organised a protest attended by over 3000 people. The protestors marched on the local police station, disrupted traffic, smashed police car windows and pushed and shoved police officers. Many historians regard the raids and the subsequent response as pivotal to gay and lesbian activism in Canada; through a demonstration of militancy and visible anger, the community was able to stave off more intrusive and perhaps more violent police action. These raids and the community's very public and aggressive response has been mythologised as a pivotal moment in the history of both the development of Toronto's gay ghetto and the maturing of Canada's gay and lesbian political movement (Kinsman 1996; Warner 2002).

The TWBC was the only overtly political queer women's organisation in Toronto at the time of the September 2000 raid, and there were realistically no other organised queer women's organisations to look to for support. In the weeks and months immediately following the Pussy Palace raid, the TWBC looked to gay male activists and prominent community organisations for assistance. One of the initial activists involved in assisting the TWBC immediately after the raids was gay activist Tim McCaskell who had coordinated the response to the 'original' 1981 bathhouse raids. At McCaskell's suggestion, a public meeting was organised at the 519 Community Centre in the heart of Toronto's Gay Village. Hundreds of people attended and fundraising discussions quickly turned into a 'spontaneous' protest march on police headquarters, an event almost identical in timing to the initial protest marches in 1981. The march halted traffic as protestors chanted 'No more shit!' and 'Fuck You 52 – pussies bite back!' and ended in a 'kiss-in' covered by local media. As Loralee explains:

> I remember going to Tim McCaskell and he was like 'Alright, this is what we do.' And he organized marshals (I think he's been organizing marshals for thirty years). He's like 'This is what you have to do Loralee, get up at the front and lead people. Here's the microphone. You can shout "No more shit!"' ... We did that and it was kind of neat because people told us those old things. And then we did things like 'What do you want? Pussy.' 'When do we want it? Now.'   (Loralee, 16 January 2004)

It is evident from this quotation that the involvement of gay male activists was seen as immensely valuable by TWBC organisers. In very practical ways, the gay male activists shared their political knowledge by coordinating march marshals, locating bullhorns, developing slogans and calling the media. The slogans that they prompted Loralee with, however, were not original. Two decades earlier gay men had shouted the very same phrases – 'No more shit!' and 'Fuck you 52' – as they, too, organised a larger 4000-person rally at Queen's Park and 52 Division, and a 2000-person sit-in in the middle of Yonge and Bloor Streets to protest the raids and call for an inquiry.

The fact that the TWBC looked to gay male activists for assistance and the imposition of the gay male bathhouse narrative on understandings of the TWBC events reflects the unequal power relations in play in locations such as Toronto's gay community. There were no other lesbian or women's organisations in Toronto able to economically and politically provide the support needed to assist the TWBC. This highlights what Heidi Nast (2002) has described as the special economic, political and social privileges enjoyed by largely gay, white, middle class males in the newly emergent urban spaces of commodification and consumerism. Groups such as the TWBC operate outside the traditional sexual and gendered identities of the normative gay and lesbian communities as they are reproduced in Toronto. Without an established political and social location within the gay and lesbian communities and with absolutely no visibility or recognition from the heteronormative state, groups such as the TWBC had limited choice but to appeal to gay men for assistance (Van Deusen 2004). In response to these appeals, gay male activists such as Tim McCaskell resorted to those normative understandings developed through the mediation of conflict between gay male activists and state authorities over several decades (Valverde and Cirak 2003).

As Van Deusen (2004, 32), drawing on Nast (2002), points out, gay male political success has meant that "gay white men, traditionally marginalized as sexual minorities, in fact benefit from the heteronormative state." In containing and soothing cultural conflict, the state can deploy versions of identity and citizenship that exclude, rework or render invisible certain identities and, as Van Deusen (2004) argues, the heteronormative state has largely been receptive to particular forms of gay, white male culture, including, in the Canadian context, a general acceptance of gay male bathhouse culture (Warner 2002; Bain and Nash forthcoming). This has the effect of marginalising other queer identities not centrally located or accepted in this mainstreamed gay male culture, ensuring that, even within subcultural groups, state mediation of sexual citizenship can position groups outside "the margins of dominant cultural identities" (Marston 2004, 12).

The queer project fostered by the TWBC was able to operate within gay male space as an event with little impact on or interest from the gay male community. As the TWBC interviewees acknowledged, alternative gendered and sexualised identities such as those encouraged at the Pussy Palace would not have been tolerated within gay male bathhouse space (or other sexualised male spaces), reflecting the hegemonic norms of the dominant sexual group. Queering certain spaces depends, in part, on the relations of power already in circulation in that space. With dominant gay male sexualities and gendered representations firmly established in urban gay neighbourhoods in North America, arguably the ability to queer that space might meet with some considerable resistance. Once the TWBC activities surfaced in the more public sphere, gay male interpretations of those activities quickly transformed the TWBC queer politic into the more traditional gay and lesbian understandings of lesbian's activities – a more 'liberationist' political argument about sexual freedom. The spatial expression of alternative sexualised and gendered expressions were effectively rendered invisible thereby making the 'gay public sphere' the normalised standard (Phelan 1997; Rushbrook 2002).

Yet despite the influence of gay male understandings, the TWBC organisers were able to insert their own style of activism that was "flamboyant, fun, and unrepentantly pro-sex" and, we would argue, presented a 'female' presence in an otherwise male proceeding (Gallant and Gillis 2001, 159). Draped in feather boas and carrying signs, they staged a 'Panty Picket' in front of police at 52 Division. Protestors shouted "Sluts can't be shamed" and hung lacy and colourful lingerie on a clothesline. Like their gay male mentors, the organisers explained: "[i]n the fashion of queer activists before us, we responded by refusing and reframing that attempt to shame and humiliate us" (Gallant and Gillis 2001, 159). Despite these modest attempts to remake the nature of these protests, the mentorship role performed by gay male activists is clearly illustrated in the following quotation:

> Tim had a broader political context around the issues that at the time I totally didn't have. I went through such a huge learning curve around these issues ... I had never written a press release before, I'd never organized a press meeting ... When we were organizing fundraisers that really targeted the gay men, they gave us names of people. It's like an old boy's club you know, not really old but middle-aged fag club. They all know the influencers, they know the important people. I didn't know these people but they got people out, they got all these fags to come and pay $150.00 for a cocktail and appetizers ... The older fags really rallied around us. (Loralee, 16 January 2004)

Not only were the older gay men politically savvy, they also had valuable social connections to wealthy people within the gay and lesbian communities who could help to sponsor the legal fund by attending fundraisers. Much like the raids two decades ago, the fundraising was based on the premise that an attack on a gay 'institution' was an attack on the community as whole, even if many of the individuals who contributed were not patrons of the bathhouses (Kinsman 1996; Warner 2002; Nash 2006).

The (re)positioning of the Pussy Palace raid as a repeat of the 1981 raids on gay male bathhouses can clearly be seen in mainstream and alternative media coverage. Between the fall of 1998, when the first Pussy Palace was held, and the fall of 2003, 80 articles and editorials have been published in Toronto's mainstream and alternative newspapers and magazines on the Pussy Palace. Initially, the Pussy Palace made headlines in the alternative 'pro-gay' media "as a sign of female sexual freedom" – a classic framing of events within a more 'liberationist' rhetoric (Giese 2000, OP1). After the police raid, the alternative media provided regular coverage of the Pussy Palace and took a strong oppositional stance to police behaviour at the bathhouse event. Consistent support for the Pussy Palace and critique of police motives can also be found within mainstream newspapers and magazines.

Of the 80 newspaper articles discussing the Pussy Palace, 16 articles make explicit reference to the 1981 gay male bathhouse raids, evenly split numerically between mainstream and alternative presses. A *Macleans* article written by the late gay male activist George Hislop (2000) nine months prior to the Pussy Palace raid describes how the media and public outrage surrounding the 1981 raids turned public opinion in favour of gay rights, and sets the stage for subsequent media coverage. That sense of public 'outrage' is elicited again in one of the first articles in the mainstream press on the Pussy Palace raid. Openly gay Toronto City Councilor Kyle Rae, who represents the ward where the bathhouse is located, called the raid an outrage to a community that had clearly established itself as a distinct cultural group, and pointed out that the police have not harassed the bathhouses in almost 20 years. Councilor Rae made it clear that, given the history that Toronto police have with gay bathhouses, the police should have known that had "no business going into private spaces where consenting adults are doing what they do" (Kennedy 2000, 10). Kyle Rae quickly became a spokesperson and advocate for the TWBC, demonstrating that, once again, members of the gay male community played, and continue to play, a central role in framing public discussions and understandings of the women's bathhouse. The most explicit construction tracing the historical lineage of the Pussy Palace raid to 1981 is a 'raiding history' or timeline in the gay community newspaper *X-tra*, where an analysis of the history of reporting on bathhouse raids illustrates how such raids in Toronto are constructed as a gay male experience (see Gallant 2001). As we have documented elsewhere (Bain and Nash forthcoming), the gendered and sexualised identities of the participants and organisers of the Pussy Palace were flattened in the press to that of lesbian women. The words bisexual, transsexual and transgender were not used in mainstream press coverage of the Pussy Palace raid, and little consideration was given to what it means to be 'queer women' and to create 'queer space', which is at the heart of the TWBC political project. Perhaps this is not surprising given that gay men in Toronto do not appear to have embraced a particularly politicised notion of queer in any of their events or spaces.

Conclusions

The activities of the TWBC represent the aspirations of a group of women committed to a queering of spaces and identities through the establishment of a space that supported not only lesbian identities but attempts to radicalise the underlying understandings of the basic binary conceptualisations of sexual and gendered identity. These queer activist interventions, in utilising existing understandings of the behaviours encouraged by the sexualised spaces of gay male bathhouses, were made in the hope that the space of the bath, by its very meaning and physical structure, would promote alternative forms of women's sexual expression. The TWBC goal was to intentionally challenge both mainstream and lesbian ideas about women's sexuality.

As a politic, attempts at 'queering' the binary understandings of gender and sexuality proved a complex task with unexpected and unforeseen results. As our interviews and anecdotal stories suggest, there is little doubt that the Pussy Palace events continue to have an impact on certain segments of lesbian life and contribute to the making of a broader queer social life in Toronto. There are more spaces labelled as 'queer' that support the presence of a range of sexual identities, including those we label heterosexual. But in terms of the bathhouses, they remain a temporary event held once or twice a year and entirely dependent on the creative energies of volunteer organisers.

Yet despite the success of the TWBC in opening up alternative sexual spaces, that success is arguably predicated on their relative invisibility and lack of impact on the hegemonic gay male culture that dominates the urban landscape of Toronto's gay neighbourhoods. Once forced into the public spotlight by the police raid, the TWBC looked to experienced gay male activists for support. In doing so, the TWBC's framing of the bathhouse events as a queering of women's and lesbians' gendered and sexualised practices and spaces was absorbed within the more comfortable and homonormative discourses negotiated between the predominantly white, middle class gay male culture in Toronto and the state over the last 20 years.

This research also suggests how the process of queering identities and space is itself gendered in that more 'traditional' gay male spaces such as the bathhouses resist or perhaps actively discourage the emergence of 'queer' identities. This raises interesting questions about how the processes of 'queering' identities and spaces play out differently in practice across differently gendered and sexualised gay and lesbian spaces. While the label and concept of queer is arguably here to stay, it occupies a precarious and perhaps ironic position outside of, yet within, homonormative gay and lesbian communities.

Chapter 14

Religion, Identity and Activism: Queer Muslim Diasporic Identities

Farhang Rouhani

Critically Considering Queer Complicities

Within the context of queer academic geographies, we are currently witnessing the simultaneous trends of a desire to radicalise the academy, on one hand, and gradual normalisation of homosexuality, on the other. Natalie Oswin argues that such a coupling of trends necessitates a theorisation of 'complicit queer futures' (Oswin 2005). She critiques geographers who locate queerness outside the spaces of capitalist logics and practices. Among other things, such a distinction unintentionally constructs difference based on authenticity, with authentic radical political action being perceived as outside the spaces of domination. "Instead of thinking complicit space as total and negative, we might reconceptualise it as ambivalent and porous, as an undetermined set of processes that simultaneously enables both resistance and capitulation" (Oswin 2005, 84).

I find a similar problematic division within queer activist circles in the US between constructions of 'gay' and 'queer', with gay representing mainstream organisations and actions and queer representing radical ones. The implications of such a distinction are that gay politics capitulate and are inauthentic, while queer politics resist and are thus authentic. Such a valorisation of queer politics is simultaneously useful and deeply problematic. While useful as a critique of the normalisation of gay and lesbian politics and identities, the politics of differentiation and authenticity in such a construction runs the risk of withholding the possibility of a critique of queer complicities.

The concept of queer complicities is particularly useful in examining the recent florescence of a transnational queer[1] Muslim[2] movement. This construct can aid in examining the complex political processes through which non-heterosexual Muslim

---

1   I follow Scott Kugle in his use of the term 'queer' rather than gay in referring to this movement in order to enable "a more descriptive and complex analysis of a variety of sexual orientations and practices that are distinct, but united in their common difference from hetero-normative sexuality" (Kugle 2003, 199). In addition, the term 'queer' can be used to identify a wider and more fluid range of political alternatives.

2   I have adopted the standard academic practice of capitalising 'Muslim' but not 'queer' in referring to queer Muslim identities and movements. In doing this, however, I do

immigrant minorities are seeking to form new spaces and spheres of engagement, the challenges and constraints they face, and the possibilities for new assimilationist and radical identity formations that create academic, activist, and spiritual connections. This chapter considers the situatedness of the transnational queer Muslim movement in relation to: scholarship on Islam and sexuality; the construction of queer and Muslim identities; the new radical-progressive Muslim activist movement; and international sexual human rights movements. I focus specifically on the formation of a queer Muslim identity and movement through Al-Fatiha, the most prominent queer Muslim group in the US.

The name Al-Fatiha is a direct Quranic reference meaning 'The Opening'. The organisation was founded in Washington DC in 1997 by Faisal Alam as an internet list-serve. It now has subscribers in over 40 countries, with chapters in eight US states, and sister organisations in the UK, Canada, South Africa and Saudi Arabia, among others. What has made Al-Fatiha particularly successful is its blending of list-serve activity with conferences, retreats, and workshops, the first of which was held in Boston in 1998, and the most recent in Atlanta in 2005.

Al-Fatiha's identity as an organisation, defined in terms of its mission and enacted through list-serves and conferences, is based on a combination of religious, academic, and political goals. There is, on one hand, the on-going effort to look within biographical histories of group members to construct new stories of spiritual empowerment. The retreats, in particular, are an overwhelmingly positive effort to build community through spiritual interactions. At the same time, social justice goals are at the centre of its mission, and Al-Fatiha actively works with other LGBTQ (lesbian, gay, bisexual, trans and queer) and progressive Muslim organisations for support and funding related to HIV, immigration and asylum, and to voice concern about national and international human rights abuses, among other things. Al-Fatiha fully embraces progressive Muslim notions of equality and justice and seeks to aid Muslim sexual minorities in integrating within their faith, families and communities. This is an ambitious goal considering the diversity of US queer Muslims in terms of class, race, ethnicity, gender and sexuality. The existence of Al-Fatiha as a group is also partly the product of the dispersed leadership and the lack of dependence on particular spaces of worship among immigrant Muslim communities (Smith 1999; Metcalf 1996a). This decentredness allows for Al-Fatiha functions to be led by its own queer-identified prayer leaders. It also allows for such things as virtual internet space and conference halls to serve as spaces of religious identification and worship.

Al-Fatiha's website is easily accessible and usable and includes extensive information on the organisation, news, events and mailing lists. On a resources page, there are listed a number of books on sexuality in Islamic societies, LGBT and progressive Muslim lists, and even a few conservative Muslim sites under the category 'opposing views'. There is much activity on the group's various list-serves, which are organised geographically on an international scale. Most of the posts are relevant articles and announcements on different religious and/or sexuality-based

---

not mean this to suggest that the Muslim portion of the identities and movements is more important than the queer portion.

issues. Very little discussion, though, occurs through these groups; they primarily operate to disseminate opinions and information. By contrast, the discussions that develop in the conferences, retreats and workshops are lively. Other email lists and groups, such as Gay Muslims, Iman, and Muslim Gay Men, are also more active in communicative exchanges and debates. The virtual and material spaces are mutually dependent and significant in creating a space that is at once not tied to any particular place and highly dependent upon spaces of public social interaction. In this sense, the queer Muslim community being built through Al-Fatiha, like other movements and communities that depend on a mutuality of virtual and material social spaces, is at once extraterritorial and recurringly territorialised (Saint-Blancat 2002). It is simultaneously removed from material space and constantly re-grounded in new ways.

My research here consists of observations based on my participation in the activities at two conferences in 2003 and 2005, selected interviews with group members, and textual analysis of the group's website and list-serves. I focus most intensively on my observations of and participation in Al-Fatiha conferences and workshops because I found them to be most dynamic and revealing in the complex politics of complicity and radicalisation. By examining the situatedness of Al-Fatiha members within academic scholarship on sexuality and Islam, the construction of Muslim and queer identities, and multiple forms of political activism, I seek to understand the complex politics of assimilation and differentiation, capitulation and resistance, within a specific segment of a transnational queer Muslim movement.

Sexuality, Scholarship and Islam

Academic research on Islam and sexuality has been overwhelmingly dominated by an emphasis on women's sexual and, often more specifically, reproductive rights. This research has focused primarily upon structural oppressions on women's bodies, and the opportunities for negotiating and challenging these oppressions. Comparatively speaking, the literature on sexual orientation, which is the primary area of interest here, is small, but dynamic and growing. This research has focused on the histories and contemporary realities of same-sex desire and non-heterosexual identities through literary, political and cultural forms of analysis (AbuKhalil 1997; Dunne 1991; Kugle 2001; 2003; Murray and Roscoe 1997; Wright and Rowson 1993; Schmidt and Sofer 1992; Tapinc 1992). Approaches include both essentialist perspectives that approach homosexuality as an already always existing identity category and more constructivist approaches that actively avoid applying modern-Western identity labels to other contexts, instead working diligently with the terminology of the context of research. More than anything, the literature emphasises the geographic and historical diversity of sexual practices and identities among Muslims, but it is also highly problematic in the ways that it sometimes uncritically conflates sexualities with religious practices, national and transnational identities, and is sometimes underlain by assumptions made based on Eurocentric categories. In this sense, the literature is entrenched in some ways in Orientalist modes of analysis.

Joseph Massad levels the most trenchant critique by considering this research to be an integral part of the agenda of the Gay International (Massad 2002). He defines the Gay International as an assemblage of the missionary tasks and organisations that constitute the problematic universalisation of western, primarily US-based, 'gay rights' discourse (Massad 2002, 361). Writing on the ramifications of this universalisation within the "Arab world", Massad implicates "an academic literature of historical, literary, and anthropological accounts, written mostly by white male European or American gay scholars, who purport to describe and explain 'homosexuality' in the past and present of the Arab and Muslim worlds" (Massad 2002, 362). Massad critiques this literature for its essentialist transhistorical and transgeographical understandings of Islam and homosexuality; the representation of Arabs and Muslims as objects, rather than subjects or audiences, of European scholarship; and a Eurocentric denigration of Muslims that fails to problematise and thus idealises a mythically egalitarian West. In this sense, this literature on sexuality and Islam is implicated within Euro-American imperialisms that seek to 'tame' and reconstruct sexual identities and rights among Muslims and Arabs. I will discuss the problematic dimensions of these processes in subsequent sections.

In a subsequent pair of responses to Massad's original piece, Arno Schmitt, one of the main Orientalist subjects of Massad's critique, points to Massad's reification of space, time and identity categories such as the "Arab world", "the medieval period", and "Arab Muslim" (Schmitt 2003, 589). As much as Massad critiques the uses of the Orientailist term 'Middle East', his own reference points of analysis are not clearly or critically defined. At the same time that he levels this critique, though, Schmitt himself replicates uncritical conceptualisations of "Western ideas" and "Western cultural products" (Schmitt 2003, 588–591). Indeed, Massad responds by claiming the "arrogance of the scope" of Schmitt's Orientalist claim of being an expert on "Eastern Mediterranean people over a 3,000-year period" (Massad 2003, 593). This engagement between Schmitt and Massad reveals some of the significant current problems in framing, conducting and writing such research, particularly the uncritical formation of time-space categories, and the conflation of research on Muslim sexual identities with research on the region alternatively termed the Middle East, the Arab World, Southwest Asia and so on. The greatest strength in Massad's work lies in getting to the question of why we need to academically compartmentalise religious and sexual worlds into space-time regions. In seeking to represent the diversity of sexual and religious practices, this literature often replicates and reaffirms the coloniser's model of dividing the world into discrete regions of analysis without attention to their relations and mutual constitution.

In response to existing limitations in the conceptualisation of sexuality among Muslims, Scott Kugle proposes a radically new critical research agenda that seeks to simultaneously reclaim and redefine the ambiguities and diversities of sexual identities, pleasure and politics within the Qur'an and its interpretations. Contrary to the perceptions of Islam as a sexually repressive religion, Kugle interprets the Prophet Muhammad's writings as sex-positive religion, a perspective within which sexuality is a field, among economic, social, and political fields, through which spirituality is worked out. As such, sexual pleasure is viewed positively not just as a source of procreation, but also as a source of spiritual gratification (Kugle 2003,

194–195). Furthermore, he argues that the acceptance and significance of diversity in physical appearance, constitution, stature and colour of human beings emphasised in the Qur'an extends to a diversity in sexuality and sexual orientations "as a natural consequence of Divine wisdom and creation" (Kugle 2003, 196; see also Wadud 2003, 274).

Kugle utilises the Story of Lut (the Quranic version of the Story of Lot) as an example of the significance of ambiguity and the dynamically changing spatial and temporal nature of interpretations of the Qur'an. While often cited by conservative clerics today as a condemnation of homosexuality, he asserts a different interpretation, argued also by some classical Islamic scholars, of the story as a condemnation of greed, miserliness, sexual oppression, and a rejection of the prophet's ethics of care (Kugle 2003, 214). This re-reading of the Story of Lut has become a significant rallying point for queer Muslim discussion and scholarship (Abdullah and Greenberg 2003; Abdullah and Wadud 2005).

Kugle's analysis advances a more constructionist perspective for Islamic sexuality studies. He clearly rejects the category of the homosexual as something that existed in classical times and continually foregrounds the significance of the diversity and ambiguities of interpretation. At the same time, Kugle is complicit in the essentialising of Islam through his own interpretation of the Qur'an as a sexually liberating source. He is among a long lineage of Islamic scholars who assert a new interpretation by returning to the original source: the Qur'an. Andrew Yip similarly identifies a multi-level process through which non-heterosexual religious scholars respond to the modern condemnation of homosexuality by highlighting the inaccuracies and socio-cultural specificities of earlier interpretations and relocating authentic interpretive authority to individualised, personal experience (Yip 2005). Thus, he identifies a constructivist approach for arguing against heteronormative interpretations, but none that would critically interrogate the religious texts themselves.

This problem is what Imam Daayiee Abdullah identified in an Al-Fatiha conference workshop as an obsession with historical justification (Daiyee and Wadud 2005). Of course, Islam is by no means the only subject of this kind of essentialising. US politicians obsessed with interpreting the original intent of the US Constitution engage in the same problematic endeavor, for example. The problem is that such an endeavor binds and limits the practitioners to the interpretation of an 'original source' that was written within a very different context. While the re-reading of the Story of Lut offers an important critique of the condemnation of homosexuality, it creates its own new essentialism.

Immigration, Identity and Queer Muslims

The majority of the research on Muslims and sexuality has been conducted in and about predominantly Islamic societies, and comparatively little exists on the expressions and politics of sexuality among Muslims in Euro-American contexts. Two notable recent exceptions are Andrew Yip's research on British non-heterosexual Muslims and the Peter Barbosa and Garrett Lenoir' documentary, *I Exist*. In examining family and kin relations among primarily non-heterosexual Muslim South Asian first and second generation immigrants in the UK, Yip critically interrogates a pervasive Muslim immigrant cultural censure of homosexuality as a 'western disease', highlights the significance placed by immigrant sexual minorities on ideas of familial respect and honour, and stresses the complexities of living between worlds (Yip 2004). Similarly, the documentary *I Exist* chronicles the narratives of gay and lesbian 'middle eastern'-identified first and second generation migrants in the US as they combat against stereotypes about their sexualities and ethnicities, share their coming out stories, and reflect on familial and societal relations (Barbosa and Lenoir 2003).

Both pieces significantly contribute to examining and understanding the particular problems facing Muslim sexual minority individuals as they negotiate multiple fronts of resistance to their sexual self-expressions. At the same time, they both operate with a very narrow definition of the role of religion within the lives of the research participants. In Yip's research, a dichotomy develops between the participants' individual expressions of sexuality, on one hand, and the structural role of social-religious obligations on the other. In this sense, Islam is reproduced as a uniformly opposing, traditionalist force, and sexual expression becomes an individualist expression of identity. In *I Exist*, too, ideas about religion in general and Islam in particular are conspicuously absent. Instead, the film chronicles the familiar Euro-American narrative territory of "coming-out and staying-in stories, with the typical repercussions that range from acceptance to resignation to physical assaults and banishment" (Morris 2005). In this sense, it replicates the common US narrative of coming out as a developmental process and thus serves as vehicle of immigrant assimilation. The absence of religion is a product of the identification of religion as a structural cause of oppression, much as in Yip's research. This is a problematic construction that prevents us from seeing how queer Muslims can actively reinterpret and reclaim their religious identities. Furthermore, this approach operates within a narrowly individualised conception of sexual identity that does not represent the realities and potential for group identity formations.

These observations are in keeping with research on contemporary immigration experiences. Contrary to popular notions of contemporary immigrants as uninterested or incapable of integration or assimilation, research in Europe and the US suggests that the overwhelming aim of immigrant organisations is to integrate and become accepted as part of the mainstream (Brubaker 2001; Kymlicka 2001; Nagel and Staeheli 2004; Statham 1999). In examining citizenship and identity among Arab migrants in the US, Nagel and Staeheli (2004), for example, argue that migrants negotiate their roles in relation to homeland and national integration through multiple forms of political and territorial attachment.

This complex form of multicultural assimilation is also evident in the only extensive scholarly piece on Al-Fatiha and the construction of 'progressive gay Muslim' identities (Minwalla et al. 2005). Minwalla et al. seek to document the experiences of Muslim gay men in North America. Their primary findings are that Muslim gay men configure their identities through ways of negotiating their religious, ethno-cultural, and colour dimensions. The researchers interpret a psycho-developmental process of relationships with religion, from earlier stages of rejection and alienation to latter stages of more inclusive, tolerant religious interpretations

(Minwalla et al. 2005, 124). Moreover, the interviewees perceive the condemnation of homosexuality to be a product of social, traditional and cultural forces and contexts rather than religion.

In my interviews with Al-Fatiha members, I have heard this disentanglement of religion from culture and society consistently. Rather than seeing the religion itself as a product of those complex social, traditional and cultural factors, many Al-Fatiha members wish to maintain a distinction between the religion of Islam and the contexts of its interpretation. Thus, in the process of politically struggling against and responding to the condemnation of homosexuality within modern Islamic interpretations, they do not want to give up on an essentialised sense of Islam. In my own participation with Al-Fatiha activities, I have found myself in a tenuous position as what is often termed a 'cultural Muslim', meaning someone who identifies with the religion in an ethnicised, rather than a spiritual or religious sense. In a debate that developed at the 2005 Atlanta conference, in fact, some members identified the term 'cultural Muslim' to be patriarchal and imperialist in its claim that religion is something to be known and studied as a part of a cultural assemblage, rather than something to be experienced on an essential, spiritual level. Others, too, responded to what they perceived as the problem of individuals like me identifying as Muslim but not taking it seriously enough in practice within everyday life. I can certainly see how for someone with a strong essential belief, my culturalist sentiments would be troubling and deeply problematic. A number of participants also expressed concerns against the cultural dilution of Islamic practices and beliefs. For example, one conference presenter identified a dissatisfaction with the "Western gay culture" of promiscuity and binge drinking and wanted her organisation to be based more on religion rather than an excuse for socialising. Sentiments like this indicate a strong concern for the maintenance of a strong sense of faith that does not get diluted through the cultural and social processes of migration. But at the same time, such comments morally narrow the scope of what a Muslim can be.

The second and third categories of Minwalla et al.'s discussion, those of ethnic and racial identity, have also been strongly present in my experiences with Al-Fatiha members. On a general level, these differences are expressed within the immigration process through a differentiation between homeland as East and hostland as West. A constructed dichotomy between East and West, with differently constructed moral, social and cultural obligations and roles within each, is commonly expressed among Al-Fatiha members. There is also, much as argued by Nagel and Staeheli, a complex form of cultural-national integration. I can express this no better than by describing the stage of the opening session of the 2005 Atlanta conference. The room was festooned with national flags of Muslim-majority countries, presumably representing the home countries of conference participants, running down two parallel walls and leading to a very large rainbow flag on the front stage. Such an approach, as well as the use of the East/West dichotomy and the appropriation of terms such as 'people of colour', indicates a process of national integration and assimilation not through homogenisation, but through an acceptance of an ideology of multiculturalism that seeks to add other cultures into the US model, rather than question the legitimacy and the salience of the model itself.

Absent from Minwalla et al.'s discussion is the category of gender. The significance of gender identity and categorisation as important dimensions of queer Muslim identity formation were revealed through the 2005 Atlanta conference, themed 'Sexism, misogyny, and gender oppression: Breaking down systems of patriarchy'. This conference provided a rare opportunity for sexual minority Muslims and their allies to discuss on critical personal and social levels their relationships to gender. At the same time, the structure of the sessions was problematic to the extent that the sessions were divided into four simultaneous retreats: men and genderqueer; women and genderqueer; trans and genderqueer; and allies. As a participant, I found it difficult to decide which retreat to attend: the retreat for Muslim gay men or for allies? Ultimately, despite ambivalence due to my 'cultural-Muslim-ness', I opted for the men's retreat and found it to be an important seldom experienced opportunity for Muslim gay men, who otherwise may only see themselves as the victims of structures of oppression, to examine their structural roles and agency within multiple US, immigrant, and sexual patriarchal systems. One of the most critically enlightening moments for me occurred at the end of the conference when the four groups came together to share what they had learned. The 'trans group' was particularly critical in pointing out flaws in the structure of the conference that prevented those that identified as 'belonging' to the identities of multiple sessions from being able to go easily from one session to another. They also pointed out the extensive ways in which throughout the course of the conference they were both subtly and overtly discriminated against through constant interpretations of their genders and of which session they should consequently be attending.

My point in detailing these conference experiences is to relate the complexities of assimilation and differentiation within the identity experiences of queer Muslims. What we need are greater opportunities to understand the complexities of queer complicities as immigrants and as sexual, ethnic, and gender majorities and minorities.

Political Activism

The queer Muslim movement is also situated within a broader progressive movement developing out of different Muslim communities. Muslim communities in the US are divided among more conservative, liberal, and radical groupings. The conservative, traditionalist groups are the least accepting of the identity of queer Muslims, have been under the greatest level of scrutiny from the US government in the post-September 11 context, and have thus lost much of their influence within Muslim communities. The mainstream, liberal Muslim groups have received the greatest amount of attention in the post-September 11 context. These are the groups who have been active in promoting tolerance and encouraging a view of Islam as a 'religion of peace', both of which are a part of conceiving Islam as a part of the multicultural religious landscape in the US. Liberal Muslims, too, have been reticent to accept the identity of queer Muslims, except within a very mainstream tone of tolerating difference. The groups that have engaged most directly with queer Muslims and who have inspired them most directly are a group of radical progressive Muslim groups, most notably the Progressive Muslim Union of North America and Muslim Wakeup.

At this point, it is also important to note that there are significant movements to reform Islam within Islamic societies. This is perhaps most evident in the influence of Islamist reform scholars such as Fatima Mernissi and Abdolkarim Soroush on the current movements within Western immigrant communities. At the same time, the more dispersed leadership among Muslim groups in the West has allowed for a greater dynamism and freedom of expression. Because groups are decentralised within localities throughout different countries, different political ideologies have emerged, women play more central role in the structure and orchestration of religious observances, and leaders are much more likely to be popularly elected or chosen at a local scale (Metcalf 1996a, 11). This flexibility is further accentuated by the fact that Muslim rituals do not require a particular sanctioned place of worship as other religions might. In different Western contexts, churches, secular community centres and people's homes have served as spaces of religious gathering. Mosques are significant sacred spaces in particular places, but for religious purposes, it is the practice that creates the space. This makes possible immigrant Muslim practices and social gatherings in locations without a mosque infrastructure (Metcalf 1996a, 6).

The radical Muslim movement is based on a set of concerns for pluralism defined in opposition to the multiculturalism evoked by mainstream Muslim reform groups. Pluralism in this sense involves active engagement, understanding, respect, and commitment to difference – rather than diversity, tolerance, and relativism, which can continue to propagate ignorance. Radical progressive Muslim activists such as Omid Safi argue that the mainstream Muslim and US perpetuation of the idea of "Islam as religion of peace" is a "hollow phrase full of apologism and hypocrisy" that "can be and has been co-opted and adopted by hegemonic powers to preserve the unjust status quo" (Safi 2003a, 24–25). The statement of Islam as religion of peace is thus unproductive, if not destructive, because it has been used by mainstream Muslim groups and the Bush administration as a license to avoid dealing with real economic, social, and political problems. The idea of pluralism in the progressive Muslim movement thus includes a simultaneous concern for economic and social justice and for justices based on identity politics, both of which are conceptualised as integrated principles. Amina Wadud, for example, addresses these within the context of the race, gender and class politics that have produced a gulf between African American and immigrant Muslim communities (Wadud 2003).

Throughout the writings produced by the progressive Muslim movement, there is a deep concern for the same kind of constructivist approach that Kugle undertakes. Safi, for example, discusses the two current ways of discussing Islam – a normative, theological way that depends on major representative leaders, on the one hand, and a more decentred, historical, analytical, people-centered way that is more common within this movement, on the other (Safi 2003a, 21). The use of the term 'progressive Muslims' instead of 'progressive Islam' is based on the goal of bringing people and interpretation to the center and avoiding the essentialising of Islam as an actor. "Islam as such", he writes, "teaches us nothing. The Prophet Muhammad does. Interpretive communities do" (Safi 2003a, 21). This decentring from Islam to Muslims is the product of the opportunities developing out of the decentring of

Muslim leadership among immigrant communities, and it enables reclamations, reformations and reinterpretations of Islam by Muslims. At the same time, this decentring is maintained within an essential view of the Qur'an as providing the inspiration for the implementation of the human rights goals of justice (*'adl*) and the realisation of goodness and beauty *(ihsan)*. Both the progressive Muslim movement, more generally, and the discussion of sexuality by Kugle, in particular, have this common factor of a tenuous balance between a constructivist, pluralistic approach and an essentialising view of human rights and goals.

So, where does the queer Muslim movement fit here? The most common configuring of non-heterosexual identities within Islam offered by progressive Muslims is as the *mustad'afun fi'l-ard*. "That is, they are among those individuals and groups mentioned in the Qur'an who, for no reason of their own, are pushed to the edges of society and live in oppression" (Islam 2004, 6). Under these circumstances and according to this interpretation of the teachings of Muhammad, Muslims have a duty to defend them; to deny justice to the marginalised is a violation of the duties of Muslims. Not all progressive Muslims, though, have been equally supportive of the rights of queer Muslims. Some have argued that other battles, such as gender justice, are more immediate and/or important. Others have criticised queer Muslims, particularly the activist group Al-Fatiha, as being too single-issue driven, and have argued that gender justice is a more vital and important goal than sexuality-based rights. These critiques signal the extent to which the identity of queer Muslims is one of the central and heated debates within the wider progressive Muslim movement.

Despite the tenuous relationships with the liberal and progressive Muslim movements, Al-Fatiha participants share similar goals of connecting Islam to other social issues that would enable the production of Muslim liberation theologies. Furthermore, perhaps because of the lack of acceptance of sexuality-based issues within the wider progressive Muslim movement, Al-Fatiha actively collaborates with other organisations, including other queer Muslim organisations in Canada, the UK and South Africa, and international human rights organisations such as Amnesty, Human Rights Watch, Human Rights Campaign, and the International Gay and Lesbian Human Rights Commission (Grundy and Smith 2005). In connection with these organisations, Al-Fatiha specifically works toward changing attitudes about homosexuality within Muslim diasporas and aiding sexual minorities in Muslim countries through, among other things, an asylum project inaugurated in 2003.

The complex domestic and international goals and responsibilities of the Al-Fatiha Foundation are evident in a recent press release statement made in response to the 19 July 2005 reporting of two 'gay' teenagers being hanged in Iran. This reporting was perpetuated by both exiled royalist, anti-Islamic republic organisations, like the National Council of Resistance of Iran, and gay rights organisations such as Human Rights Campaign. The British organisation Outrage! went as far as to call for the European Union to institute trade sanctions against Iran for its persecution of sexual minorities. After this news, accompanied by the final images of the teenagers before being hanged, was disseminated through the international gay media, it became increasingly unclear whether the boys were hanged for 'being gay' or the officially claimed charge of rape. Other organisations, such as Human Rights Watch, thus took a much more careful critical stance. In any case, while the details of the case are unclear, this case, according to one press release, "only fed to the growing Islamophobia and hatred toward Muslims and the Islamic world" (Alam 2005). Based on the circumstances, Alam identifies two important strategic goals: the imperative that Western gay human rights organisations need to build stronger ties with on-the-ground NGOs working within countries like Iran, and the imperative that the gay and lesbian liberation movement needs to be viewed within a broader lens of social justice. Both these goals, in essence, require the much greater and careful incorporation of the queer Muslim movement within other political movements.

The circumstances surrounding this event reveal the complex and tenuous set of relations through which the Al-Fatiha Foundation, in particular, and the diasporic queer Muslim movement, in general, engage with diasporic exiled Muslim, Euro-American gay human rights, and other social justice organisations. Alam seeks to differentiate al-Fatiha and the voices of queer Muslims from conservative diasporic communities and Islamophobic gay media and human rights organisations in a vein similar to Massad's critique of the Orientialist Gay International (Massad 2002). At the same time, Al-Fatiha members seek to identify with sexual minorities in Muslim countries as similarly 'gay' and similarly 'Muslim'. This sense of relatedness is what enables and mobilises efforts to support sexual minorities in predominantly Muslim countries in different ways, but it is also what Massad identifies as the most problematic element of the Gay International: the imposition of Euro-American notions of identity on sexual diversity in Muslim countries, which ultimately has the effect of marginalising and making the lives of sexual minorities all the more vulnerable. As a burgeoning diasporic community, queer Muslims must deal with the question of how to support sexual minorities in predominantly Muslim countries according to the ways that those minorities want to be supported.

The queer Muslim movement, in this sense, is the product of a complex set of struggles, identities, experiences and politics through a unique critical combination of religious, activist and academic pursuits. The combination of these relations and pursuits places the movement simultaneously within and in resistance to larger cultural and political systems. The future efficacy of a transnational queer Muslim movement depends upon a critical awareness of the multiple forms of complicity and resistance, assimilation and differentiation that exist in the spaces that diasporic identities inhabit.

Chapter 15

HIV+ Bodyspace: AIDS and the Queer Politics of Future Negation in Aotearoa/ New Zealand

Matthew Sothern

What we have to offer in our lesbian and gay families is exactly what children need to grow into healthy, happy, well-adjusted children. (Martin 1993, 25)

As an adult gaybie (isn't that condescending?!) or 'queerspawn' (doesn't that sound delicious?!), I'm so fucking sick of all you queer Uncle Toms whitewashing our experiences in heteroculture ... We're all faggots. Did our parent's genital proclivities turn us into homosexuals? No. So what makes us, the grown children of queer parents, faggots? You do, gay parents ... Even if we grow up to be straight, we don't grow up to be normal... (Black 2005)

'Queerspawn'

These two quotes frame the following discussion of contemporary HIV politics in three ways. First, they underscore how much contemporary gay and lesbian political organising is focused on issues of marriage and the family, and, thus, they are suggestive of a troubling silence about the status of HIV/AIDS ('isn't that a South East Asian/African disease now?'). The second frame these extracts provide my discussion concerns the kinds of politics each invoke. The first quote is a liberal invocation of normalcy, a kind of non-queer benignity where gays and lesbians underscore their significant contributions to hegemonic projects of national domesticity (what Lisa Duggan (2002) calls 'the new homonormativity'); whereas the second promotes a queer politics of deviance, a radical politics of subversion and an angry celebration of abnormality. These two framings are obvious enough, of course, so in this chapter, I draw on the provocative work of Lee Edelman (2004) to suggest the need for a third and even queerer frame that uses the opposition between rhetorical figures of 'The Child' haunting both quotations and the queer figure of HIV+ bodyspace (the living-dead that William Harver (1996) calls the space of future AIDS corpses). This opposition between The Child and the queer, as Edelman stresses, serves to frame 'The Political' in terms of life/death, the world with a future/ the world without a future, the worthy/the unworthy, innocence/evil. In other words, I am interested in the opposition between the image of the happy and healthy child lurking on the side of liberal gay and lesbian politics of homonormativity (in whose name the future of the nation is assured) and that of the space of the HIV+ body, a figure overdetermined by its subjection to a lack of futurity and therefore excluded a priori from the promise of the future afforded by The Child.

Of course, Edelman is trying to be somewhat provocative with this opposition, and, certainly, the apparent audacity of his attack on 'the politics of the child' goes a long way towards proving his point that the figure of The Child is so sacred within current discourses of the political. As he goes on to argue, the kind of moralism that attends the sacred space of The Child, a figure produced at the expense of the reality of actual children's lives, ensures that a politics articulated toward a future that is to be realised in the name of The Child is necessarily a politics orientated toward a replication of the same. For Edelman, the childish politics of futurity is a politics of radical continuity, as the *Gay and Lesbian Parenting Handbook* so clearly celebrates; the happy, healthy, well adjusted child is the child for whom the status quo presents little difficultly. It is hardly surprising, therefore, that the space of the HIV+ body has emerged as an especially important space for imagining a politics without The Child, which is to say a politics quite literally without the future as we know it. As many queer theorists now argue, queer political futures cannot be complicit with what Munoz (1999) calls the liberal logic of identity because queer is not a basis for solidifying identity claims, but rather the space beyond which identity exhausts itself (cf. Oswin 2005 on queer complicity). The future heralded by queer politics is not the whitewash imagined by the *Lesbian and Gay Parenting Handbook*, but rather it is the negation of that future offered by 'queerspawn'.

Empirically this chapter uses these rhetorical figures of 'the queer' and 'The Child' as a device to interrogate the relationship between the liberal politics of futurity and what might be called a queer politics of future negation in a recent HIV/AIDS prevention campaign in Aotearoa/New Zealand. 'Negative Role Model', the specific campaign this chapter considers, was launched by the New Zealand AIDS Foundation (Te Tuuaapapa Mate Aaraikore o Aotearoa, hereafter NZAF) in 2003. Like many developed countries, New Zealand has witnessed dramatic increases in HIV infection rates in the millennium, and in 2003 there were more new HIV infections in New Zealand than at any time previously. Most disturbing for NZAF is that a disproportionate number of these new infections have been located in communities of gay men, the group which had historically responded best to safer sex education. However, 'Negative Role Model' was somewhat different to NZAF's traditional prevention campaigns. NZAF, like other AIDS service organisations (ASOs), had historically relied on images of the HIV+ body as diseased, threatening, sinister, and always and already dead (Crimp 1989; 2002). These visual logics of HIV prevention campaigns worked to rhetorically produce the space of the HIV+ body as so thoroughly and completely Other. This othering, emerging from the steady flow of media images of disease and death, positioned the HIV+ body as a kind of visceral 'warning' that mobilised fear as a way of motivating changes in sexual behavior. Such representations have been central to safer sex messages emerging from the AIDS crisis (Borowitdz 2003) and were relatively easy to engender given the spectacular and comparatively rapid nature of AIDS deaths earlier in the pandemic. This othering caused little difficulty because HIV diagnosis assured a rapid death; now, in contradistinction, biomedical advances in the treatment of HIV mean that people are living longer with HIV, and the sense of urgency surrounding the AIDS crisis has waned. As a result, in what might be termed an AIDS-as-post-crisis moment, a new set of political arguments, centred on support and advocacy, has emerged around the space of the HIV+ body, and persons living with AIDS are challenging messages that see them as sites of disease, contagion and death (Crimp 2002).

In current configurations of AIDS politics, the HIV+ body emerges as a particularly dense and contradictory site. On the one hand, the logics of HIV prevention continue to imagine the HIV+ body as a site for the diffusion of the HIV virus, a site of contagion and death: it is a space of the Other to be avoided (if only by latex). On the other hand, within the logics of advocacy and support, the HIV+ body is seen as a space in which political claims to inclusion and equality can be fixed. The HIV+ body is and is not Other simultaneously. This contradiction in the representations of the space of the HIV+ body raises important questions for contemporary HIV/AIDS politics. 'Negative Role Model' was NZAF's response to this contradiction. It was an attempt to shift the representational burdens of prevention work away from the HIV+ body. Rather than rely on images that explicitly other the HIV+ body, 'Negative Role Model' sought to provide a positive image of the benefits of remaining HIV-, thereby avoiding the potential conflict between advocacy and prevention.

This positive role model came in the form of Nathan Brown, a 25 year old university student. 'Negative Role Model' was to be structured around Nathan's struggle to avoid HIV infection. The press release announcing the campaign tells us that Nathan is a healthy, attractive and vibrant young gay man; he is a good son from a typical Kiwi family with deep connections to his local community and university. While Nathan was an attempt not to blame 'bad' HIV+ people for spreading the disease, I argue that this campaign ultimately fails in this attempt. This failure, I suggest, has to do with the ways in which politics demands a vision of the future that the HIV+ body is incapable of sustaining. The HIV+ body is not happy, healthy and normal but is instead more akin to 'queerspawn', a space irreconcilable with the liberal politics of futurity.

The chapter is divided into three sections. The first section outlines Edelman's provocative arguments about the politics of futurity caught between the opposition of the figures of 'The Child' and 'the queer'. The second section pushes this conceptualisation a little further to interrogate the ways in which queer political theory has understood the AIDS crisis. This section pays particular attention to the ways that HIV/AIDS has engendered new understandings of the temporality of the political. The final section returns to reconsider the spatial and politic logigs of the 'Negative Role Model' campaign. I conclude that what HIV/AIDS ultimately reveals is that arguments about gay benignity, such as those of the *Lesbian and Gay Parenting Handbook*, are untenable and are rather complicit with what Lisa Duggan (2002) calls the new 'homonormativity'.

'The Child' and the 'Fascism' of Futurity

While the figure of The Child has always served as a backdrop to much of the politics of sexuality, it has taken on a particular purchase within the new politics of homonormativity. As Edelman brilliantly shows, when the political is orientated toward an always yet to arrive future (a future, of course, that can never arrive), The Child is invoked both as the symbol in whose name we must fight for the future and as that future's ultimate bearer. Nowhere is the centrality of the figure The Child more evident than in the current debates over gay marriage: the rhetorical plea 'won't someone please think of the children?' is mobilised to argue both in defence of 'traditional' regimes of the family and in recognition of the families gay parents already make. In such debates, The Child is a privileged signifier that frames the contours of contemporary gay and lesbian politics – and, indeed, of the political more generally – by disciplining it to what Edelman calls the logic of 'reproductive futurism'. The Child polices identity: it shows how happy and normal gays, lesbians and their families can be, even while it is the figure that proves a priori the absurdity of such claims (see Sedgwick 1993). As the overdetermined signifier of futurity, The Child, in whose name we must always struggle to save the body politic, becomes a kind of fascist ontology that renders 'not fighting for the children' unthinkable.

Fascism is a loaded term, and surely attaching it to the benign and smiling face of The Child is likely to provoke sustained criticism. Yet to call this politics of The Child fascist is not in any way to use the term lightly, and nor does it dilute the real horrors for which the label is reverently reserved. Rather, I suspect that Edelman's use of the term fascist – however irreverent it may be – is warranted for it is precisely through the very everydayness, commonplaceness and taken for granted-ness of The Child that the hegemony of the reproductive politics of futurity is constituted. It is in the seeming insignificance or political irrelevance of reproductive futurism that Edelman's critique uncovers the establishment of the very impossibility of imagining a politics without the future. Fascism reigns by admitting no possibility of an alternative; insofar, therefore, as understandings of The Political require a conception of The Future and insofar as The Child remains central to the possibility of imagining the future, then the figurative Child remains the fascist core of the political. The figure of The Child functions as a moralistic intervention that quashes more radical political imageries by subtending them to a logic of an always yet to arrive future – a phantasmic projection on whose horizon The Child perpetually sits:

> The Child has come to embody for us the telos of the social order and come to be seen as the one for whom that order is held in perpetual trust ... We are no more able to conceive of a politics without a fantasy of the future than we are able to conceive of a future without the figure of The Child ... The Child is the obligatory token of futurity.
>
> (Edelman 2004, 11–12)

The powerful symbol of The Child should not be underestimated; indeed, the profound innocence of The Child is even seen as washing clean the filth of sex, at least insofar as it bestows a potentially redemptive quality on heterogenital relations (a redemption even the Pope willingly bestows). The Child is a figure that installs "generational succession, temporality, and narrative sequence, [organised] not toward the end of enabling change, but, instead, of perpetuating sameness, of turning back time to assure repetition – or to assure a logic of resemblance (more precisely: a logic of metaphoricity) in the service of representation and by extension, of desire"

(Edelman 2004, 60). The Child legitimates the social expression of desire – not the asocial narcissism of fucking, but rather the proof that Love ensures that 'tomorrow, tomorrow you're only a day away'.

Insofar as The Child carries all this representational weight on its shoulders, it should come as no surprise that Edelman argues that the Other of The Child is the future negating queer, those who refuse to assume their proper place in the Oedipal drama of the symbolic. The queer's aberrant sexuality tarnishes the iridescent promise of the future in the favour of the immediacy of self-destructive *jouissance* (is there anything more sexy than 'the death drive'? (Crimp 2002, 106; Bataille 1986)). While Edelman understands this *jouissance* in typically Lacanian terms, Judith Halberstam's (2005, 6) new work explores the subcultural configurations of various queer modes of world-making that refigure "the temporal frames of bourgeois reproduction and family, longevity, risk/safety, and inheritance." Halberstam examines the emergent postmodern formations of 'What If' where AIDS (and other forms of sanctioned violence against queers) "squeeze new possibilities out of the time at hand ... [where AIDS] is not only about compression and annihilation; it is also about the potentiality of a life unscripted by the conventions of family, inheritance and child rearing" (2005, 5). These new queer geographies of 'What If' are not a politics organised around an assumption of stable identity but instead a politics of liminal non-identity that offer potential subversions to neoliberal formations of time and space. As work by Lisa Duggan (2002), Barbara Cruickshank (1999) and Nikolas Rose (1999) demonstrates, as political ideology, neoliberalism reproduces conservative political structures. As the state abdicates its responsibility for social service provisions and scales back its commitment to social justice, it does so by seeking to govern through an investment in a conservative cultural politics of family and self-responsibility where the ideals of the long, productive and respectable life buttress the logics of flexible capital accumulation to ensure that 'everything has a time and place'.

For Leo Bersani, queer modes of 'What If' are experimental orderings of desire, pleasure and of the body that challenge the governmental regimes of liberalism because in queer formations of time and space "[we] risk [our] boundaries, risk not knowing where [we end] and the other begins. This is lawless pederasty – not because it violates statutes that legislates our sexual behavior, but because it rejects personhood, a status the law needs in order to discipline us" (Bersani 1995, 128–9). AIDS disrupts the assumed naturalness of time and place scripted by the assumption of what counts as the 'normal' life subject to the laws of intelligibility. The power of 'queer' (and of course its anti-social danger) is that it reveals the heteronormative precondition of the laws by which the normal is constituted. Much recent writing in queer theory, and even more so in other forms of queer cultural production, are profoundly political because they are attempts to explore the potentialities of queer senses of time and space – they are imaginaries of 'What If' that confound the limits of political representation policed by the figure of The Child (Butler 2002; 2004; Halberstam 2005; Edelman 2004; Harver 1996; Brown 2001).

The politics of queer time and space are not orientated toward the future, to longevity and respectability: in other words, these are not imaginaries where lesbian and gay families bring up healthy and happy, well adjusted children – where the measure of their success is the degree to which they shore up the symbolic status quo – but rather imaginaries where various 'queerspawn' grow up to be deliciously abnormal. Certainly these queer reformulations of time and space must take the risk of the radically undetermined nature of the human: in lieu of *his* determination, his place, there is only the posthuman body (Halberstam and Livingstone 1995), a Deleuzian-style 'enfleshed materialism' (Braidotti 2002) where the certainty of authenticity, identity, gender, justice and even the future are anything but certain. The queer body deploys these terms – terms that are so central to the lexicon of The Political as to render thinking about The Political without them an impossibility – and circulates them in complex new ways that confuse the (hetero/homo)normative systems of production, consumption and desire in which The Political trades. In the politics of queer time and space, the queer body "constitutes a loss, a disorientation, but one where the human stands a chance of coming into being anew" (Butler 2004, 38–9).

The best work in queer politics challenges the norms by which the human is intelligible; queer politics is not The Child but the destruction of the future The Child promises. 'Queer' marks a space beyond conventional politics because it is so thoroughly saturated with challenging the stable economy of identities through which The Political is constituted. Therefore, queer is not so much an oppositional category (for opposition is contained within the logic of sameness and difference, the logic of Self/Other on which identity feeds) so much as it is a deconstructive move. One is tempted to say that this deconstructive move is one where The Child is finally slain, or at the very least malnourished – where the category of The Human itself is undone, and just maybe, hopefully, performatively made anew (see Irigary 1993; Butler 2004; Ordover 2003; Haraway 1997). In this vein, it is easy to see why AIDS has become a privileged site for exploring this kind of queer politics. This is because AIDS quite literally demands a reckoning of The Political without subtending it to the tyranny of futurity, to "the fascism of the baby's face" (Edelman 2004, 151) or to what Jeffery Nealon (1998, 20) somewhat less provocatively calls "the reassurance of a deferred future."

AIDS and the Future Negating Queer

As decades of writing in queer studies have argued, AIDS has a double face. AIDS has been a catastrophe for the 'gay community'; in certain places, it has decimated entire generations of gay men. However, it is misleading to describe AIDS solely in terms of loss because AIDS has also provided the mechanism for constituting the 'gay community' through the political efforts mobilised to respond to entrenched homophobia (Altman 1988; Crimp 2002; Duttmann 1996; Yingling 1997; O'Neil 2001). To a certain extent, AIDS transcended issues of class, race and geography to reveal the commonality of gay men. AIDS, we are repeatedly told, respects no boundaries. This paradox between the losses and productivities wrought by AIDS, as William Harver (1996) explains, revolves around the ways in which AIDS deaths have been politicised as a symbol through which a particular liberal gay identity politics centred on loss and mourning is constituted (on liberalism and mourning see Eng and Kazanjian 2003; Agamben 2002). Such a politics works to produce a symbolic AIDS death around which a call for inclusion can be mobilised. Hence the culmination of Tony Kushner's *Angels in America* (Kushner 1993) – "we won't die silent deaths anymore" – stresses the centrality of naming AIDS deaths as part of the projects of gay inclusion in hegemonic narratives of American nationalism and the full benefits of citizenship these narratives protect. Similarly, ACT-UP's famous slogan 'Silence = Death' serves to constitute the category of AIDS deaths: not random deaths from obscure cancers, for example, but deaths from institutionalised homophobia – death wrought from the privilege of unmarked heteronormativity, race and class.

While it is important not to overgeneralise these literatures, consideration of the history of NZAF certainly underscores the usefulness of these ideas to analyzing the New Zealand context. NZAF began in the mid 1980s as a political agitation group that hoped to embarrass the New Zealand government into taking action over the emerging AIDS crisis. From the outset, NZAF has seen AIDS in a broad political context and viewed its job as facilitating gay community building and arguing for institutional and legislative change across the spectrum of sexual politics issues. In recent years, NZAF have publicly lobbied the government on issues of gay marriage, prostitution, hate speech and the like. This placing of AIDS within the broader agenda of gay identity politics in New Zealand has served to constitute NZAF as a very important site for consolidating gay identity as a political category within the mainstream political landscape of New Zealand.

Central to NZAF's emergence as an identity politics organisation has been the production of a symbolic AIDS death. Political organising around the space that Harver (1996) calls the AIDS corpse becomes a site of transcendence that retroactively produces the categories of contemporary gay identity politics. The AIDS corpse therefore becomes not the corpse of the individual but rather the site at which the individual is collapsed into the identity category 'gay'. This collapse allows for the production of 'gay' identity as an ideal category because "the possession of others is possible only when they are dead; only then is nothing opposed to our image of them" (Bersani 1990, 7). Gay identity politics thus emerges as a site of loss, what Wendy Brown (1995) calls a 'state of injury'. Brown suggests that a common feature of identity politics movements in modern liberal democracies is a claim to the political significance of particular groups' marginalisation or exclusion. The gay identity politics emerging from the AIDS crisis revolved around the way mounting AIDS deaths revealed the commonality of gay identity as a site marked by its exclusion from the category of the fully human, those whose lives are worth saving. Thusly constituted as an identity emerging from exclusion, gay becomes a site from which to mount a campaign for inclusion.

Such analyses of AIDS politics were commonplace under the conditions of AIDS-as-crisis where the assured rapidity of AIDS death provided both the urgency and mechanism for gay identity politics centred on the production of this symbolic death. Paradoxically, it is from the ashes of the political mourning of AIDS deaths that a politics of the affirmation of gay life emerged. But do we now exist in an AIDS-as-post-crisis moment?

[T]he so called phenomena of AIDS has become very much part of the texture of the quotidian, central to our commonsense perceptions of the way the world is, and thereby to our sense of commonality. For example, many of our undergraduates have never known and perhaps will never know, sex without latex; we are now being urged to think of HIV seropositivity, and indeed of 'AIDS itself', as a chronic condition on the order of diabetes; we are, in short, becoming persuaded that AIDS belongs to the normative rather than the extraordinary, that AIDS is chronic rather than a crisis. We have erected, perhaps in place of other erections, entire structures of intelligibility and comprehensibility on and around the pandemic, structures that themselves render AIDS normative and routine: the business of AIDS, constructed and carried on around an impossible object, has become – like genocide, nuclear terror, racism, misogyny, and heteronormativity (or what I would prefer to call orthosexuality) – business as usual.   (Harver 1996, 2–3)

Biomedical advances in the treatment of HIV have rescripted the temporality of AIDS such that it is no longer (for the privileged few at least) the immediate death sentence it once was. The result has been both a decline in AIDS deaths and a simultaneous increase in infections as the sense of urgency and visibility of the pandemic has waned in most Western countries. This lack of urgency might, for want of a better term, be called AIDS-as-post-crisis. In this AIDS-as-post-crisis moment, the political space of the HIV+ body, those facing the real struggle of surviving with HIV, has emerged as a significant new challenge for ASOs. Western ASOs now strive to balance the ongoing need for HIV prevention work with the new mission of advocating on behalf of those living long-term with the disease. The HIV+ body emerges, then, as a particularly dense site of contradiction: on the one hand prevention campaigns focused on advocating safer sex necessarily produce an othering of the HIV+ body (as a space to be avoided, if only by latex); whereas, on the other hand, advocacy on behalf of those living with HIV engenders a set of arguments for inclusion and equivalence (where the HIV+ body is just as normal, productive and useful as other bodies and so should not be discriminated against).

This contradiction is perhaps not terribly surprising insofar as the living HIV+ body does not allow for the production of a symbolic death, and yet the HIV+ body is not a space around which arguments about the assuredness of the future are particularly stable. How, for instance, is one to argue for the politics of the HIV+ body when the only guarantee is that it is the body that is defined by its subjection to the lack of futurity. Moreover, the HIV+ body further complicates the simple politics of homonormativity, where gays and lesbians argue that they are normal and productive, because it is a space produced at the intersection of flesh and technology. The HIV+ body is a cyborg space in which flesh and biomedical technology become indistinguishable and is thus a space that is precisely not-normal, not fully human (or, rather, where the category of The Human is renegotiated – see Squier 2004). The HIV+ body confounds liberal arguments about the benignity of liberal gay politics because the HIV+ body materially confounds the murderous politics of normalcy that are made in the name of realising a future from which it is necessarily excluded. Above all, therefore, HIV+ bodyspace is an important new kind of political space, albeit one overdetermined by the familiar and tired circuits of race, class and homophobia. It is a site that requires a politics after The Child because the logics of reproductive futurism are woefully inadequate as response to the conditions of AIDS-as-post-crisis. The seemingly simple question 'what does it mean to be *living with AIDS?*' is, as Garcia Duttman (1996) argues, anything but simple. What happens to the subject – the subject who is defined by their subjection to the inevitability of death – when they claim the status of 'living with AIDS'? Can there be a stable identifiable 'I' within the identity claim of Person Living with AIDS (PLWA), and can the HIV+ body be a stable container to which that 'I' is supposed to be self-identical (Haraway 1991; Martin 1994)? For many queer theorists the answer is no. Thomas Yingling (1997), for instance, challenges us to think about AIDS not simply as a physical undoing of the self, but as producing a no-longer-self through which to think the political anew – a politics that Australian cultural critic Eric Michael's (1994) calls 'unbecoming':

> That is the *thing* of AIDS, it is the signifier through which we understand the cancer of being, the oncology of ontology – not only in its threat to our being, its announcement that we are moving toward non-being, indeed are already inscribed with it, in it … But also that it itself is deeply not identical, never quite the same, appearing under different guises, none of which is a disguise, following circuitous routes into visibility and action. It is the disease that announces the end of identity.   (Yingling 1997, 15)

While the cultural politics of AIDS have been understood largely in terms of desire – excessive, deviant, othered desires as well as the regressive, homophobic, racist desires of the conservative cultural right – Yingling insists we pay attention to the cultural politics of non-identity:

> [W]e might ask – through Lacan and others – about the mirror stage and the relation of the 'I' to its body, for we encounter in living with AIDS the production of non-subjects, people for whom the mirroring illusions of discourse are broken: the host body in this case continually reminds its subject – with every glance in the mirror – of the distance between the 'I' and its lesions, and of the fact that the lesions may not be subsumed into any transcendence.   (Yingling 1997, 16)

The queer struggle of and for AIDS is a struggle for a cultural politics of non-identity, and with it new scriptings of what counts as a life worth living. This is the struggle for a representation that is adequate to the experience of the radical reterritorialisation of the subject that AIDS has wrought. This is because AIDS renders mute – or at least mutable – the territorialised boundaries on which modes of liberal interpolation are based: the territories of Self/Other, for example (Haraway 1991; Martin 1994); territories of Life/Death (Harver 1996); territories of Science/Culture (Martin 1994); and ultimately, of course, the territories of Politics/Sex, which is perhaps only the boundary imagined as Public/Private (Watney 1994; Warner and Berlant 2002; Delaney 1999). Thus AIDS "foreground[s] the impossibility of speaking the condition of loss being written into (and onto) the body. So it is not desire that is in question, but identity: the whole problem of a disappearing body, of a body quite literally shitting itself away. That is AIDS" (Yingling 1997, 16).

HIV+ bodyspace therefore becomes a kind of limit point – what Agamben (1998) would call a threshold figure (Norris 2005a) – within the biopolitical regimes of liberal modernity because it is at the limit of the acceptable and thus delimits and produces the acceptable solely in the productive act of its exclusion. For Butler (1993), this is the political function of 'abjection'. Agamben (2005) prefers to call this the act of 'abandonment'. HIV+ bodyspace is a space that cannot be reconciled with the politics of The Child because it is a space that exudes a radically different non-future (Butler 1993, 27–55). This opposition of The Child as the signifier of the future (the phantasmic projection that is the stuff of the political) and the future negating queer (epitomised by the figure of the living-dead HIV+ body – the body that spreads death) causes major problems for liberal gay and lesbian politics of homonormativity and their arguments about gay benignity. In the final section of this chapter I want to return to the 'Negative Role Model Campaign' to explore the consequences of this difficulty for the cultural politics of HIV/AIDS in Aotearoa/ New Zealand.

The Child vs The Queer in Aotearoa/New Zealand

Aotearoa/New Zealand, like other Western countries, has seen a dramatic increase in new HIV infections under the conditions of AIDS-as-post-crisis. In 2003, 183 new HIV infections were recorded. This was the largest number of new diagnoses in 12 years. In response, NZAF launched a new media campaign to combat what it called "an undue optimism around HIV/AIDS." Importantly, NZAF viewed this optimism as emerging largely from the AIDS culture in Sydney and North America (especially San Francisco). What interests me about this new campaign is the way that it sought to overcome the routine images of AIDS-as-post-crisis by redoubling NZAF's education efforts and shifting the focus from 'how to avoid AIDS' onto 'why to avoid AIDS'. As they stressed in their 16 June 2004 press release announcing the 'End the Silence Campaign':

> [T]he 'End the Silence' campaign will be making a strategic shift from telling people *how* to prevent HIV infection to *why* it should be prevented … The success of HIV treatments at delaying the onset of AIDS; the continued prejudice and judgment in our communities against men living with HIV; and a glossing over of the often very difficult and unpleasant consequences of HIV infection and treatment have contributed to a silence and/or an undue optimism around HIV/AIDS that has to end if we are to stem this record rise in new HIV infections.   (NZAF 2004)

Much debate ensued over how ending this silence would be best approached. Initially NZAF talked about producing a series of widespread media images aimed at gay men that would show the 'awful truth' of living with – and dying from – AIDS. This campaign was to prominently feature images of emaciated and 'obviously sick' PLWA as a way of providing a visceral counterpoint to the logics of AIDS-as-post-crisis. The politics of this approach to visceral truth telling – particularly with respect to the possible reaction from New Zealand branches of BodyPositive, an HIV/AIDS support and advocacy group – meant that NZAF decided to take a different tack. In place of the initial approach, NZAF decided to advocate a 'Negative Role Model', which was embodied in 25-year-old university student Nathan Brown. The choice of Nathan, who sits on the elastic threshold between The Child in its innocence and the fully narcissistic queer, is telling. For Nathan occupies a kind of ambivalent position between the politics of innocence that is The Child and blameworthiness that can legitimately be attached to the adult. The fact he is always shown with his parents (Figure 15. 1) underscores the degree to which we are encouraged to read him as so thoroughly on the side of The Child and thereby central to the politics of homonormativity and the projects of nationalism it supports.

Figure 15.1   The Promise of the Future?

By permission of the Out There! A New Zealand Queer Youth Development Project by Rainbow Youth Inc (Auckland) and the New Zealand AIDS Foundation / Te Tuuaapapa Mate Aaraikore o Aotearoa. The image contains Nathan Brown and his parents Jaye and Russell.

Nathan and his parents are intended to provide a positive 'negative' role model as a way of promoting "living free of HIV as a desirable lifestyle." But what exactly is at stake in the images of this attractive, young, healthy gay man and his supportive middle-class parents, who, of course, are members of Parents and Friends of Lesbians and Gays (PFLAG)? Nathan is intended as a liberal body whose vitality is supposed to make the case for remaining HIV-. Consequently the campaign has become structured around the production of an ongoing narrative of his struggle to avoid HIV. This is intended as a move that does not reproduce the demonisation and scapegoating of PLWA. On the surface, therefore, the use of Nathan as a figure moves the representational burdens of prevention work away from HIV+ bodies, thereby avoiding potential conflict with PLWA.

Hovering just below the image of liberal Brown familial domesticity, however, is the failure of the 'Negative Role Model' campaign to shift the focus from illiberal HIV+ bodies, leaving these bodies as the campaign's repressed invisible centre. Let's consider what we are told about Nathan: he is 25 and a Film and Media major at Dunedin's Otago University (where his Father also did his Veterinary training). He is president of UNIQ, Otago University's LGBT students' association. We are also assured that despite the pressures of having a gay son, his family is "a typical Kiwi family with strong values around work, family and community" who have lived in Dunedin for 23 years. His parents, who are quoted liberally in the NZAF press release, fully support and love their son – to the extent that they are willing expose themselves to 'negative comments' and be part of this national campaign. As Nathan's mother Jaye explains, "I worry about my gay son, just like I worry about my straight son, but the higher presence of HIV in the gay community does mean it figures larger for me when I think about Nathan, his health and his future" (NZAF 2004).

There are several issues that warrant mention here. First, the campaign seems to rest upon and reinscribe the problematic logics of risk categories (and thus of the exceptionalist arguments of minoritorising discourses of sexuality identified by Sedgwick (1990)). Further, these logics not only assume a 'normal', non-gay public that is not at risk, but places a rhetorical blame on those gay youth who do seroconvert while simultaneously obfuscating the kinds of race and class privilege Nathan enjoys. And, of course, it should be noted Nathan and his family are hardly poster-children for gay alienation and familial homophobia. In short, Nathan was never likely to be counted amongst the burgeoning number of new infections in Aotearoa/New Zealand, which were overwhelmingly located in minority populations in the transnational spaces of urban Auckland. Moreover, lurking at the centre of this image – perhaps just beyond the suburban gate at whose threshold Nathan, but not his parents, stand – is the dirty world of sex, populated by bodies that are vectors of HIV, as a threatening constitutive Other to the familial rhetoric in which this image trades. Thus, the only way in which this image can become effective as a mechanism of prevention is through a negation of representations of the HIV+ body that nonetheless provide its modus operandi. While this premier image does not make explicit reference to PLWA, it does work to position the HIV+ body and queer sex as a kind of silent diffusing Other whose antidote seems to be the nuclear, heterosexual family, whose own position is produced, in part, through the exclusionary logics of nationalism. In this respect it is perhaps no coincidence that the Browns live in the regional town of Dunedin and not the urban centres of Auckland or Wellington where the kinds of suburban spatial logics of this image might be undermined by circuits of immigration and modes of transnationalism wrought within the context of increasingly important processes of globalisation. The heteronormative logics of nationalism operative here also demand that he is shown with his parents and not a partner (or fuck-buddy), friends or other support-network (such as NZAF outreach staff), and is shown at home rather than at a bar, club, cruising-ground or bathhouse – where presumably the struggle to remain HIV- is most urgent.

The 'Negative Role Model' campaign, in its current form, may do little to displace AIDS-as-post-crisis because it merely reproduces a regressive liberal political logic of family, nation and self-responsibility – the same set of logics through which AIDS itself is structured. As an example of the new politics of homonormativity, 'Negative Role Model' denies the ongoing veracity of social structures of race, class, homophobia and displacement and places a rhetorical blame both on HIV+ bodies and those stupid enough to use their bodies in illiberal fashion; it individualises the moment of seroconversion, divorcing it of its social content and thus demands nothing of the future but a reproduction of the same. HIV+ bodyspace proves the inadequacy of this non-demand. Above all, however, this campaign is ultimately disturbing precisely because of its apparent benignity, its (or, perhaps, is it Nathan's?) claim to innocence. It does not succeed in avoiding a demonisation the HIV+ body. Instead, it more thoroughly and surreptitiously places rhetorical blame on the HIV+ body, which is constituted as ever present in its threatening absence. Thusly constituted as absence, there is no place within 'Negative Role Model' for PLWA to claim any space of political engagement; it may not, in this regard, cause friction with BodyPositive but it should be viewed with some scepticism nonetheless.

'Negative Role Model' betrays the representational difficulties of the political space of the HIV+ body within the liberal framework of prevention. The difficulty here, as I have tried to show, is that an attempt to shift the representational burdens of prevention work away from HIV+ bodies only more thoroughly implicates them at the centre of this prevention work. The queer HIV+ body threatens to rupture the liberal political logics of gay identity – the promised normal, healthy life Nathan's remaining HIV- portends – and thus is only useful to the liberal politics of homonormativity insofar as it dies and becomes a site that transcends itself into ideal realm of identity. The living HIV+ body, however, is capable of performing no such transcendence. Thus, as Douglas Crimp (1989; 2002) argues, the traditional mode of representing people living with AIDS as always-and-already dead rests upon our ultimate fear of queer modes of being, queer formulations of time and space, and, of course, of queer sex. The HIV+ body may be likened to something like the enfleshed space of queer. The HIV+ body refuses to be redeemed by the future that is the promise of The Child:

> [T]here is a deeper explanation for portrayals of PWA [Persons With AIDS], and especially of gay men with AIDS, as desperately ill, as either grotesquely disfigured or as having wasted to fleshless ethereal bodies. These are not images that are intended to overcome our fear of disease and death, as is sometimes claimed. Nor are they intended to reinforce the status of the PWA as victim or pariah, as we often charge. Rather, they are, precisely, *phobic* images, images of the terror at imagining the person with AIDS as still sexual.
>
> (Crimp 2002, 106)

Terror, in other words, that HIV+ bodyspace should be a container for a person at all. NZAF's 'Negative Role Model' campaign, as an example of the liberal politics of homonormativity, therefore, is so invested in the politics of The Child that HIV+ bodyspace is, like all those queerspawn who refuse the privilege of normality, negated in favour of the future it cannot bring. What the logic of AIDS-as-post-crisis openly acknowledge[s] is that the (entirely fantasmatic) clean and proper body politic is maintained only in the process of the exclusion of an expendable social surplus comprised of people and peoples of color, sex workers, IV drug users, and queers; the exclusion, that is to say, of all those of us whose bodies are encoded as excessively and preternaturally erotic, all those of us who are said to be excessively devoted to the practices of pleasure; all those of us who are held to be therefore incapable of distinguishing with any eroto-epistemological surety between self and other. Our very corporeality is taken to be at once the locus of immoderate pleasures and therefore that of contagion, impurity and death.

(Harver 1996, 9)

The negation in the representations of the space of the HIV+ body within 'Negative Role Model' betrays the untenability of the claims of the benignity of gay and lesbian politics. The HIV+ body refuses the liberal political logic of an always deferred future and instead forces a confrontation with the hegemony of heteronormative constructs of time and space instantiated by the figure of The Child. What the HIV+ body mandates is a whole new conception of The Political itself. Such a demand is not the conciliatory and assimilationist liberal politics of the *Gay and Lesbian Parenting Handbook* but, rather, the radical challenge made by 'queerspawn'.

Chapter 16

Autonomy, Affinity and Play in the Spaces of Radical Queer Activism

Gavin Brown

To be autonomous is not to be alone or to act in any way one chooses – a law unto oneself – but to act with regard for others, to feel responsibility for others. This is the crux of autonomy, an ethic of responsibility and reciprocity that comes through recognition that others both desire and are capable of autonomy too.   (Notes from Nowhere 2003, 110)

The Notes from Nowhere collective (2003) has suggested that the current growing appeal of autonomy is a direct result of the crisis of faith and trust in representative democracy. For a growing minority the current political climate has posed more profound questions about the meaning of democracy and popular involvement in the decision-making process. As a result, "participation, deliberation, consensus, and direct democracy are emerging from the margins and, in many instances, are being reaffirmed as the centre of gravity for communities the world over." (Notes from Nowhere 2003, 108).

These developments often go hand in hand with a broader questioning of societal values and a re-evaluation of how life should be lived. Many of the most vibrant forms of contemporary radicalism are beginning to move beyond purely oppositional politics and are attempting to reconstruct society around a different set of norms (for example, co-operation, consensus and non-hierarchical horizontal structures), often through localised experiments.

In this chapter, I want to discuss the political practices that constitute and are constituted through radical queer experiments in social autonomy. Specifically, I examine the spaces of international Queeruption gatherings. Like 'Reclaim The Streets' and 'Food Not Bombs' (Day 2005; Ferrell 2001), Queeruption events are what Day (2004) has called 'non-branded tactics' – a model of activist work that can be adopted and adapted by anyone who chooses to work with it. Queeruption is a specifically queer tactic of constructive direct action – a space where radical queer activists from different countries can come together to share information, skills and community for a short period.

The queer activist networks that create the Queeruption gatherings (and there have now been nine of them since 1998 – two in London, and others in New York, San Francisco, Berlin, Amsterdam, Sydney, Barcelona and Tel Aviv) have a complex and non-linear heritage, drawing on many strands of activism. Their precursors include ecological and post-anarchist movements such as Reclaim the Streets (Jordan 1998), Earth First! (Plows 1998) and the Zapatistas (Marcos 2001); as well as early Gay Liberation (Cant and Hemmings 1988), the Cockettes (Weissman and Weber 2002), the women's peace camp at Greenham Common (Roseneil 2000), and the queercore and riot grrl music scenes (Spencer 2005).

Their political goals, structures and methods distinguish them from both mainstream reformist gay political organisations (like Stonewall in the UK or the National Gay and Lesbian Task Force in the US), which seek equality with normative heterosexuality, and also from more radical gay groups like Outrage! or Queer Nation, who have been prepared to use civil disobedience to further an agenda of demands that is not that different from the mainstream (even though they initially offered something very different).

The first wave of queer politics in the early 1990s offered the promise of a move beyond a politics of representation for minority sexual groups towards a more thorough resistance to "regimes of the normal" (Warner 1993, *xxvi*). In both Britain and the US, this political development grew out of the involvement of lesbian and gay activists in AIDS activism and civil disobedience organised through ACT-UP (the AIDS Coalition to Unleash Power). In the US, at least, this occurred partly because ACT-UP had been so successful in building a coalition of those affected by HIV/AIDS that it partially lost its earlier focus on fighting the homophobia that underpinned the neglect of many people with AIDS (Crimp 1993). Queer Nation and Outrage! organised many media-focused stunts designed to highlight official homophobia and to challenge the invisibility of queer people in pubic space. However, as Cohen (1997) has argued, the militancy of the ACT-UP groups in many major US cities (but particularly New York and San Francisco) was founded on the 'class dislocation' that many upper middle class gay men experienced as a result of AIDS: not only were their friendship networks being decimated by the disease, their (sexual) citizenship was undermined and their ability to consume as 'normal' was disrupted by the cost of health-care. As a result, although their tactics of non-violent direct action appeared radical, the politics that underpinned them were more complex and contradictory. There was always a danger that in employing transgressive tactics to expose the ubiquity of heterosexual privilege, these groups actually served to draw attention to and, ultimately, reinforce the homo/hetero binary (Heckert 2004).

In contrast to these expressions of gay political organising, contemporary radical queer politics creatively experiments on the basis of affinity and autonomy to find means of fostering non-hierarchical queer community. These diffuse activist networks are avowedly anti-assimilationist and sex positive. The right to bear arms, join the clergy and marry are not rights that overly concern them. They would far rather celebrate and defend the diversity of people who are attempting to live outside the confines of heteronormativity, rather than muscle in on the institutions that sustain it. All of the events I describe in this chapter were consciously promoted as being 'for queers of all sexualities and genders.'

While these activist spaces endeavour to be inclusive of bisexuals and transgendered people, queer is still more than simply an umbrella term for all those who are 'othered' by normative heterosexuality. Indeed, 'queer' in these spaces is as opposed to homonormativity as it is to heteronormativity. That is, they oppose and contest the complacent politics of mainstream gay politicians who actively work to win gay people's compliance to a depoliticised culture based on domesticity and privatised consumption. Queer celebrates gender and sexual fluidity and consciously blurs binaries. It is more of a relational process than a simple identity category. As Heckert (2004) suggests, a truly radical politics of sexuality must move beyond simple transgression, and incorporate its ethical goals (for example, co-operative, non-hierarchical, sex-positive relationships) into its mode of operation. This politics is infused with a creative 'do-it-yourself' ethos that prefers charity shop glamour, which shuttles back and forth between masculine and feminine signifiers (Hennen 2004), over the latest designer labels. Queer revels in its otherness, difference and distance from mainstream society (gay or straight), even as it recognises that this distance is always incomplete. In other words, by aligning myself with these networks, I am using 'queer' neither as an umbrella term and a synonym for 'LGBT', nor simply as a marker of alterity.

In the metropolitan centres of the Global North, the gay scene is saturated by the commodity. Through it, people consume products and experiences that confirm their identity as 'gay'. This investment in consumption means that people no longer relate to each other as active participants in the creation of society, but as the owners (or not) of things that are divorced from the processes by which they came into being. The social relations of production, of 'doing', are converted into 'being' – in this case, *being gay*. For Holloway (2002) capitalism is precisely that: the separating of people from their own doing. This is the crux of Queeruption's anti-capitalist critique of hegemonic gay identities, culture and politics. The queeruptors are not interested in perpetuating a situation where sexuality is reduced to the acquisition of commodities that have been separated from the conditions of their production and from the experiences of those that produced them. They are not interested in engaging in a demands-based politics that is oriented towards the state. Such politics perpetuates other separations – the separating of leaders from led and 'serious' political activity from 'frivolous' personal activity (Holloway 2002). In contrast to this experience of life on the receiving end of an anti-social 'power-over', the queeruptors are interested in making modest, low-key attempts to re-engage their 'power-to-do', which is always part of a social process of doing with others. And so I return to my earlier point that 'queer' within these networks functions more as a relational process, rather than a simple identity category. A queer positionality, in this context, is produced through the very process of working collectively to create a less alienated and more empowered space in which to explore a multiplicity of sexual and gendered potentialities.

In the rest of this chapter I will contextualise the Queeruption gatherings in relation to the reassertion of a politics of affinity and autonomy over the last decade or so. I will then recount some tales from Queeruption IV – the last to be held in London – by way of illustration. Part of my motivation in writing this chapter has been to tentatively explore the connections between recent geographical writing on the transnational activist networks of the 'grassroots globalisation' movement (Routledge 2003) and the work undertaken (mostly) by other geographers on performativity and affect. I explore these potential connections in the penultimate section of the chapter. In doing so, I hope to expand the political frames of reference for geographers of sexualities.

Autonomy and Affinity

As I chart the return of the politics of autonomy and affinity as a context for Queeruption, I hope that you will hold in the back of your mind what these modes of organising reveal about the "the wide array of affective networks that [often] underpin activism" (Cvetkovich 2003a, 173), the emotions and visceral feelings that can inspire (or hinder) political action, or the many ways in which friendships and romances are sustained through, around and despite activist practice.

As Chatterton (2005, 547) has explained in his study of neighbourhood-level experiments with autonomy in Argentina, the idea of autonomy stems from "the *desire* for freedom, self-organisation and mutual aid," (emphasis added). Whether in Chiapas, Buenos Aires or a large warehouse on the banks of the Ij in Amsterdam (squatted as the location for a Queeruption gathering) autonomy is embedded in social relations *through* daily practice. In part, collective autonomy is a refusal to engage in mainstream circuits of capitalism; but, simultaneously it creatively experiments with alternatives in the here and now, rather than slipping into a nihilist stupor or postponing all dreams until some 'post-revolutionary' future (Gibson-Graham 2006). Autonomy's refusals are acts of creation. Where other anti-capitalist political traditions have bogged themselves down in polemic, critique and endless analysis, autonomy creates the tools and strategies for changing the world through its creative experimentation. Autonomy is both a means and an end.

Figure 16.1   Queer Bloc on Anti-Fascist Action in den Haag, June 2004

Chatterton (2005, 547) suggests that "autonomy implies a different notion of self" and self-belief. For him, "it is a radical imaginary, the urge to imagine an 'other' society" based on the collective realisation that in so many ways we are already "co-authors of the world." To work with and towards autonomy is to do politics differently – it is about working consciously with respect, dignity and purpose with others and without hierarchies or permanent leaders to help our (individual and collective) selves. In the next section of this chapter, I shall examine some of the social and spatial practices that constitute queer autonomous space.

However, there is another important aspect of the creation of Queeruption gatherings, and that is the convergence of activists from many disparate social and cultural contexts. In conjunction with the growth of local experiments in autonomy across the globe over the last decade has been a shift in the motivation and emphasis of the relations between otherwise physically, socially and culturally distant social movements. There has been a shift from the altruistic solidarity of Minority World activists towards those in the Majority World (e.g. solidarity with South African anti-apartheid activists in the 1970s and 80s) towards a more horizontal, two-way process of mutual solidarity based on a common, but not necessarily unitary, social critique – typified, for example, by the relationship between the Zapatistas and the European social centre networks (Olesen 2005, 107).

Although the radical queer networks that come together at Queeruption gatherings are most concentrated in Western Europe and North America, conscious attempts have been made to work with groups in the Majority World. To this end, work has been undertaken with radical queer groups in Argentina (Espacio ReSaCa – Space for a Sexual and Anti-capitalist Revolution), Israel/Palestine (the Black Laundry, a queer group opposed to the Israeli occupation of Palestinian land), Serbia (the organisers of the aborted Belgrade Pride) and Turkey (KAOS-GL, a lesbian and gay anarchist group). These are all organisations that, to varying degrees, are not simply attempting to create a space in their respective cultures for American and Western European style gay rights, consumption and 'lifestyles', but are experimenting in their local contexts with different anti-capitalist means of living autonomous queer lives.

In most cases, the European and American activists have initially responded to calls for solidarity and assistance from these Majority World organisations, but there has been an attempt for this work to be undertaken in a spirit of mutual learning and development. Nonetheless, there is clearly a latent danger of a colonising impulse coming from the more privileged activists based in the Minority World, as ideas about what it means to be a 'radical queer' seem to travel more in one direction than the other.

I should also point out that in addition to developing networks of affinity and mutual aid with other queer anti-capitalist activists, in creating and hosting Queeruption gatherings local queer radical groups have often been able to draw on the skills and resources of broader networks of autonomist and non-hierarchical activists in their localities. Individual queer activists are often involved in these broader activist networks in any number of different ways.

Day (2004, 717) has contrasted this new politics of autonomy and affinity with earlier political cultures which operated within a logic of hegemony. He suggests that a provisional definition of the logic of affinity would be "that which always already undermines hegemony." In other words, these newest social movements are not attempting to gain influence and power-over within the system of states and corporations, but are focused instead on "the possibilities offered by the *displacement* and *replacement* of this system" (Day 2004, 719). In the words of Amin and Thrift (2005, 226), in their recent consideration of the future for Left geography, "what these challenges add up to are powerful new geographies of organisation, belonging and attachment, which are literally redefining the spaces of what it is to be political."

Although the queer activists that I am writing about here are, individually and collectively, often involved in the 'spectacular activism' of disrupting meetings of the G8, for example, much of their activism involves the more 'mundane' business of experimenting with alternative modes of being and relating to each other. They attempt to incorporate the ethics that inspire their vision of the future (for example, co-operative, non-hierarchical, sex-positive relationships) into their political work in the here and now. In order to examine this point in more detail, I turn now to an analysis of the creation and performance of queer autonomy at a recent Queeruption gathering.

Experimenting with Queertopia

The fourth Queeruption gathering took place in a squatted block of flats in east London for five days in mid-March 2002. The publicity for the event promised that it would be:

an international DIY [do-it-yourself] gathering of queer folk of all sexualities with an interest in:

- learning from and sharing with each other
- forging anti-commercial queer community
- fucking with gender
- building alternatives and living our dreams (Queeruption London 2003, 7).

Nearly 500 people participated in this free gathering, with 150 of them staying for the duration in one of the sixty-four apartments situated around the central courtyard in a former hostel for the homeless. On entering the building, participants were greeted by one of the volunteer hosts, who explained the ethos of the gathering and showed them a map of the building, revealing which of the apartments were still vacant as accommodation and which had already been claimed for themed spaces. Once newcomers had settled into their flats and had become acquainted with the layout of the building, the tendency was for them to find somewhere to sit and chat with new and old friends. For much of the event, the small reception area, as much as the café, became a relaxed space in which people could gather to share news and stories.

Elsewhere in the building, space was found for a vegan café, the waste from which was composted. In other spaces, a wide range of discussions and skill-sharing workshops took place, covering everything from how to make your own sex toys to queer self-defence or healing with chocolate. Creativity was unleashed with life modelling, body painting and creative (erotic) writing workshops. There were also political discussions, many of them focusing around health-related issues, such as fertility rights and AIDS dissident theories – the suggestion that the link between HIV and AIDS has yet to be adequately proven and a plethora of other critiques of established medical orthodoxy. As Marinella, a Shiatsu practitioner and the main impetus behind the health space, reflected afterwards:

> The health space at Queeruption was my first attempt at combining issues of health, complimentary therapies, autonomy and radical politics. The radical queer space was perfect for this, because many people were already open to connections between health and politics, and the necessity to take our health – and maybe, the collective health of radical movements – into our own hands.   (Queeruption London 2003, 18)

She continues,

> The most important thing for me is that people used the [health] space in a very autonomous way. I was there some of the time – sometimes I just wanted to slob out in the caff, or cook, or go to a workshop even – and I did not feel territorial about the space at [all] (read: I needed lots of breaks), so it was great to walk back in and find folks chilling out, swapping treatments, doing their thing. That's when I realised the space was really working after all. … [F]rom the minute people started coming in, the health space, like all of Queeruption, became everybody's work (and play), that's why it turned out to be as good as it did.   (Queeruption London 2003, 18)

For the most part, this claim to a lack of territoriality is genuine – my experience of these queer spaces is that a small group of people will take responsibility for ensuring that something they are interested in, or that they recognise is essential, takes place. After that, the success or failure of the space largely relies on who turns up and the part they play in making the space 'work'.

Within the venue, there was also a room set aside as a dedicated dress-up space – drag, in all its many gender-fucking forms, is a central aspect of the spaces created by the Queeruption network. But, as Hennen (2004) has argued, in the context of Radical Faerie gay men's 'New Age' spiritual gatherings (Thompson 1987) in the US, this is thrift shop drag. It also tends not to be a form of drag that is motivated by the desire to 'pass' as the opposite sex. For the biological men (as distinct from female-to-male transsexuals), at least, dragging up tends to involve a playful combination of traditional masculine and feminine signifiers, or the artful transcendence of established gender categories towards the performance of some kind of trans-human entity. It is also a space where high femme performances by queer women are appreciated, although these often take the form more of 'faux queen' performances that again blur gender roles, rather than more archetypal 'lipstick lesbian' performances (Cefai 2004).

The evenings at the gathering were filled with home-made entertainment – one night a drag king band played ska; another was filled with poetry and performance art; there were dance parties; and, perhaps inevitably, the gathering culminated in a sex party. The following review of the cabaret that opened the sex party gives a flavour not just of that evening, but of the entertainment throughout the gathering (and illustrates my earlier point about the deployment of drag within this milieu).

Picture this: a moustached girl in drag holds the microphone in front of a boy who is performing homoerotic poems in the nude; a female night nurse struggles with the temptations incited by a semi-unconscious bisexual female patient played by a boy in a skimpy miniskirt; an alluring singing lady turns into a butch boy, then becomes a woman again and performs a naked ritual of martial arts with a big wooden stick; two camp boys and a girl with a strap-on dildo enact on a chaise longue the logistics and troubles of executing double anal penetration during a gay male threesome; a girl wearing only a translucent sack of onions plays a clarinet.   (Queeruption London 2003, 52)

For me, this description also emphasises the point that this is very much a *queer* and not a (lesbian and) gay space – the boundaries between sexualities and gender identities are blurred, porous and, ultimately, not particularly important.

For five days, the squat pulsated with what one participant described as "a functional anarchic system" (Lechat 2002). Another conceded that although it had been fun "queertopia was exhausting" (Lechat 2002). The value of these queer political actions is not found in their transgression of heteronormative sexual mores but in their modest steps towards the development of alternative sexual *and* social values. Their politics of affinity and ethical relationships allow for the exploration and development of ways of being that gently challenge the social divisions that result from a politics of identity.

An Affective Politics of Possibilities

This focus on the importance of spontaneity and the eventful intensification of experience has some resonance with recent post-Deleuzian interventions and debates in geography around *affect* (Lim, Chapter 4, this volume; McCormack 2003; Thrift 2004). In particular, the creation of these autonomous queer spaces resonates with Thrift's call for a politics of 'emotional liberty' (2004, 69) that broadens the realm of play and seeks pleasure rather than simply 'averting pain' (2004, 70). These gatherings deploy a politics of joyful creativity and experimentation rather than righteous indignation and rage, although those emotions still surface from time to time and have their place in political organising – even if they are limiting when they exist as the sole affective paradigm for activists.

The approach to community building within the Queeruption network also has parallels with the emphasis McCormack (2003, 495) places on ethical attachments that "emerge through the cultivation of the affective dimensions of sensibility" and require "openness to the uncertain affective potentiality of the eventful encounter" (2003, 503). I am increasingly of the view that one of the politically most important aspects of Queeruption events is the way that they increase the intensity of affective attachment, creativity and connectivity. As one participant reflected after Queeruption IV,

having entered the premises through the green CEITEX™ door once, gravity was different. With every moment it got harder to believe that there was an outside world. Periods of recovery and regaining a sense for the outside world's reality took its time. However, things will never be the same again.   (Queeruption London 2003, 109)

This should also serve as a reminder that the functions of queer autonomous spaces are not limited simply to being play spaces that offer a certain degree of safety for sexual dissidents and gender outlaws. Freed from the sexual and gender constraints of the quotidian world, participants in these queer autonomous spaces often find themselves questioning the social relations that normally restrict the free expression of their desires. Immersion in these autonomous spaces can be an intensely emotional and cathartic experience, with the consequence that when they end many participants experience a real 'come down'.

There is considerable room for scholarship on activist movements that examines both the affects and emotions that are generated by political organising. It should engage with the excitement and camaraderie of activism, as well as the ambivalences and disagreements. More than this, it should consider the ways in which affects serve as an "index of how social life is felt" (Cvetkovich 2003a, 48) and become the raw material for cultural and political formations that seek both to refuse and to transcend neo-liberalism in varied and unpredictable ways.

The modernist political movements of the twentieth century held the pretension that the future of the world could be carefully and rationally planned. In contrast, I would suggest that the grassroots globalisation movements of the last decade or so are becoming more comfortable with the realisation that, as Thrift (2003, 2021) has put it, "uncertain outcomes built upon partial knowledges are a constant of human life." As a result, they are engaged in a re-imagining of political practice that is revelatory, rather than programmatic.

To quote Thrift (2003, 2021–22) again, for those who

> study the fragile and temporary construction of the social aggregates and whose main concern is modes of development … politics is a precious thing, a fragile form of life and one of the chief means through which society is achieved. It is also necessarily a hesitant entity as it is highly performative.

These performative approaches to politics are attempting to "expand the existing pool of alternatives and corresponding forms of dissent" (Thrift 2003, 2021). To relate this back to Queeruption gatherings and the networks of activists who attend them, the chief concern is not to achieve equality within a flawed and tightly bounded system, nor to simply present a hollow critique of the existing commercial gay scene and the reified identities that it fosters. Rather, the concern is to attempt to create (and experience in a very embodied, sensual way) alternatives that are built from the bottom up, where the process of experimentation is consistent with the desired end results – even if those cannot fully be known or articulated.

Latham with Conradson (2003, 1903) have argued that as geographers and other social scientists have come to acknowledge the "event-ness of the world" and the "profound importance of affect in the unfolding of this event-ness", so they have come to appreciate the openness and multiple dimensions of social life and the world at large. For them, this "demands modes of thought that acknowledge both the openness of practice and the importance of emotion, desire, intuition, and belief in the unfolding of our worlds," (Latham with Conradson 2003, 1902). Although I suspect some theorists of the growth of collective autonomy would contest a view of it as an 'event' rather than a "process of affirmation of self-belief that comes through self-organising together" (Chatterton 2005, 559), I think there are still echoes of this desire for a lack of universalising principles in the work of these autonomous movements. For sure, universal discourses such as autonomy and dignity are articulated by these movements, but there is no desire for them to be applied in universal or hegemonic ways – the ideas and inspiration may spread across multi-scalar and transnational networks of affinity, but what they mean in practice is left to the imagination of activists in each locality. The ideas travel as inspiration and invitation, rather than command.

Concerning Sustainability

While the invitation to liberate oneself through collective experimentation has inspired the activists involved in hosting and attending the international Queeruption gatherings, there is a danger that with each gathering that takes place the events become more formulaic and repetitive, failing to fully engage with the distinctiveness of the locations in which they occur. It is also certainly true that, for the most part, those people who have been inspired to experiment with queer autonomy are activists who are already in the orbit of the autonomous strands of the global justice movement of movements. The attendees at these events are predominantly white, well-educated, relatively privileged and under thirty. Although there are many activists in their 30s, 40s and beyond within the network, several of them find themselves less than inspired by the prospect of spending a week sleeping on the floor of cold and unsanitary squats. If the movement is to become sustainable across the generations and to inspire and involve broader layers of queer people who are open to exploring alternative ways of being, then a broader range of experiments will need to be enacted.

In this vein, criticisms are beginning to arise from within these radical queer networks that the international gatherings are becoming little more than an annual circuit party for nomadic queer punks, with political content becoming increasingly marginalised within the events. My own reaction to this critique is somewhat mixed. On the one hand, it does seem that overt political discussions are becoming less central to the gatherings, yet, on the other hand, discussions about 'no borders' actions for the free-flow of migrants across national borders occurred at both the Sydney and Barcelona gatherings in 2005, and plans for the 'queer barrio' (encampment) at the rural convergence centre against the Gleneagles meeting of the G8 (in July 2005) were concretised as a result of discussions in Barcelona. Furthermore, I still maintain that the very process of non-hierarchical organising that takes place before and during the gatherings is inherently political, especially as it fosters networking between activists from different national cultures, encourages horizontal skill sharing (albeit, often in little more than mass vegan catering and knitting) and blurs the boundaries between work and play. As a result, I question whether those who are raising the claim of a lack of politics at the gatherings are stuck in an understanding of politics that sees it as a discursive activity that is in some ways quite distinct from other aspects of (social) life – politics as a discrete activity (for example, a protest or a debate) or a set of demands, rather than a process. In his discussion of the development of these newest social movements, Day (2005) has articulated a distinction between the *politics of the act* and the *politics of the demand.* For him, only a politics of the act can break out of the cycle by which every demand, in anticipating a response, *perpetuates* structures of discipline and control. And, as Chatterton (2005, 559) has argued, "[a]utonomous movements are made and remade in emergent, complex adaptive behaviours based upon participative self-organising and collective intelligence that can lead to higher order outcomes which are difficult to control and predict."

Of course, the logic and ethics of participative self-organisation throws the onus for creating 'more political' gatherings back into the hands of those who raised the criticisms in the first place: these, so the argument goes, are participative, do-it-yourself events, so, if you think there is something missing or unaddressed, it is your duty to make it happen or to work with others who have an affinity with your critique to create alternative events that address these issues. Although that might sound harsh to some, it is an integral aspect of the politics of affinity and autonomy from which these gatherings originate and of the "new geographies of affect, obligation and commitment to distant others" (Amin and Thrift 2005, 234) that have been developing over the last dozen years or so. In this respect, I want to close with the words of Richard Day (2004, 741), who has suggested that, "if anarchist-influenced groups look disorganised, this is perhaps because the ways in which they are organized cannot be understood from within the common sense maintained by the hegemony of hegemony. Perhaps [then] a new, *uncommon* sense is needed."

And, what could be queerer than that? Despite the concerns raised in these final paragraphs, these radical queer spaces are important because they provide a constructive and practical attempt to offer a non-hierarchical, participatory alternative to a gay scene that has become saturated by the commodity. They offer more than empty transgression. They are experimental spaces in which new forms of ethical relationships and encounters based on co-operation, respect and dignity can be developed. The queerness of these spaces is constituted as much through the process of building relationships on this basis, as it is from any attachment to specific sexual or gender identities.

Chapter 17

Counting on Queer Geography

Michael Brown

"What's happened to Michael?!"

The quote in the header above comes from Dr Hester Parr, a critical cultural geographer who has been an outstanding colleague in the field of health and body geographies. Her exclamation was uttered in reaction to an equation on a slide I displayed. It explained how location quotients produced an index of concentration of same-sex couples in Gates and Ost's (2004) *The Gay and Lesbian Atlas*. My paper used regional-science techniques to ask whether the maps of same-sex households in the atlas reproduced heteronormativity, and re-closeted certain gays, lesbians and other queer folk in the city (Brown and Knopp 2006). Although my critique was intellectually framed as an extension of Foucault's increasingly popular concept of governmentality – and thus as extending poststructural geography – she was surprised by my central use of GIS (Geographical Information Systems) and by my discussion of the equation used to calculate the statistics for the map. Such artifacts of quantitative geography and spatial science were hardly what she expected from the likes of me, given my theoretical orientations (Brown 1999), my typical use of qualitative methods (Brown 1997), and that my early work was explicitly critical of spatial science (Brown 1995). Had I 'sold out'? 'Gone quantoid'? Or, in the terms of my home at the University of Washington, had I become a 'space cadet' (Morrill 2002; Barnes 1998)?!

Whether a brief detour or emerging trend, I cannot say. But I acknowledge that the use of realist epistemology, quantitative techniques and GIS technologies have become an integral part of my work over the past five years or so. Such moves stand in tension with the general suspicion of quantification and GIS found within the various intellectual movements and communities into which I became a geographer: poststructuralism (Doel 2001), the new cultural geography (Anderson et al. 2003), but especially (and more intensely) in the context of queer geography and sexuality and space studies (Bell and Valentine 1995; Binnie 1997). In these critical traditions there has been a certain and steadfast opposition and sometimes a hostility to quantitative techniques and technologies and the epistemologies that typically saturate them. Binnie (1997), for instance, has argued quite compellingly that positivism itself is squarely and irretrievably heteronormative, grounding and conservatising by foreclosing debate on the nature of what 'is'. His resolve certainly echoes across papers I have read, referees' reports I have received, and indignant students' (both graduate and undergraduate) feelings that if they do queer theory, they shouldn't deign to have to learn quantitative analyses, GIS or the like. Thus, while

I admit my own perspective to be situated and partial, I nonetheless have a general sense of a queer geography where there is a not-so-subtle disciplining against *any* use of quantitative techniques and analyses or of GIS. And, of course, these echoes are wider. Queer theory more generally comes out of the arts and humanities rather than the social sciences, thereby foregrounding the analysis of the text in scholarship, rather than any empirical object under analysis (see Rosneau 1991). And within the American academy at least, there has been a longstanding, if sometimes internecine, epistemological-cum-methodological division between those who do 'gay and lesbian studies' and those who do queer studies (Jagose 1996a; Sullivan 2003).[1] So what's up with this apparent shift in my scholarship?

My brief from the editors for this chapter was to think reflectively and critically about "the various advantages and problems of the mapping strategies and techniques you deployed in your study, and how thinking about these strategies and techniques has developed over the years." So I take this generous opportunity to explicate the motivations behind my own 'quantitative turn', but not for the sake of a solipsistic intellectual biography. Rather, my aim is to suggest why the limited and careful use of quantitative techniques and data and GIS might fruitfully energise queer geography by providing an internal critique to scientific studies of sexuality, by claiming space and ontology against those who would ignore or erase our very presence in their world, and to avoid the trap of being the 'others' who are spoken of and about, but never heard themselves.

To be sure, the arguments I make about the possibility of such ecumenical scholarly community (or the inevitable tensions in it) here are by no means original. I draw guidance and insight from several colleagues in critical geography generally (Barnes 1997; Philo 2005; Dixon and Jones 1998; Plummer and Sheppard 2001; Sheppard 2001; Johnston et. al. 2003; Johnston 2006) and feminist geography in particular (Hanson 1994; Lawson 1995; McLafferty 1995; Staeheli and Lawson 1995; Kwan 2002; 2004; Hickey and Lawson 2006) who have considered the relationship between scientific epistemologies and techniques and critical theoretical work. The contribution I hope to make here is to describe my own position with the specific aim of counting on queer geography to also find a place for scientific scholarship so as not to foreclose important scholarly and political projects.

Orientations

As a synecdoche for poststructuralism and postmodernism, queer theory typically destabilises assumed fixed dualisms around sex and gender but also around other kinds of representations (Jagose 1996a; Sullivan 2005). Its métier is to expose the structural/structuring assumption behind the evidence and the criteria we use to validate such evidence (e.g. Scott 1990). It treats the world we study as a text, but also treats our evidence from it as a text. Queer theory demands that we – never

---

1   For example, it has been a fairly 'open secret' that research explicitly grounded in queer theory was looked upon unfavorably in the competitions for the only sexuality graduate fellowship in the US. Geography students, meanwhile, often are excluded from more humanities-based sources because they are housed in social-science departments.

unidimensional or essentialised subjects – shine light on the dark corners of our doxa and allegiances. So, in a very curious and paradoxical sense, scholars and activists have worried about the extent to which the theoretical domain's political allegiance to sexual freedom and equality is belied by an intellectual enthusiasm for sheer indefatigable critique (e.g. Sothern 2004). Nevertheless, queer geography has demanded such endless, relentless critique – both towards geographers outside our ken, but also amongst ourselves (e.g. Bell and Binnie 2000). Ultimately, the point is to achieve incisive scholarship that tells important and weighty truths about the world, along with a resistance to the accretions of power and privilege endemic to all fixities, dualisms and shared assumptions.

While I recognise that there is a cultural tendency in the social sciences to treat the definition of science like an accordion, expanding and contracting as the debate requires, I tend to define 'science' quite narrowly as a method (Gower 1997; cf. Chalmers 1999). It is a means of answering questions about the world truthfully, based in rationalism and empiricism, and operationalised in a highly formalised set of ordered steps meant to minimise bias and error (Popkin and Stroll 1993). Typically implicit (though not always) in that algorithm is a modernist epistemology that presumes a knowable, law-following universe independent of our knowledge of it. Quantification need not necessarily be an element of scientific method, though it commonly is.

Yet it was on a proudly queer path that my original foray into queer geography was tread, with research about the local responses to AIDS in Vancouver, Canada (Brown 1997). My critique (or at least half of it) was fairly simple: that implicit – and sometimes abjectly explicit – in this medical geography was the conceptualisation of HIV-positive bodies and spaces as vectors of diffusion (Brown 1995). Elided a priori in such a conceptualisation was any means to consider that such bodies and spaces might also *block* the spread of the virus and syndrome. The story of a medical geography like a 'wine stain spreading across a tablecloth' (Gould 1999) closeted a political and cultural geography of HIV/AIDS prevention and response. It was a story of shock-horror, gay men, IV drug users, and sex workers, all of whom were in fact *cleaning up* the wine stain and caring for the sick and dying, all under our very noses. My intellectual inspiration came from new directions in political theory (Mouffe 1992), but also from the cultural theory/activism around AIDS in North America (Crimp and Rawlston 1990), which strategically and tactically destabilised every assumption about something so biologically foundational and difficult to imagine as a virus that turns the body against itself and leaves it defenceless (e.g. Patton 1990).

At that stage, my stand was that geography urgently needed that critique and others like it. We needed it politically: to resist powerful scientific discourses in society that seemed to speak about – rather than for or from – gay men. We needed it intellectually: to think better about the syndrome, about nature-society, about place and space. AIDS wasn't just a medical geography. It was a political geography. Thus, I had hoped my intervention would help us tell more truths about the world and would reign in our arrogant hubris about what we thought we could know and what we thought we needed to know. 'We' were: sexuality-and-space scholars, cultural geographers, political geographers, health geographers. My reputation as a critic of spatial science was firmly established.

I certainly won't argue that my suspicion about the spatial science of AIDS was ill-founded. But I will claim that I was (and have been) far more sympathetic to something broadly called science than I have been read as being. By friends and foes alike, 'Ironies of distance' (Brown 1995) has been read and cited for its front-end rather than its rear end, and that has always nonplussed me. Both critics and allies alike read the piece as a rout of spatial science. For instance, one very sympathetic referee asked that I cut out the latter part of the essay for publication and let the spatial scientists make their own arguments against poststructural geography. I was ceding them too much ground, s/he tactically warned. From the other end of the court, the obvious example is Peter Gould's (1999) imperial assault on my paper at St. Andrews, Scotland, later published as 'Beyond cathartic geography'.[2]

To my mind, there was something too instrumental and, frankly, dishonest about these partial readings. That's because in the second half of that essay I had tried (ironically, I'd hoped) to show that poststructural epistemologies and qualitative methods did not solve the problem of the social distance between researcher and researched, even if they did address the concerns I had regarding scientific and quantitative tacks. Poststructural epistemology and qualitative techniques of data collection and analysis were no simple 'solution' to the dilemmas of how to write a geography of AIDS. They were not unproblematic in terms of truth or politics. My experiential authority – as Clifford (1988) would call it – was pained by the inevitable social distances at work in my own ethnography. "Ghoulish grist for the ethnographic mill," in the poignant words of Judith Stacey (1988, 22). My work was certainly not going to be used as evidence to leverage state funds for the care of people who were suffering, or to prevent others from becoming infected; spatial science's had been and would be.

Moreover, I had come to admit the not-so-subtle practices of science that were inherent in my scholarship. I presented positive, empirical evidence through direct quotes and thick description, marshalling a sign system and presuming/declaring a fixity between signifier and signified that was stable between me and my interlocutors. My research project was framed as an inductive exercise in theory construction, as opposed to deductive theory-testing, but not necessarily as unscientific (Glaser and Strauss 1967). I relied on the twin epistemological pillars of science – empiricism and rationalism – to produce my scholarship. Through empiricism I showed a thickly-described political geography of Vancouver at work, while through rationalism I used reason and logic to convince my reader that this was radical democracy and also worth imagining geographically. My sampling strategy was deliberate and appropriate to the research question. Academic debate and career ladders may have positioned me within one camp and against another, and I might have willingly gone along, but I knew better. Part of what I had done was 'scientific' – epistemologically and even methodologically.

---

2   It must be noted that I was not Gould's only target. He assailed my supervisor, Geraldine Pratt; my committee member Derek Gregory; as well as my fellow graduate student and brilliant current colleague Matt Sparke – all under the malign of 'the Western Canada Center for Political Correctness'. I was a virus that was part of a larger syndrome fatally infecting geography, it seemed.

Arguments for a Queer Science

The simple answer to Hester Parr's question above, then, is that I have always been a closet scientist, albeit in drag (in the sense that I lacked the white lab coat and/or its more subtle symbolic equivalents). In this section, however, I want to elaborate on this performative in order to explain how and why I think these ironically scientific moves need to be made. Rosneau (1991) has helpfully identified scepticism as a central feature of postmodernism's ontology in the social sciences. By this she means that postmodernism is not so much a rejection of modernist science, but rather a relentless scepticism about its promise that if we follow it, we will arrive at 'the truth'.[3] More philosophically, this standpoint is akin to Lyotard's (1984) definition of postmodernism as 'incredulity to metanarrative'. A postmodern relation to knowledge-production, then, is not some pre-modern superstitious rejection; nor is it some hermetic counter position. It is rather a moment of honest troubling. It is scepticism. If doing science only works so long as we assume that there is a mirror correspondence between our signs and an external nature, what if we know that that's just not true? What do we do when we catch ourselves suspending disbelief in real time? Isn't it bad science not to worry about such things when they so obviously reek of falsity and obfuscation (Baynes, Bohman and McCarthy 1987)?! In other words, postmodernism is not a rejection of metanarrative, but incredulity to it, and an acknowledgement of its plurality. A sense of scepticism, wonderment, unease, and qualification towards science's premises, products and effects is how I have always conceptualised the term. That's not a rejection of science. To me, it is a call for a *very* difficult plurality of scholarship, but one where there is productive tension and debate, or at the very least one where there is awareness of power dynamics in the academy.[4]

It is on these philosophical grounds that I agree with Barnes (1999) on the power of internal critique. For example, in my and Knopp's (2006) critique of location quotients we relied on the very criticisms within regional science about their partiality, limits, and confusions. Simply put, as ratios of ratios location quotients are always scale-dependent. Should 'we' not agree on the proper contextual scale, then mapping relative concentrations becomes quite arbitrary, even though cloaked in the guise of scientific calculation and authoritative GIS representation (Schurman 2004). By no means were the Atlas authors (Gates and Ost 2004) ignorant of this crisis of representation. Still, they trivialised it to a technical rather than an epistemological dilemma and went ahead anyway. We replicated their results (a wholly valid, if disrespected, scientific practice) to show just how governmentalised their geographies were, a governmentalisation which reduced places to polygons.

Likewise, in a recent sex-panic in Seattle over allegedly rising HIV infections amongst gay men the local gay media made the egregious scientific mistake of treating an indicator as a material process, and conflating probability with

---

3   She does, however, acknowledge other forms of postmodernism, some of which reject the very possibility of scientific epistemology and the utility of its concurrent methods.

4   And on this score, it has always puzzled me how irrational and emotional many dyed-in-the-wool scientists become when the spectre of postmodern critique is raised. Science itself (in whatever particular form it may take) is somehow off limits to critique.

documentation. A short-term rise in new HIV infections alongside long-term rising rates of gonorrhea, chlamydia, and syphilis amongst men who have sex with men in King County, Washington was used as evidence that immoral gays were fucking without conscience, illicitly spreading HIV. A closer and longer term reading of epidemiological data showed, however, that the HIV blip was an anomaly. The local-media narrative that gay men had stopped practicing 'safe sex' was an erroneous proposition predicated on a faulty understanding of sampling error. Epidemiological evidence shows that it is easier to transmit gonorrhea, Chlamydia, and even syphilis than it is to transmit HIV (Brown 2006). Gay men and their partners, it seemed, were practicing safe sex to prevent HIV, just not other sexually transmitted infections. These data can certainly be treated as texts, but they also tell us politically important stories, and quantitative techniques help us read them thusly. The science behind this politics of shame faltered on the very scientific principles of validity and reliability. Illogicality is bad science.

Rather more politically, I am motivated by being an inevitably conscripted foot soldier in the American culture wars (Hunter 1991). In this frame, an a priori rejection of scholarship (both scientific and 'unscientific') operates in concert with a social conservativism in American culture that is – by its own rhetoric – determined to eradicate all things 'homosexual' – never mind *queer* – from the homeland. A major epistemological attack by the religious right in the US has been in the form of deciding that all 'homosexuality is a choice' and is thus different from other social identities like race (e.g. Baier 2006). Consequently, it is taken as a valid ground for moral discrimination (Simon 2006) as well as for biomedical attempts at conversion therapy. In a recent study, however, multivariate analyses suggested – in contrast to qualitative analysis – that such claims were not persuasive in an urban electoral conflict (Brown, Knopp and Morrill 2005). My point here is simply that by ceding science in whatever form it takes (its epistemology, its techniques, its modes of representation), queer geography risks not addressing the political-cultural geographies of the culture wars here in the US.

A second argument I would offer is that there is a politics of strategic positivism that may work against a powerful form of heteronormativity: the closet. Now to be sure, queer theory has been quite suspicious of the 'evidence of experience' when it is essentialised and facilely treated as theory neutral (Scott 1992). Nonetheless, if the closet refers to the concealment, denial, and erasure of queer folk, claiming positive presence, documenting it, and disseminating that geography works against that oppression (Brown 2000). It is for this reason that my volunteer work for the past three years has been with the Northwest Lesbian and Gay History Museum Project. This Seattle organisation collects, analyzes and disseminates local queer histories. The fragile nature of gay and lesbian history has been well-acknowledged. In this work, I have learned basic GIS and produced an expandable database of major sites of significance in local queer history. I plotted and designed a widely disseminated and popular map of major sites of queer history in Seattle from the 1930s–1990s (see http://lgbthistorynw.org/). This artifact of spatial science documents nearly 250 sites including bars, bathhouses, bookstores, cruising areas, and religious and social organisations. The organisation now uses it in its research, presentations, walking tours and publicity. Now to be sure, the map is not without its epistemological and ontological paradoxes, dilemmas and silences, but by offering such a conventional historical geography (the map) based in and on GIScience, we have literally used positivism to resist the closet of history and to claim space in the city.

This politics also has academic dimensions. Like many American geography departments, our undergraduates are, by and large, interested in GIS. Many go so far as to declare that they are 'majoring in GIS' rather than Geography. They are deeply suspicious of critical-theoretical work – not because they are conservative, but because it seems useless to their (neoliberal) instrumental rationality of training themselves with rare and valuable skills. And so it queers these students to no end to see my own use of GIS in my community-activism work with the Northwest Lesbian and Gay History Museum Project. Thus having a 'theory-head' do GIS productively destabilises rigid hierarchies in our undergraduate curriculum. It is also, it needs to be said, an important means of contesting the closet of queer scholarship and the historical experiences of gay Seattle. Science has helped me queer my department's own internal divides and encampments.

My third argument is one of political tactics and opposition. Simply stated: if we don't do it, someone else will. When I published my critique of the geography of AIDS (Brown 1995), a decade into the pandemic's wrenching of North America, the discipline had almost exclusively framed that spatial story as one of a non-human virus diffusing through various vectors and at various scales across the planet. This story had slotted unproblematically into the longstanding and honourable tradition of medical geography's investigation into the spatial epidemiology of disease. Following in the row of cholera, malaria, and tuberculosis, HIV/AIDS was fresh grist for the mill of spatial epidemiology. But there was something imperialistic about all these data points and extrapolated models when it was your friends and family who were being represented by them; when proudly heterosexual colleagues put up dot maps of HIV+ people in a metropolitan area, and your own dying brother couldn't tell his family that he was gay and positive because he was so afraid of that publicity; when you, yourself, have had to go through the ignominy of epidemiological interviews and regular HIV testing, with all their attendant governmentalising and heteronormalising. There is benefit and truth in the science, but there is also harm and cost in the very real and ubiquitous sense of being spoken for and about rather than doing the representing.

Moreover, quantitative social science in the US has ever-increasing access to data sets that involve queer folk. These data are out there and are being used. The example that opened this paper, for example, used the 2000 US census data. It was the first census in which one could reliably detect the presence of same-sex couples. There has been a small but growing use of these data in the discipline by heterosexual geographers (Cooke 2005). Furthermore, public health data, for example, remain an important resource through which sexualities can be represented – and also interrogated. If we are unable to read or analyze these data for ourselves, we run the risk of being represented by those who use us as just a new data source.

Queering the Equation; Not Solving It

My goal in this chapter has been to specify some of the concerns I have with the relationship between queer geography and science. I've tried to do this by specifying and elaborating on my own relation to these two forms of scholarship in the discipline. To summarise, I have three concerns about the relationship. First, internal critiques of an argument can be every bit as powerful – sometimes even *more* powerful – than external ones. If queer geography is to offer such powerful critiques of scientific ways of knowing and researching sexualities, it must have some room for those who are internally conversant in scientific epistemology, methods, and discourse. Second, if closeting is a particularly apposite form of oppression, then there is a powerful politics of positivism: 'We are here!' is a quintessentially positive statement that would be irresponsible not to enact. Third, if we don't use these techniques and methods, others will. Armed with even the most benevolent and progressive of intentions, their representations of us are not our own. They come to speak of and for the other.

I want to close off this chapter by taking the very postmodern tack of saying what it is not. My essay has not been an argument that we should stop working in a poststructural vein, or that we should stop using qualitative methods of data collection or analysis. Nor am I am implying that quantitative and positivist work should be immune from critique. Let's never ignore or downplay the masculinism of complex mathematics and quantification. Let's never forget our geographies are always partial and situated – especially when our techniques and methods have historically implied otherwise. Let's not ignore the powerful discourses that work through scientific epistemologies and techniques. Let's always readily deconstruct their representations. Indeed, I have no doubt that outstanding work in this vein will continue. This essay is also not an argument for a return to some simulacra of unassailable scientific or quantitative geography. There is no going back. The bell cannot be unrung! The point is to work with all of the insights, epistemologies and methods geography and geographers have gleaned – to use the techniques and heed the critiques. This chapter is also not a call for some quick or easy rapprochement between poststructural geography and spatial science. As I have written elsewhere (Brown and Colton 2001), I think the tensions themselves are worth exploring.

For me, being able to critique, to deconstruct, and to represent inevitably involves scholarship aware of – if not necessarily practicing – science. I am inevitably a scientist, but never just that. It would be insufficient for the kinds of truths I think are important to tell and for the scholarly and pedagogic interventions that must be made. Mine is not some clarion call for an epistemological or methodological shift in queer geography. But on its own terms, I do think there is space in queer studies for pluralising our epistemologies and methodologies. And I'm counting on queer geography to do so, albeit always critically.

Acknowledgements

Thanks to Gavin Brown, Larry Knopp, Vicky Lawson, Dick Morrill, Ron Johnston and Suzanne Withers for their help and comments.

Conclusions and Future Directions, or Our Hopes for Geographies of Sexualities (and Queer Geographies)

Jason Lim, Gavin Brown and Kath Browne

It was my (Kath's) idea not to write a 'normal' conclusion for what I believe is a very queer book. Although we will speak of directions for the future – where we see geographies of sexualities going – I wanted to finish the book with something that wasn't coherent, that didn't offer a single conclusion, or even a number of conclusions. Rather, I wanted to finish messy – to have something that highlighted not just the variety of views we the editors have on queer theory, queer geographies and geographies of sexualities, but also something that showed how co-writing both enriches texts and has an excess where individualities are necessarily co-opted into a broader frame. In this part of the conclusions I will consider my hopes for the future of geographies of sexualities and queer geographies.

When I initially engaged with *Mapping Desire* during my undergraduate degree, I was riveted; hiding the book under my pillow when I went home (to my parental home in Ireland) reflected some of the themes of closeting in its pages. Practices and experiences of 'being' lesbian, gay, bisexual and trans continue to need recognition and exploration, even where these identities are questioned (or questionable?) (see Matejskova, Chapter 11, this volume). Geographies of sexualities that record lesbian and gay histories of cities, such as Larry Knopp and Michael Brown's work on Seattle, are not only important methodologically and theoretically but also have impacts beyond the realm of academic endeavours. Continuing the projects of exploring spatial practices and of mapping lesbian and gay places and spaces, whilst simultaneously contesting the heterosexist and masculinist processes that constitute what count as legitimate 'practices', 'mapping' and 'territories' (Peace 2001), perhaps at times requires a particular reification of identities, bodies and boundaries. Moreover, in diversifying geographies of sexualities beyond specific minority world contexts, geographies of sexualities should incorporate identities, lives and relationships that do not fit and may indeed challenge the globalising discourses of 'lesbian' and 'gay' (See G. Brown, Hacker, Matejskova this volume).

The reification of specific identities in minority world contexts could allow for those who continue to be under-explored in this dominant literature to be included; as many authors beyond geographies have noted, the move to queer has often led to a dearth of academic explorations of gendered power relations and to a focus on men. As women's studies struggles to survive in the current climate of gender theory and genderqueer, explorations of gendered power relations are, for me, central. These need to account not just for the plethora of diverse masculinities, hegemonic or otherwise, but also for powerful femininities, which are not necessarily located on biologically specific bodies. However, geographies of sexualities can and should continue to explore the geographies lived by lesbian, bisexual and queer 'women' (including transmen, bois and other genderqueers), as well as exploring female heterosexualities.

When we were planning this book, Jason and I had a long email discussion regarding the inclusion of women among those who would write the 'reflective' pieces. I was nervous of tokenistic attempts at 'inclusion' when the criteria we had set had asked for 'reflections' from those who had 'begun' the geographical investigations into sexualities in the 1980s and 1990s and who continue to work within the sub-discipline in some way. Jason, on the other hand, wanted to foster a "gender reparative book that would challenge the male dominance" of geographical work on sexualities. We both recognised our "place within the canon" and agreed that "[d]isciplines themselves can be a modality of power and conservatism, so I [Jason] think we should not be in thrall to their extant formation, but should actively seek to reshape them." What we were too late to do, however, was to consider how the more 'established' figures could be made to "practice their politics like outsiders again." We asked a number of women to reflect as David, Larry, Michael and Phil have done in this volume; unfortunately, even though two agreed they had to pull out. Although, through invited contributions, we ensured there would be a degree of balance between male and female contributors to the book (the relative lack of trans people and those who position themselves publicly between the boundaries of man and woman is problematic), it remains the case that those who are more 'established' within geographies of sexualities more broadly are men. This might be a reflection of the overall gender composition within geographies of sexualities and queer geographies (including the relative absence of trans individuals), or perhaps the absence of established female authors is due to intersection of the gender discriminatory structuring of academic life with the overwork, differing priorities or other commitments of female scholars. In reshaping geographies of sexualities, hopefully this book will facilitate (and celebrate) the establishment of more female authors within the sub-discipline. It also incorporates a call to advance trans and genderqueer participation in geographies of sexualities (even if this is because there is annoyance at 'our' considerations of trans and genderqueer geographies).

Discussions of trans lives, gender transgressions and genderqueer within geographical sub-disciplines is interestingly located within gender geographies or queer geographies; the latter can overlap with geographies of sexualities but not necessarily so. The consideration of genderqueer, trans and gender transgressive lives, identities and contestations within queer geographies involves a questioning of understandings of queer geographies as synonymous with lesbian and gay lives. Yet, the increasing use of the acronym LGBT (lesbian, gay, bisexual and trans) often locates gendered contestations within geographies of sexualities. Entitling this book *Geographies of Sexualities* while opening up discussions about the possibilities of queer and the diversity of its theoretical, empirical and political uses, will (hopefully) help to develop the potentialities of queer geographies – a project that might both engage with and take its distance from *Geographies of Sexualities* the book and the sub-discipline (see Knopp, Chapter 1, this volume). Such moves would not have to be antagonistic or oppositional, especially given that the position of both queer geographies and geographies of sexualities – both intellectually and in terms of the materialities of discrimination, prejudice and exclusion – can be precarious, particularly in the contemporary political climate in nations such as the United States. Queer geographies might, however, challenge the gendered boundaries of geographies of sexualities and its synonymy with lesbian and gay, and in this way offer further insights and critiques beyond what has been possible in this book or even beyond what has more broadly been offered to date. In arguing not for a total movement towards queer, but instead for a recognition of the multiplicity of theories, practices and politics, I hope that mutually and constructively critical discussions might occur.

Within these spaces, further explorations would follow Phil Hubbard's call to queer heteronormativity (Chapter 12). Such explorations would also critique homonormativity, but not only from the margins, as Gavin Brown and Mark Casey do in this volume, but also from the 'centre'. The contradictions and complexities of being both 'normal' and 'dissident' have yet to be fully recognised. How does the idea of homonormativity complicate our understandings of 'power' in a world that continues to be dominated by heterosexuality? What are the spatial implications of this 'power' in the production of 'gay ghettos' as well as in the use of everyday space? As previously marginalised sexualities become socially, culturally and politically 'accepted' and 'celebrated', there is a need to explore how hegemonies, hierarchies and elites are being established and recreated, and not simply on the basis of exclusion and marginalisation from these new homonormativities.

For me, these homonormativites are geographically specific, both within the national and local context within which I work and live (Brighton, UK) and within international contexts. I am acutely aware of the specificity of my writing and of my white, middle class, educated, able-bodied, employed in a permanent job in a British institution etc. status that informs this. I think it is important that this privilege is not just acknowledged (Vanderbeck 2005), but is explored and used. My hope is that geographies of sexualities and queer geographies foster diverse critiques that offer and effect change, not only for the discipline of geographies and its plethora of sub-disciplines, or only for the communities, nations and contexts in which we work – but also for 'us', those who continue to create the (sub-)discipline.

Kath

Earlier in this web of concluding remarks, Kath suggested that "queer has often led to a dearth of academic explorations of gendered power relations and to a focus on men." I (Gavin) think that that observation needs complicating somewhat. While such an observation might, to a certain extent, accord with what is happening in the context of academic theorisations of queer, it is at odds with my experience of 'second wave' queer activist networks and the social spaces that are associated with them. At least, in London, where I am based, and a number of other UK and northern European cities with which I am familiar, transmen and 'bio-women' play such a prominent role in the organisation and population of many 'queer' spaces that I have increasingly heard men ask 'where have all the queer boys gone?' Of course, my observation is largely anecdotal. It evokes the need for new research and investigation. It also prompts me to question whether the feminist critique of 'queer' – that it concentrates on men and makes women's lives invisible – itself undermines, normalises and de-queers the lived experiences of many queer women, drag kings and transmen.

From a different perspective, I think there is a need for closer geographic attention to be paid to the full range of non-hegemonic performances of masculinity. Despite the emblematic centrality of the drag queen to early queer explorations of performativity, within geographic scholarship little attention has been paid recently to the spatial expression of what we might call male femininities. Much as I like the hyper-masculine gay skinhead that appeared within many British sexual geographies in the 1990s (Bell *et al.* 1994; Bell 1997; Bell and Binnie 1998; Binnie 1995), maybe it's time to encounter his fey sister-brothers (Hennen 2004).

More than this, I would like to reinforce Kath's observation that "the contradictions and complexities of being both 'normal' and 'dissident' have yet to be fully recognised," and her recognition of the need for a greater exploration of the operation of power amongst those who are marginalised by 'the new homonormativity' (Duggan 2002). I have been surprised by the number of contributors to this book who have engaged in a critique of 'homonormativity'. It seems we are all anti-homonormative now. Yet, I think we might do well to pause and reflect upon the potential pitfalls as well as the potential benefits of taking such a stance. How far does a critique of homonormativity reinforce our collective 'critical' credentials while eliding the difference and creativity that might still go on within the spaces, settings and relations that we have characterised as homonormative? To what extent does a critique of homonormativity rely upon our analyses being several degrees removed from the everyday lives and aspirations of so many urban lesbians and gay men in the Global North? What problems might such a gap present? I may not choose to socialise much on the commercial gay scene, I may not choose the same consumer products as so many gay men of my generation and income levels, and I may question the rush to marriage (and civil partnerships), but these things give real meaning to many. I offer kudos to Kath for being the only geographer I have heard so far to acknowledge just how 'homonormative' her life is (indeed, it's been the source of long-running banter between the two of us throughout this project).

In part, my questioning of the potential for overusing the term 'homonormativity' arises from a concern that it might suggest that it is in some way possible to be an 'authentic queer' whose life is untainted by homonormative social relations. As Oswin (2005) has recently reminded us, we cannot escape our complicity in these practices. I am certainly prepared to acknowledge that despite my involvement in radical queer activist networks and a certain distance from the commercial gay scene, it would be very easy for many observers to mistake my partner and me for any other professional gay couple rapidly approaching our forties and enjoying everything that city centre living has to offer. My point is that most of our lives are marked by compromise and contradiction. I believe there is a danger in our rush to critically queer contemporary social relations that we make nothing more than an empty critique unless we offer the prospect of alternative ways of being. Happily, I believe several authors in this collection are committed to exploring alternatives to the present state of being. It seems invidious to single out one or two authors at this point (not least of all because I suspect I would end up highlighting those whose visions I find appealing); what is important to me is that we keep open a dialogue about what could be.

This brings me to my suggestions for some strands of work that are, as yet, under-developed within the geographies of sexualities literature. When writing about urban gay life we collectively tend either to write about the commercial gay scene or about non-commodified spaces of public homosex. There are a huge number of other sites and spaces of lesbian, gay and queer socialisation that exist between these two poles – for example, the queer autonomous experiments I have written about (in this book and Brown forthcoming) or Farhang Rouhani's work on transnational queer Muslim organisations (Chapter 14, this volume). There are countless other examples. I would like to see more work that analyses the social, economic and political geographies of post-capitalist (Gibson-Graham 2006) and 'alternative' lesbian, gay and queer spaces and practices. There is, of course, a growing body of work by political and economic geographers on autonomous spaces (Chatterton 2005), utopian communities (Pinder 2005), protest sites (Chatterton 2006; Routledge 1997; 2003), and alternative economic practices (Gibson-Graham 2006; North 1999). Quite apart from specifically queer experiments of these kinds, such sites and practices have a sexual geography of their own that needs to be studied.

There is another, increasingly important, trend in current (British) social and cultural geography that I think needs queering, or at least (re)connecting with some sense of the sexual. That is the 'non-representational turn'. As Jon Binnie highlights in Chapter 2, this methodological turn was, in part, influenced by Butler's queer writings on performativity. And yet, somewhere in the process, non-representational theory has rendered the body abstract and not particularly queer. Despite this criticism, I think that the 'more-than-representational' register (Lorimer 2005) has much potential for engaging with the messy materialities of sexual body. It is time to move beyond non-representational writings on dance and yoga, and to get to grips with the more-than-representational geographies of sex(ual) practices – cruising and caressing, flirting and fucking. In Chapter 7, Doug Herman offers an interesting insight into the queer spaces of the (predominantly heterosexual) BDSM scene in North America. To me, this chapter opens up a whole range of possibilities for future research that might contemplate what a sensuous embodied geography of thinking, feeling and doing BDSM would look like. Geographers' attention to the material culture of non-human objects and the hybrid geographies of human and non-human relations (Whatmore 2003) could be extended into the sphere of sexual geographies. When will we read a geography of the dildo, the leather harness or the enrolment of the ceramic of the urinal into a cruising scene? Similarly, it seems strange that so little of the geographic work that is concerned with affect and emotion has, so far, considered the geographies of desire, lust, unrequited love or being jilted.

In these respects, I am optimistic that there is still much innovative work to be done within the sub-discipline of sexual geographies, without simply rehashing tried and tested themes in new combinations. And yet, we have certainly not exhausted the possibilities for expanding the knowledge base in the areas that we already do well – as Mark Casey and Tatiana Matejskova's chapters (10 and 11, respectively) attest, there is still a need for subtle explorations of local gay vernaculars that pay greater attention to the regional as well as national similarities and differences in sexual cultures.

Gavin

One of the characteristics that marks most of the contributions to this book is the engagement geographers of sexualities have made with queer theory in its various guises. While some of the contributors have developed something of an arms length relationship with queer theory, others have placed it centre stage in their discussions. Yet all have had to engage with queer theory in one way or another. In common with the engagement with and adoption of queer theories by scholars elsewhere in the social sciences, this process has involved translating a set of ideas developed primarily in the humanities into ideas that might productively be used in the social sciences to think about the material, institutional, practical and spatial dimensions of sexualities. In this context, I (Jason) think it is interesting that some, such as Jon Binnie in Chapter 2 and Gavin earlier in these concluding remarks, have averred that a similar challenge is now posed by 'non-representational theory', the challenge being to find a way to use what are sometimes rather abstract ideas to attend to the materiality, embodiment and affect of contemporary sexualities. What is particularly interesting is that this challenge arises despite non-representational theory's avowed interest in precisely these dimensions of everyday practice.

What strikes me is that it might be thought that both queer theory and non-representational theory are marked by a gap between what they promise in theory and what they offer in practice. For queer theory, this gap takes many forms. For instance, Catherine Nash and Alison Bain (Chapter 13) suggest that the queer ideals that underpinned the Toronto Women's Bathhouse Committee's (TWBC) 'Pussy Palace' events were in some ways belied by the practical necessities of having to defend the organisers of these events against prosecution for liquor licence violations. Becoming embroiled within the institutionalised politics of a legal battle and so turning to more experienced gay male activists for support, the TWBC's queer philosophy became subsumed to the expediency of framing of the legal battle within a homonormative engagement with rights-based politics. Although this may be, in part, a matter of time and experience, such work raises the question of where the current limits of the efficacy of queer politics lie when operating within a broader polity. The gap between queer theorisations and queer practice, however, does not only apply within the sphere of activists' queer lives. This gap between queer ideals and queer practice also exists at the level of theory itself. Jon Binnie (Chapter 2), for example, points out how despite its professed epistemological humility and its professed determination to undo minority world domination of sexual discourse, much queer scholarship remains in practice a project dominated by privileged Western thinkers. This is a critique that, unfortunately, could fairly be levelled at this book, too. Jin Haritaworn (Chapter 8), building on the arguments of Prosser (1998), draws our attention to how the theoretical transgressiveness of gender, race and sexuality categories offered in Judith Butler's seminal queer theorisation of Jenny Livingston's film *Paris is Burning* relies upon a covert objectification of the bodies of minoritised Others – in this case those of the transsexual women of colour Octavia St. Laurent and Venus Xtravaganza. Maybe these gaps between the theoretical promises of queer and what it actually delivers are suggestive of fertile grounds for attempts at resolution – attempts to make theory and practice more consistent with one another. Or maybe they are suggestive of fundamental problems with queer theorisations, problems that might mean queer theorisations need to be reworked.

One such problem, which appears to be relevant to at least the issues discussed by Catherine and Alison and by Jin, might be the tensions that arise between, on the one hand, queer emphases on conceiving and fostering new identities and, on the other, queer practices of fostering new material and affective conditions for sexual desire and pleasure.

Non-representational theory might also be thought to be marked by a gap between its theoretical promises and what it delivers in practice. Despite its appeal to materiality, embodiment, affect and practice, many scholars not well versed in the non-representational canon take its treatments of these subjects to be abstruse and overly theoretical. The feeling amongst these scholars is that the recondite complexity of non-representational theory hinders its usefulness in thinking about material and embodied practice. While some of these misunderstandings arise because those who work with non-representational theories and those who do not tend to be asking different questions, practitioners of non-representational theory might like to consider whether it is possible and desirable for their ideas to be transplanted into other contexts where other kinds of questions are asked.

Working both on geographies of sexualities and with what others have called non-representational theories, I have a rather ambivalent relationship to these disjunctures: I both understand them and attempt to work beyond them. One of my regrets about my contribution to this collection is that I did not write about the possible intersections between non-representational theory and geographies of sexualities in a less theoretical way. Given that few of the contributions focus specifically on the racialised and ethnicised politics and relations of sexuality, I also regret not writing explicitly about this aspect of my research. In addition to the discussion Kath and I had about the gender composition of the contributors of the reflective chapters, Kath, Gavin and I also had long discussions about how to include more contributions regarding the racialisation and ethnicisation of sexual politics and relations, and the geographies of sexualities in the majority world. Despite our best efforts, our attempts did not bear the amount of fruit we had hoped for.

Nevertheless, I think within the collection there is still a huge breadth of ideas relevant to engagements with many contemporary changes in geographies of sexualities around the world. In the context of the current intersection of, on the one hand, the globalisation of Western and modernist ideas about sexuality (see, for example, Hanna Hacker's contribution, Chapter 5) and, on the other hand, the heightened tensions surrounding the spread of purportedly new modern forms of religion and religious politics, Farhang Rouhani's contribution to this book (Chapter 14) warns us against the dangers of positioning Muslims as either the Others to sexual freedom or the victims of sexual oppression. Indeed, I would hope that Farhang's chapter invigorates broader discussions of the intersections between religion and sexuality (see also Althaus-Reid 2001; 2003), especially given the increasing importance of the transnationalisation of religious discourse and institutions. Among other trends, we are witnessing the spread via the media and via migration of conservative Christianity from the US and from Africa into Europe; the consolidation of 'liberal' Anglicanism in England and in North America; and the intersections and disjunctures among ideas of modern consumerist sexualities, queer diasporas (see Gopinath 2005) and nationhood understood, in part, through religion in South and South East Asia.

As somebody whose research within geographies of sexualities is concerned primarily with interrogating heterosexuality, I am glad that there are contributions to this collection that also directly address heterosexuality (in particular, Doug Herman's and Phil Hubbard's chapters – Chapters 7 and 12, respectively – but also, in part, Vincent Del Casino Jr's and Jin Haritaworn's – Chapters 3 and 8, respectively). Of course, there is still a huge amount of room for more work to be done on this front within the sub-discipline. Both Doug's and Phil's contributions raise the question of whether this potential research would be most productively framed in terms of thinking about the marginality of various heterosexual practices. Certainly, I think there is much merit to Phil's problematisation of the presumed normality of heterosexuality, given the polymorphous forms that heterosexual practices actually take. I don't think, however, that this precludes an attentiveness to the multiplicity of mundane heterosexualities, and such an attentiveness might be an opportunity for the development of a non-representational register in the study of sexual practices. Here, the value of such a register might arise from it providing an important counterpoint to the existing narratives of many psychoanalytic considerations of heterosexual practice and desire.

It must also be asked how useful queer theories are in the project of interrogating heterosexualities, or, rather, what do these theories do in relation to the heterosexual? While we might think of queer as an unsettling movement – one that disrupts the hidden assumptions and exclusions of everyday practice and institutions, and that thereby disrupts the normative operation of identity and therefore otherness – we might also consider how this unsettling of the taken-for-granted might proceed through other means than the validation of transgression (see Jin Haritaworn's contribution, Chapter 8, for a critique of the deployment of transgression in queer theory). I am suggesting that queer unsettling should (and indeed does) involve a reworking of mundane, commonplace practices of sex and sexual relationships through a process of decomposition and recomposition. Here, a distinction needs to be made between the 'normal' (commonplace) practices of heterosexual desire, and the hidden assumptions, exclusions and injunctions that enforce sexual norms and render other kinds of sexual desire and practice marginal. A queer reworking of heterosexualities, then, might be deconstructive of the techniques that enforce heterosexual norms, but also nurturing of a more productive experimentation with different practices and desires. In many ways, these latter processes of experimentation have always been a part of the queering of LGBT practices and desires, as several chapters in this volume are testimony to. What I am hoping to suggest is that such experimentation might not only be pursued as a proliferation of the margins, but also as a recomposition of the centres, both heterosexual and LGBT. Alongside a critical project in which transgressiveness produces new sexualities, queer theories, practices and politics might also come together as a reparative project for cultivating and fostering our desires.

Jason

Bibliography

Abdullah, I. and Gutenberg, S. (2003), 'Thematic and Contextual Reading of the Story of Lut (Lot)', workshop presented at *Our Individual Lives, Our Collective Journey* conference, Al-Fatiha Foundation, New York. 9 August.

Abdullah, I. and Wadud, A. (2005), 'Islam and Homosexuality', workshop presented at *Sexism, Misogyny and Gender Oppression: Breaking Down Systems of Patriarchy* conference, Al-Fatiha Foundation, Atlanta. 3 September.

Abelove, H., Barale, M.A. and Halperin, D.M. (eds.) (1993), *The Lesbian and Gay Studies Reader*, (London: Routledge).

AbuKhalil, A. (1997), 'Gender Boundaries and Sexual Categories in the Arab World', *Feminist Issues* 15:1–2, 91–104.

Achilles, N. (1967), 'The Development of the Homosexual Bar as an Institution', reprinted in P. Nardi and B. Schneider, (eds) (1998).

Adams, V. and Pigg, S.L. (eds) (2005), *Sex in Development: Science, Sexuality, and Morality in Global Perspective*, (Durham: Duke University Press).

Agamben, G. (1998), *Homo Sacer: Sovereign Power and Bare Life* (Stanford: Stanford University Press).

— (2002) *Remnants of Auschwitz: The Witness and the Archive* (New York: Zone Books).

— (2005) *State of Exception* (Chicago: Chicago University Press).

Ahmed, S. (1997), '"It's Just a Sun Tan, Isn't It?": Auto-biography as an Identificatory Practice', in H. Mirza (ed.).

Aitcheson, C., MacLeod, N.E. and Shaw, S.J. (2000), *Leisure and Tourism Landscapes. Social and Cultural Geographies.* (New York: Routledge).

Aitken, S.C. (2000), 'Fathering and Faltering: "Sorry, but you don't have the necessary accoutrements"', *Environment and Planning A* 32, 581–598.

Alam, F. (2005), 'Gay Media's Failure to Accurately Report Stories Adds to Growing Islamophobia and Hatred Toward Islamic World', press release, Al-Faitha Foundation, 1 August.

Allison, D. (1996), *Two or Three Things I Know for Sure* (New York: Plume).

Allman, K.M. (1996), '(Un)Natural Boundaries: Mixed Race, Gender, and Sexuality', in M.P.P. Root (ed.).

'Alt.com, World's Largest BDSM & Alternative Lifestyle Personals' Various, Inc. 2005 <http://www.alt.com/>.

Althaus-Reid, M. (2001), *Indecent Theology: Theological Perversions in Sex, Gender, and Politics*, (New York: Routledge).

— (2003), *The Queer God*, (New York: Routledge).

Altman, D. (1988), 'Legitimation through Disaster: AIDS and the Gay Movement', in E. Fee and D. Fox (eds.).

— (2001), *Global Sex* (Crows Nest, NSW: Allen and Unwin).

Amin, A. and Thrift, N. (2005), 'What's Left? Just the Future', *Antipode* 37: 2, 220–238.

Anderson B. (2004), 'Time-stilled Space Slowed: How Boredom Matters', *Geoforum*, 35, 739–754.

— (2005), 'Practices of Judgement and Domestic Geographies of Affect', *Social and Cultural Geography* 6: 5, 645–660.

— (2006), 'Becoming and being hopeful: towards a theory of affect', *Environment and Planning D: Society and Space* 24: 5, 733–752.

Anderson, J. and Duncan, S. (eds) (1983), *Redundant Spaces in Cities and Regions?*, (London: Academic Press).

Anderson, K., Domosh, M., Pile, S. and Thrift, N. (eds) (2003), *Handbook of Cultural Geography* (London: Sage).

Anderson, P. and Kitchen, R. (2000), 'Disability, Space and Sexuality: Access to Family Planning Services', *Social Science & Medicine* 51, 1163–1173.

Angerer, M-L., Peters, K. and Sofoulis, Z. (eds) (2002), *Future Bodies: Zur Visualisierung von Körpern in Science und Fiction*, (Vienna: Springer Verlag).

Anzaldúa, G. (1987), *Borderlands/La Frontera* (San Francisco: Aunt Lute).

Appadurai, A. (1996), *Modernity at Large. Cultural Dimensions of Globalization* (Minneapolis: University of Minnesota Press).

Attitude Magazine, (2003), Youth Special, February 2003.

Austin, P. (2002), 'Femme-Inism: Lessons of My Mother', in D. Hernández and B. Rehman (eds.).

Badgett, L. (2001) *Money, Myths and Change: The Economic Lives of Lesbians and Gay Men* (Chicago: University of Chicago Press).

Baer, B.J. (2002), 'Russian Gays/Western Gaze: Mapping (Homo)Sexual Desire in Post-Soviet Russia', *GLQ- A Journal of Lesbian and Gay Studies* 8, 499–521.

Baier, E. (2006), 'Meetings Debate Whether Homosexuality is a Choice", *South Florida Sun-Sentinel*, 7 May, 1, (Ft. Lauderdale, Florida)

Bain, A.L. and Nash, C. J. (2006), 'Undressing the Researcher: Feminism, Embodiment and Sexuality at a Queer Bathhouse Event', *Area* 38:1, 99-106.

— (forthcoming), 'The Toronto Women's Bathhouse Raid: Querying Queer Identities in the Courtroom', *Antipode*.

Bakhtin, M.M. (1965), *Rabelais and His World*, trans. H. Iswolsky, 1984, (Bloomington: Indiana University Press).

Balls, A. (2003), 'Ask me Bollox', *Free! Magazine for the Irish Gay Scene*, 12, 6 (Dublin).

Barbosa, P. and Lenoir, G. (2003), *I Exist*, (San Francisco: Eyebite Productions).

Barnard, S. (2000), 'Construction and Corporeality: Theoretical Psychology and Biomedical Technologies', *Theory & Psychology* 10, 669–688.

Barnes, T. (1995), *Logics of Dislocation* (New York: Guilford).

— (1998), 'A History of Regression: Actors, Networks, Machines, and Numbers', *Environment and Planning A* 30, 203–223.

Barton, L. (ed.) (1996), *Disability and Society: Emerging Issues and Insights*, (London: Longman).

Bataille, G. (1986), *Erotism: Death & Sensuality* (San Francisco: City Light Books).

Bawer, B. (1993), A Place at the Table: The Gay Individual in American Society (New York: Poseidan Press).

Baynes, K., Bohman, J. and McCarthy, T. (eds) (1987), *After Philosophy: End or Transformation?* (Cambridge, MA: MIT Press).

BBC Capital of Culture (2005), [website] <http://www.bbc.co.uk/capitalofculture/newcastlegateshead/background.shtml>, accessed 29 July 2005.

Bech, H. (1997), *When Men Meet: Homosexuality and Modernity* (Cambridge: Polity Press).

Beckerman, N. (2002), 'Couples Coping with Discordant HIV Status', *AIDS Patient Care and STDs*, 16: 2, 55–59.

Beemyn, B. (ed.) (1997), *Creating a Place for Ourselves: Lesbian, Gay, and Bisexual Community Histories*, (London: Routledge).

Beemyn, B. and Eliason, M. (eds) (1996), *Queer Studies*, (New York: New York University Press).

Bell, D. (1991), 'Insignificant Others: Lesbian and gay geographies', *Area* 23: 4, 323–329.

— (1994a), 'In Bed with the State: Political Geography and Sexual Politics', *Geoforum* 25: 445–452.

— (1994b), 'Bi-sexuality: A Place on the Margins', in S. Whittle (ed.).

— (1995a), 'Guest Editorial: [*Screw*]ing Geography: Censor's Version', *Environment & Planning D: Society & Space* 13: 127–131.

— (1995b), 'Perverse Dynamics, Sexual Citizenship and the Transformation of Intimacy' in D. Bell and G. Valentine (eds).

— (1997), 'One-Handed Geographies: An Archaeology of Public Sex' in G.B. Ingram, *et al.* (eds).

— (2001), 'Fragments for a Queer City', in D. Bell et al. (eds).

— (2006), 'Bodies, Technologies, Spaces: On "Dogging"', *Sexualities*, 9:4, 387–407.

— (forthcoming), 'Fade to Grey: Some Reflections on Policy and Mundanity', *Environment and Planning A*.

Bell, D. and Binnie, J. (1998), 'Theatres of Cruelty, Rivers of Desire: The Erotics of the Street' in N.R. Fyfe (ed.).

— (2000), *The Sexual Citizen: Queer Politics and Beyond*, (Cambridge: Polity Press).

— (2004), 'Authenticating Queer Space: Citizenship, Urbanism and Governance', *Urban Studies*, 41: 1807-1820.

— (eds) (2006), 'Geographies as Sexual Citizenship', *Political Geography*, 28:8.

Bell, D., Binnie, J., Cream, J. and Valentine, G. (1994), 'All Hyped Up and No Place To Go', *Gender, Place and Society*, 1: 31–47.

Bell, D., Binnie, J., Holliday, R., Longhurst, R., and Peace, R. (eds) (2001), *Pleasure Zones: Bodies, Cities, Spaces*, (Syracuse: Syracuse University Press).

Bell, D. and Holliday, R. (2000), 'Naked as Nature Intended', *Body & Society*, 6:3–4, 127–40.

Bell, D. and Valentine, G. (eds) (1995), *Mapping Desire: Geographies of Sexualities* (London: Routledge).

Benjamin, W. (1999), *The Arcades Project* (Cambridge: Belknap Press) [*Das Passagenwerk* (1983)].

Berlant, L. and Warner, M. (1998), 'Sex in Public', *Critical Inquiry* 24:2 (Winter), 547–566.

Berry, C., Martin, F. and Yue, A. (eds) (2003), *Mobile Cultures. New Media in Queer Asia*, (Durham: Duke University Press).

Bersani, L. (1990), *The Culture of Redemption* (Cambridge, MA: Harvard University Press).

— (1995), *Homos* (Cambridge, MA: Harvard University Press).

Bhabha, J. and Shutter, S. (1994), *Women's Movement* (Stoke-on-Trent: Trentham Books).

Bibbings, L. and Alldridge, P. (1993), 'Sexual Expression, Body Alteration and the Defence of Consent', *Journal of Law and Society* 20: 356–370.

Billinge, M. and Gregory, D. (eds) (1984), *Recollections of a Revolution*, (London: Macmillan).

Binnie, J. (1995), 'Trading Places: Consumption, Sexuality and the Production of Queer Space', in D. Bell and G. Valentine (eds).

— (1997), 'Coming Out of Geography: Towards a Queer Epistemology', *Environment & Planning D: Society & Space*, 15, 223–237.

— (2004), *The Globalization of Sexuality* (London: Sage).

Binnie, J. and Valentine, G. (1999), 'Geographies of Sexualities – A Review of Progress', Progress in Human Geography 23:2, 175–187.

Black, J.C. (1994), '24 Frames', in S. Lim-Hing (ed.).

Black, N. (2005), 'Gay Parents: You're Not Normal and Neither Are Your Kids', *The Stranger*, Seattle, 23 June.

Blackwood, E. (2005), 'Transnational Sexualities in One Place – Indonesian Readings', *Gender and Society* 19:2, 221–242.

Blasius, M. (ed.) (2001), *Sexual Identities, Queer Politics*, (Oxford: Princeton University Press).

Blum, V. and Nast, H. (1996), 'Where's the Difference?: The Heterosexualization of Alterity in Henri Lefebvre and Jacques Lacan', *Environment and Planning D: Society and Space* 14, 559–580.

Boellstorff, T. (2003), 'The Perfect Path: Gay Men, Marriage, Indonesia', in R.J. Corber and S. Valocchi (eds).

Bolton, R. (1995), 'Tricks, Friends and Lovers: Erotic Encounters in the Field', in D. Kulick and M. Wilson (eds).

Bondi, L. (1998), 'Sexing the City', in R. Fincher and J.M. Jacobs (eds).

Boone, J.A., Dupuis, M., Meeker, M. and Quimby, K. (eds)(2000), *Queer Frontiers: Millenial Geographies, Genders, and Generations*, (Wisconsin: University of Wisconsin Press).

Bouthillette, A.M. (1994), 'Gentrification by Gay Male Communities: A Case Study of Toronto's Cabbagetown', in S.Whittle (ed.).

Braidotti, R. (1994), *Nomadic Subjects. Embodiment and Sexual Difference in Contemporary Feminist Thought* (New York: Columbia University Press).

— (2002), *Metamorphoses: Toward a Materialist Theory of Becoming* (Cambridge: Polity Press).

Brickell, C. (2000), 'Heroes and Invaders: Gay and Lesbian Pride Parades and Public/Private Distinction in New Zealand Media Accounts', *Gender, Place and Culture* 7: 2, 163–178.

Brook, J., Carlsson, C. and Peters, N.J. (eds) (1998), *Reclaiming San Francisco: History, Politics, Culture*, (San Francisco: City Light Books).

Brown, G. (forthcoming), 'Mutinous Eruptions: Autonomous Spaces of Radical Queer Activism', *Environment and Planning A*.

Brown, M. (1994), 'The Work of City Politics: Citizenship Through Employment in the Local Response to AIDS', *Environment and Planning A*, 26, 873–894.

— (1995), 'Ironies of Distance: An On-going Critique of the Geographies of AIDS', *Environment and Planning D: Society and Space*, 13, 159–183.

— (1997), *RePlacing Citizenship: AIDS Activism & Radical Democracy* (New York: Guilford).

— (1999), 'Reconceptualizing Public and Private in Urban Regime Theory', *International Journal of Urban and Regional Research* 23, 45–63.

— (2000), *Closet Space: Geographies of Metaphor from the Body to the Globe* (London: Routledge).

— (2006), 'Sexual Citizenship, Political Obligation and Disease Ecology in Gay Seattle', *Political Geography*, 25: 8, 874–898.

Brown, M. and Colton, T. (2001), 'Dying Epistemologies: An Analysis of Home Death and its Critique', *Environment & Planning A*, 33, 799–821.

Brown, M. and Knopp, L. (2003), 'Queer Cultural Geographies – We're Here! We're Queer! We're Over There , Too!', in Kay Anderson et al. (eds).

— (2006), 'Places or Polygons: Governmentality, Scale, and the Census in *The Gay and Lesbian Atlas*', *Population, Space & Place*, 12, in press.

Brown, M., Knopp, L. and Morrill, R. (2005), 'The "Culture Wars" and Urban Electoral Politics: Sexuality, Race, and Class in Tacoma, Washington', *Political Geography* 24:267–221.

Brown, W. (1995), *States of Injury: Power and Freedom in Late Modernity* (Princeton: Princeton University Press).

— (2001) *Politics Out of History* (Princeton: Princeton University Press).

Browne, K. (2004), 'Genderism and the Bathroom Problem: (Re)materialising Sexed Sites, (Re)creating Sexed Bodies', *Gender, Place and Culture* 11:331–346.

— (2005), 'Stages and Streets: Reading and Misreading Female Masculinities', in, B. Van Hoven and K. Horschelmann (eds), *Spaces of Masculinities*. London: Routledge, pp. 237–248.

— (forthcoming), 'A Party with Politics? (Re)making LGBTQ Pride spaces in Dublin and Brighton', Social and Cultural Geography.

Brubaker, R. (2001), 'The Return of Assimilation? Changing Perspectives on Immigration and its Sequels in France, Germany, and the United States', *Ethnic and Racial Studies* 24, 531–548.

Brunel-Cohen, P. (2001), *Perceiving a Pinker Triangle: Investigating the Effects of an Expanded and Improved Gay Scene in Newcastle*, Unpublished Undergraduate Dissertation (University of Newcastle: Newcastle upon Tyne).

Buchanan, I. (1997), 'The Problem of the Body in Deleuze and Guattari, Or, What Can a Body Do?' *Body and Society* 3:3, 73–91.

Burchell, G., Gordon, C. and Miller, P. (1991), *The Foucault Effect: Studies in Governmentality* (Chicago: University of Chicago Press).

Butler, J. (1990), *Gender Trouble: Feminism and the Subversion of Identity* (London: Routledge).

— (1993a), *Bodies That Matter: On the Discursive Limits of "Sex"* (London: Routledge).

— (1993b), 'Gender is Burning', in J. Butler.

— (1996), 'Gender as performance', in P. Osborne (ed.).

— (1997), *Excitable Speech: A Politics of the Performative* (London: Routledge).

— (1999), 'Revisiting Bodies and Pleasures', *Theory, Culture and Society*, 16:2, 11–20.

— (2002), *Antigione's Claim*, (New York: Columbia University Press).

— (2004), *Undoing Gender*, Routledge: London.

Butler, R. (1999), 'Double the Trouble or Twice the Fun? Disabled Bodies in the Gay Community', in R. Butler and H. Parr (eds).

Butler, R. and Parr, H. (eds) (1999), *Mind and Body Spaces: Geographies of Illness, Impairment, and Disability* (London and New York: Routledge).

Byrne, R. (2003), 'Setting the Boundaries – Tackling Public Sex Environments in Country Parks', paper presented at the Planning Research Conference, Oxford Brookes University, April 8–10.

Califia, P. (1979), 'The Secret Side of Lesbian Sexuality', *The Advocate*, Issue 287 (December 27, 1979), Reprinted in Weinberg (ed.).

Cant, B. and Hemmings, S. (eds) (1988), *Radical Records: Thirty Years of Lesbian and Gay History* (London: Routledge).

Carter, S. and McCormack, D.P. (2006), 'Film, Geopolitics and the Affective Logics of Intervention', *Political Geography*, 25:2, 228–245.

Casey, M. (2004), 'De-dyking Queer Spaces: Heterosexual Female Visibility in Gay and Lesbian Spaces', *Sexualities*, 7:4, 446–461.

— (2005), 'Bent on Normality?', *Attitude Magazine*, March: 38.

— (eds) (2006), *Intersections Between Queer and Feminist Theory*, (Oxford: Palgrave Macmillan).

Castells, M. (1983), *The City and the Grassroots* (Berkley: University of California Press).

Castronovo, R. and Nelson, D. (eds) (2002), *Materializing Democracy: Toward a Revitalized Cultural Politics* (Durham, NC: Duke University Press).

CDC (2001), *Tracking the Hidden Epidemics: Trends in STDs in the United States, 2000* (Atlanta, GA: Centers for Disease Control and Prevention).

Cefai, S. (2004), 'Negotiating Silences, Disavowing Femininity and the Construction of Lesbian Identities', in WGSG.

Cerullo, M. and Hammonds, E. (1987), 'AIDS and Africa: The Western Imagination and the Dark Continent', *Radical America* 21: 2–3, 17–23.

Chalmers, A. (1999), *What is this Thing Called Science?*, 3rd Edition (Cambridge: Hackett).

Chatterton, P. (2005), 'Making Autonomous Geographies: Argentina's Popular Uprising and the 'Movimiento de Trabajadores Desocupados' (Unemployed Workers Movement)', *Geoforum* 25: 5, 545–561.

— (2006), '"Give Up Activism" and Change the World in Unknown Ways: Or, Learning to Walk with Others on Uncommon Ground', *Antipode* 38: 2, 259–281.

Chatterton, P. and Hollands, R. (2001), *Changing our Toon: Youth Nightlife and Urban Change in Newcastle* (Newcastle upon Tyne: University of Newcastle Press).

— (2003), *Urban Nightscapes* (London: Routledge).

Chouinard, V. and Grant, A. (1996), 'On Being Not Even Anywhere Near 'The Project': Ways of Putting Ourselves in the Picture', in N. Duncan (ed.).

Clifford, J. (1988), *The Predicament of Culture* (Berkeley: University of California Press).

Cloke, P. and Little, J. (eds) (1997), *Contested Countryside Cultures: Otherness, Marginalisation and Rurality*, (London: Routledge).

Cloke, P.J., Crang, P. and Goodwin, M. (2005), *Introducing Human Geographies*, (London: Hodder Arnold).

Cocks, H.G. and Houlbrook, M. (2005), *The Modern History of Sexuality* (London: Palgrave).

Cohen, C.J. (2001), 'Punks, Bulldaggers, and Welfare Queens: The Radical Potential of Queer Politics?', in Blasius, M. (ed.).

Cohen, P.F. (1997), '"All They Needed": AIDS, Consumption, and the Politics of Class', *Journal of the History of Sexuality* 8:1, 86–115.

Collins, D. (2005), 'Identity and Urban Place-making – Exploring Gay Life in Manila', *Gender and Society*, 19:2, 180–198.

Collins, P.H. (1990), *Black Feminist Thought*, (Boston, MA: Unwin Hyman).

Colomina, B., Dollens, D., Sedgwick, E., Urbach, H. and Wigley, M. (1994), 'Something about Space is Queer', Art and Text 49, 83–4.

Colwell, C. (1997), 'Deleuze and Foucault: Series, Event, Genealogy', *Theory and Event*, 1, 2. <http://muse.jhu.edu/journals/theory_and_event/v001/1.2colwell.html>.

Conlon, D. (2004), 'Productive Bodies, Performative Spaces: Everyday Life in Christopher Park', *Sexualities*, 7:4, 462–479.

Connell, J. (ed.) (2000), *Sydney: The Emergence of a World City* (Melbourne: Oxford University Press).

Connolly, W.E. (2002), *Neuropolitics: Thinking, Culture, Speed*, (Minneapolis: University of Minnesota Press).

Cooke, T. (2005), 'Migration of Same-sex Couples', *Population, Space, and Place*, 11, 401–409.

Cooper, D. (1994), *Sexing the City: Lesbian and Gay Politics Within the Activist State*, (London: Rivers Oram).

Corber, R.J. and Valocchi, S. (eds) (2003a), *Queer Studies: An Interdisciplinary Reader*, (Oxford: Blackwell Publishing).

Corber, R.J. and Valocchi, S. (2003b), 'Introduction', in Corber, R.J. and Valocchi, S. (eds).

Cosgrove, D. (1984), *Symbolic Formation and Symbolic Landscapes*, (London: Croom Helm).

Cowen, D. (2003), 'From the American Lebensraum to the American Living Room: Class, Sexuality, and the Scaled Production of "Domestic" Intimacy', *Environment and Planning D: Society and Space* 22, 755–771.

Craddock, S. (2000a), *City of Plagues: Disease, Poverty, and Deviance in San Francisco* (Minneapolis: University of Minnesota Press).

Craddock, S. (2000b), 'Disease, Social Identity, and Risk: Rethinking the Geography of AIDS', *Transactions of the Institute of British Geographers* 25: 2, 153–169.

Crang, M. (1998), *Cultural Geography* (London: Routledge).

Cream, J. (1995), 'Re-solving Riddles: the Sexed Body', in D. Bell and G. Valentine (eds).

Crimp, D. (1993), 'Right On, Girlfriend!', in M. Warner (ed.).

— (2002), *Melancholia and Moralism: Essays on AIDS and Queer Politics* (Cambridge MA: MIT Press).

— (2003), 'Melancholia and Moralism' in D.L. Eng and D. Kazanjian (eds).

Crimp, D. and Rolston, A. (1990), *AIDS Demo Graphics* (Boston: Bay Press).

Crone, J. (1995), 'Lesbians: The Lavender Women of Ireland', in I. O'Carroll and E. Collins (eds).

Cruikshank, B. (1999), *The Will to Empower: Democratic Citizens and Other Subjects*, (Ithaca, NY: Cornell University Press).

Crysler, G. (2003), *Writing Spaces: Discourses of Architecture, Urbanism, and the Built Environment, 1960–2000* (London: Routledge).

Cvetkovich, A. (2003), *An Archive of Feelings: Trauma, Sexuality and Lesbian Public Cultures* (Durham, NC: Duke University Press).

D'Emilio, J. (1993), 'Capitalism and Gay Identity', in H. Abelove (eds).

Dahles, H. (1998), 'Of Birds and Fish. Streetguides, tourists and sexual encounters in Yogyarkarta, Indonesia', in M. Oppermann (ed.).

Dandridge, D. (2006), 'Voices from the Margins: Sadomasochism and Sexual Citizenship', *Citizenship Studies*, 10, 373–389.

Dangerous Bedfellows (eds.) (1996), *Policing Public Sex*, (Boston: South End Press).

Davidson, A.I. (ed.) (1997), *Foucault and his Interlocutors*, (Chicago: The University of Chicago Press).

Day, R.J.F. (2004), 'From Hegemony to Affinity. The Political Logic of the Newest Social Movements', *Cultural Studies* 18: 5, 716–748.

— (2005), *Gramsci is Dead: Anarchist Currents in the Newest Social Movements* (London: Pluto Press).

Delaney, S. (1999), *Times Square Red, Times Square Blue* (New York: New York University Press).

DeLauretis, T. (1991), 'Queer Theory: Lesbian and Gay Sexualities: An Introduction', *Differences* 4:1–10.

Del Casino Jr., V., (2006), 'NGOs and the Reorganization of 'Community Development': Mediating the Flows of People Living with HIV and AIDS', in A. Sleigh et al. (eds).

Del Casino Jr., V.J. (2007), 'Flaccid Theory and the Geographies of Sexual Health in the Age of Viagra', Health and Place, doi: 10.1016/J.healthplace.2007.10.003.

Del Casino Jr., V., Kochems, L. and Fisher, D. (2004), 'Discourses of Responsibility: Dividing HIV Prevention Efforts Between the "Positive" and the "Negative"', Abstract XV International AIDS Conference, Bangkok, Thailand.

Deleuze G. (1988), *Spinoza: Practical Philosophy*, (San Francisco: City Light Books).

— (1990a), *Expressionism in Philosophy: Spinoza*, (New York: Zone Books).

— (1990b), *The Logic of Sense*, (London and New York: Continuum).

— (1997), 'Desire and Pleasure', trans. Daniel W. Smith, in A.I. Davidson (ed.).

Deleuze G. and Guattari F. (1984), *Anti-Oedipus: Capitalism and Schizophrenia*, (London: The Athlone Press).

— (1988), *A Thousand Plateaus: Capitalism and Schizophrenia*, (London: The Athlone Press).

Devor, H. (1989), *Gender Blending: Confronting the Limits of Duality*, (Bloomington: Indiana University Press).

Dewsbury, J-D. (2000), 'Performativity and the Event: Enacting a Philosophy of Difference', *Environment and Planning D: Society and Space*, 18: 4, 473–496.

Divall, C. and Revell, G. (2005), 'Cultures of Transport: Representation, Practice and Technology', *Journal of Transport History* 26, 99–111.

Dixon, D. and Jones, J.P. (1998), 'My Dinner with Derrida, or Spatial Analysis and Poststructuralism do Lunch', *Environment and Planning A*, 30, 247–260.

Doel, M.(1999), *Poststructuralist Geographies: The Diabolical Art of Spatial Science*, (New York: Rowman & Littlefield).

Domosh, M. (1999), 'Sexing Feminist Geography', *Progress in Human Geography* 23:3, 429–436.

Domosh, M. and Seager, J. (2001), *Putting Women in Place: Feminist Geographers Make Sense of the World* (New York: Guilford Press).

Dorn, M. and Laws, G. (1994), 'Social Theory, Body Politics, and Medical Geography: Extending Kearn's Invitation', *The Professional Geographer* 46:1, 106–110.

Dreyfus, H. and Rabinow, P. (1982), *Michel Foucault: Beyond Structuralism and Hermeneutics* (Chicago: University of Chicago Press).

Dublin Pride (2006), [website] <http://www.dublinpride.org/contact.html>, accessed 11 December 2006.

Duggan, L. (2002), 'The New Homonormativity: The Sexual Politics of Neoliberalism', in R. Castronovo and D. Nelson (eds).

— (2003), *The Twilight of Equality? Neoliberalism, Cultural Politics, and the Attack on Democracy* (Boston: Beacon Press).

Duggan, L and Hunter, N.D. (1995), *Sex Wars* (New York: Routledge).

Duncan, J. and Ley, D. (eds) (1993), *Place/Culture/Representation* (London and New York: Routledge).

Duncan, N. (ed.) (1996a), *BodySpace: Destabilizing geographies of gender and sexuality* (New York: Routledge).

— (1996b), 'Renegotiating Gender and Sexuality in Public and Private Spaces', in N. Duncan (ed.).

Dunne, B. (1990), 'Homosexuality in the Middle East: An Agenda for Historical Research', *Arab Studies Quarterly* 12:3–4, 55–82.

Duttmann, G.A. (1996), *At Odds With Aids: Thinking and Talking About a Virus*, (Stanford: Stanford University Press).

Edelman, L. (2004), *No Future: Queer Theory and the Death Drive*, (Durham, NC: Duke University Press).

Elder, G. (1999), '"Queerying" Boundaries in the Geography Classroom', *Journal of Geography in Higher Education*, 23:1, 86–93.

— (2002), 'Response to "Queer Patriarchies, Queer Racisms, International"', *Antipode* 34, 988–991.

Elwood, S.A. (2000), 'Lesbian Living Spaces: Multiple Meanings of Home', in G. Valentine (ed.).

Eng, D. and Kazanjian, D. (eds) (2003), *Loss: The Politics of Mourning*, (Berkley: University of California Press).

Eng, D., Halberstam, J. and Munoz, J.E. (2005), 'Introduction: What's Queer about Queer Studies Now?' *Social Text*, 23:3–4, 1–17.

England, K. (1993), 'Suburban Pink Collar Ghettos: The Spatial Entrapment of Women?', *Annals of the Association of American Geographers* 83, 225–242.

— (1994a), 'Getting Personal: Reflexivity, Positionality and Feminist Research', *Professional Geographer* 46, 80–89.

— (1994b), 'From "Social Justice and the City" to Women-Friendly Cities? Feminist Theory and Politics', *Urban Geography* 15, 628–643.

Escobar, A. (1995), *Encountering Development. The Making and Unmaking of the Third World* (Princeton: Princeton University Press).

Essig, L. (1999), *Queer in Russia. A Story of Sex, Self, and the Other* (Durham: Duke University Press).

Esteva, G. (1992), 'Hilfe und Entwicklung stoppen! Eine Antwort auf den Hunger', in Gustavo Esteva, *Fiesta – jenseits von Entwicklung, Hilfe und Politik* (Frankfurt/Main: Brandes und Apsel; Südwind Verlag).

Ettorre, E.M. (1978), 'Women, Urban Social Movements and the Lesbian Ghetto', *International Journal of Urban and Regional Research*, 2, 499–520.

Evans, D. (1993), *Sexual Citizenship: the Material Construction of Sexualities* (London and New York: Routledge).

Eves, A. (2004), 'Queer Theory, Butch/Femme Identities and Lesbian Space', *Sexualities*, 7:4, 480–496.

Faderman, L. (1981), *Surpassing the Love of Men: Romantic Friendship and Love Between Women from the Renaissance to the Present* (London: Women's Press).

Fannin, M. (2003), 'Domesticating Birth in the Hospital: "Family-Centered" Birth and the Emergence of "Homelike" Birthing Rooms', *Antipode* 35:3, 513–535.

Fanon, F. (1986), *Black Skin, White Masks* (London: Pluto Press).

Ferguson, J. (1994), *The Anti-Politics-Machine: "Development", Depoliticization, and Bureaucractic Power in Lesotho* (Minneapolis: University of Minnesota Press).

Farmer, P. (1992), *AIDS and Accusation: Haiti and the Geography of Blame* (Berkeley: University of California Press).

Fee, E. and Fox, D. (eds) (1988), *AIDS: The Burdens of History* (Berkley: University of California Press).

Ferrell, J. (2001), *Tearing Down the Streets: Adventures in Urban Anarchy* (New York: Palgrave Macmillan).

Fincher, R. and Jacobs, J.M. (eds) (1998), *Cities of Difference* (New York: Guilford).

Flintoft, J.M. (2004), 'Bathroom Reading', PolicyMatters 1(1), <http://www.policymatters.org> (home page), accessed 11June 2005.

Forge Forward (2004), 'Transgender Sexual Violence Project: Early Themes', *Forge Forward*, [website] (September 20, 2005), <http://forge-forward.org/transviolence/early_themes.php>.

Foucault, M. (1978), *The History of Sexuality: An Introduction* (New York: Vintage Books).

— (1979), *The History of Sexuality: Volume One, An Introduction*, (London: Penguin).

— (1982), 'The Subject and Power', in H. Dreyfus and P. Rabinow (eds).

— (1985), *The Uses of Pleasure: The History of Sexuality Volume Two* (New York: Vintage Books).

— (1986), *The Care of the Self: The History of Sexuality Volume Three* (New York: Vintage Books).

— (1991), 'Governmentality', in G. Burchell et al. (eds).

Frankenberg, R. (1993), *White Women, Race Matters*, (New York and London: Routledge).

Fraser, N. (1995), 'From redistribution to recognition? Dilemmas of justice in a "post-socialist" age', *New Left Review*, 212, 68–94.

Fuss, D. (ed.) (1991), *Inside/Out: Lesbian Theories, Gay Theories*, (New York: Routledge).

Fyfe, N.R (ed.) (1998), *Images of the Street: Planning, Identity and Control in Public Space*, (London: Routledge).

Gaard, G. (1997), 'Toward a Queer Ecofeminism', *Hypatia* 12, 114–137.

Gabriel, Y. and Yang, T. (1995), *The Unmanageable Consumer: Contemporary Consumption and its Discontents* (London: Sage).

Gallant, C. and Gillis, L. (2001), 'Pussies Bite Back: The Story of the Women's Bathhouse Raid', *Journal of the Canadian Lesbian and Gay Studies Association* 3, 152–167.

Gallant, P. (2001), 'Since '81: Technicalities and Politics Protect the Bathhouses From More Raids', *X-tra*, 25 January.

Gates, G. and Ost, J. (2004), *The Gay and Lesbian Atlas*, (Washington, DC: The Urban Institute).

Gattrell, A. (2002), *Geographies of Health: An Introduction* (Malden, MA and Oxford: Blackwell Publishing).

Geltmaker, T. (1992), 'The Queer Nation Acts Up: Health Care, Politics, and Sexual Diversity in the County of Angels', *Environment & Planning D: Society and Space* 10, 609–650.

Gesler, W. (1992), 'Therapeutic Landscapes: Medical Issues in the Light of New Cultural Geographies', *Social Science and Medicine* 34:7, 735–746.

Gesler, W. and Kearns, R. (2002), *Culture/Place/Health* (London and New York: Routledge).

Gibson-Graham, J.K. (1996), *The End of Capitalism (As We Knew It): A Feminist Critique of Political Economy* (Oxford: Basil Blackwell).

— (2006), *Post-Capitalist Politics* (Minneapolis: University of Minnesota Press).

Giddens, A. (1984), *The Constitution of Society: Outline of the Theory of Structuration* (Cambridge: Polity Press).

Giese, R. (2000), 'Police "Panty raid" Was All About Sex, Not Crime', *Toronto Star*, OP1, 12 October.

Giffney, N. (2004), 'Denormatizing Queer Theory: More than (Simply) Lesbian and Gay Studies', *Feminist Theory* 5:1, 73–78.

Glaser, N. and Strauss, A. (1967), *The Discovery of Grounded Theory* (Hawthorne, NY: Aldine de Gruyter).

Gleeson, B. (1999), *Geographies of Disability* (London and New York: Routledge).

Glick Schiller, N., Basch, L. and Szanton Blanc, C. (1995), 'From Immigrant to Transmigrant: Theorizing Transnational Migration', *Anthropological Quarterly* 68:1, 48–63.

GMHC (2006), 'Gay Men's Health Crisis HIV/AIDS Timeline', *Gay Men's Health Crisis*. (updated 2006) <http://www.ghmc.org/about/timeline.html>, accessed 8 February 2006.

Goldstein, R. (2002), *The Attack Queers: Liberal Society and the Gay Right* (London: Verso).

Gopinath, G. (2005), *Impossible Desires: Queer Diasporas and South Asian Public Cultures*, (Durham, NC: Duke University Press).

Gorman-Murray, A. (2006), 'Homeboys: Uses of Home by Australian Gay Men', *Social and Cultural Geography*, 7:1, 53–69.

Goss, J. (1993), '"The Magic of the Mall": An Analysis of Form, Function, and Meaning in the Contemporary Retail Built Environment', *Annals of the Association of American Geographers* 83, 18–47.

Gould, P. (1993), *The Slow Plague: A Geography of the AIDS Pandemic* (Oxford: Blackwell).

Gould, P. (1999), *Becoming a Geographer* (Syracuse: Syracuse University Press).

Gould, P. and Pitts, F. (eds) (2002), *Geographical Voices*, (Syracuse: Syracuse University Press).

Gower, B. (ed.) (1997), *Scientific Method: An Historical and Philosophical Introduction* (London: Routledge).

Gregory, D. (1978), *Ideology, Science and Human Geography* (London: Hutchinson).

— (1981), 'Human Agency and Human Geography', *Transactions of the Institute of British Geographers* 6, 1–18.

Gregson, N. and Lowe, M. (1994), *Servicing the middle classes: class, gender and waged domestic work in contemporary Britain*, (London: Routledge).

— (1995), '"Home"-Making: On the Spatiality of Daily Social Reproduction in Contemporary Middle-class Britain', *Transactions of the Institute of British Geographers*, 20, 224–235.

Gregson, N. and Rose, G. (2000), 'Taking Butler Elsewhere: Performativities, Spatialities and Subjectivities', *Environment and Planning D – Society and Space* 18, 433–452.

Grossberg, L. and Radway, J. (eds) (1992), *Cultural Studies* (New York: Routledge).

Grosz, E. (1994), *Volatile Bodies* (Bloomington and Indianapolis: Indiana University Press).

— (2005a), *Space, Time, and Perversion*, (London: Routledge).

— (2005b), *Time Travels: Feminism, Nature, Power*, (Durham NC and London: Duke University Press).

Grundfest Schoepf, B. (2004), 'AIDS, History, and Struggles over Meaning', in E. Kalipeni et al. (eds).

Grundy, J. and Smith, M. (2005), 'The Politics of Multi-scalar Citizenship: The Case of Lesbian and Gay Organizing in Canada', *Citizenship Studies* 9:4, 389–404.

Gunther, S. (2005), '*Alors*, Are We "Queer" Yet?', *The Gay & Lesbian Review*, May–June 2005, 23–25.

Hacker, H. (2002), 'Zum Begriff der Transgression. Historische Ansätze und Überschreitung', *L'Homme. Zeitschrift für Feministische Geschichtswissenschaft* 13: 2, 224–238.

—(2003), 'Über Sex und Unter Entwicklung: Transnationale lesbische Interventionen', *Ihrsinn. Eine radikalfeministische Lesbenzeitschrift* 13: 27, 59–68.

Halberstam, J. (1998), *Female Masculinity*, (Durham, NC: Duke University Press).

— (2005), *In a Queer Time & Place: Transgender Bodies, Subcultural Lives*, (New York and London: New York University Press).

— (2006), 'Boys Will be Bois ...? Or Transgendered Feminism and the Forgetful Fish?', in Richardson, D. et al.

Halberstam, J. and Livingstone, I. (eds) (1995), *Posthuman Bodies*, (Bloomington: Indiana University Press).

Hall, C.M. (1994), 'Gender and Economic Interests in Tourism Prostitution: The Nature, Development and Implications of Sex Tourism in South-east Asia', in V. Kinnaird, and D. Hall (eds).

Halperin, D. (1996), 'A Response to Dennis Altman', *Australian Humanities Review* <http://www.lib.latrobe.edu.au/AHR/emuse/Globalqueering/halperin.html>, accessed 12 November 2006.

Hammonds, E. (1990), 'Missing Persons: African American Women, AIDS and the History of Disease', *Radical America* 24: 2, 7–23.

Hanson, S. (1994), 'Never Question the Assumptions, and Other Scenes from the Revolution', *Urban Geography*, 14, 552–556.

Haraway, D. (1991), *Simians, Cyborgs, and Women: The Reinvention of Nature*, (New York: Routledge).

— (1992), 'The Promises of Monsters. A Regenerative Politics of Inappropriate/d Others', in: L. Grossberg et al. (eds).

—(1997), *Modest_Witness@Second_Millenium.FemaleMan©_Meets_OncoMouse™*, (New York: Routledge).Harding, S. (1991), *Whose Science? Whose Knowledge?* (Milton Keynes: Open University Press).

Haritaworn, J. (2005), *Thai Multiracialities in Britain and Germany: An Intersectional Study*, Unpublished PhD thesis, London South Bank University.

Haritaworn, J., Kleese, C. and Lin, C-J. (forthcoming), 'Poly/logue: An Introduction to Polyamory', *Sexualities*.

Harrison, S., Pile, S. and Thrift, N. (eds) (2004), *Patterned Ground: Entanglements of Nature and Culture*, (London: Reaktion).

Hart, A. (1995), '(Re)constructing a Spanish Red-Light District: Prostitution, Space and Power', in D. Bell and G. Valentine (eds).

Harver, W. (1996), *The Body of this Death: Historicity and Sociality in the time of AIDS*, (Stanford: Stanford University Press).

Hayes, J. (2000), *Queer Nations. Marginal Sexualities in the Maghreb*, (Chicago: University of Chicago Press).

Hayslip, L.L. and Wurts, C.J. (1989), *When Heaven and Earth Changed Places*, (New York: Doubleday).

Heckert, J. (2004), 'Sexuality/Identity/Politics' in Purkis and Bowen (eds).

Hemmings, C. (1995), 'Locating Bisexual Identities: Discourses of Bisexuality and Contemporary Feminist Theory", in D. Bell and G. Valentine (eds).

— (1997), 'From Landmarks to Spaces: Mapping the Territory of a Bisexual Genealogy", in G.B. Ingram, et al. (eds).

— (2002), *Bisexual Spaces: A Geography of Sexuality and Gender*, (London: Routledge).

— (2005), 'Invoking Affect: Cultural Theory and the Ontological Turn', *Cultural Studies*, 19: 5, 548–567.

Hennen, P. (2004), 'Fae spirits and gender trouble', *Journal of Contemporary Ethnography* 33: 5, 499–533.

Herman D (1994), *Rights of Passage: Struggles for Lesbian and Gay Legal Equality*, (Toronto: University of Toronto Press).

Hernández, D. and Rehman, B. (eds) (2002), *Colonize This! Young Women of Color on Today's Feminism*, (New York: Seal Press).

Higgs, D. (ed.) (1999), *Queer Sites: Gay Urban History Since 1600*, (London: Sage).

Hindle, P. (1994), 'Gay Communities and Gay Spaces in the City', in S. Whittle (ed.).

Hird, M. (2000), 'Gender's Nature: Intersexuality, Transsexualism and the "Sex"/ Gender Binary', *Feminist Theory* 1, 347–364.

Hislop, G. (2000), 'The Bathhouse Raids were a Turning Point,' *Maclean's*, 112, 126.

Holliday, R. (1999), 'The Comfort of Identity', *Sexualities*, 2: 4, 475–491.

Holliday, R. and Hassard, J. (eds) (2001), *Contested Bodies*, (New York: Routledge).

Holloway, J. (2002), 'Twelve Theses on Changing the World without Taking Power', *The Commoner*, 4 (May), <http://www.commoner.org..uk/>.

hooks, b. (1981), *Ain't I a Woman: Black Women and Feminism*, (London: Pluto).

— (1990), 'Postmodern Blackness', *Postmodern Culture* 1: 1.

— (1992), *Black Looks*, (London: Turnaround).

Howell, P. (2000a), 'Prostitution and Racialised Sexuality: The Regulation of Prostitution in Britain and the British Empire Before the Contagious Diseases Acts', *Environment and Planning D: Society and Space,* 18: 3, 321–339.

— (2000b), 'A Private Contagious Diseases Act: Prostitution and Public Space in Victorian Cambridge', *Journal of Historical Geography,* 26: 3, 376–402.

— (2003), 'Venereal Disease and the Politics of Prostitution in the Irish Free State', *Irish Historical Studies,* 33, 320–341.

— (2004), 'Race, Space and the Regulation of Prostitution in Colonial Hong Kong', *Urban History*, 31: 2, 229–248.

Hubbard, P. (1998), 'Sexuality, Immorality and the City: Red-light Districts and the Marginalisation of Female Street Prostitutes', *Gender, Place and Culture*, 5: 1, 55–76.

— (1999), 'Researching Female Sex Work: Reflections on Geographical Exclusion, Critical Methodologies and "Useful" Knowledge', *Area* 31: 1, 229–237.

— (2000), 'Desire/Disgust: Mapping the Moral Contours of Heterosexuality', *Progress in Human Geography* 24, 191–217.

— (2001), 'Sex Zones: Intimacy, Citizenship and Public Space', *Sexualities* 4: 1, 51–71.

— (2002), 'Sexing the Self: Geographies of Engagement and Encounter', *Social and Cultural Geographies*, 3: 4, 365–381.

Hubbard, P. and Saunders, T. (2003), 'Making Space for Sex Work: Female Street Prostitution and the Production of Urban Space', *International Journal of Urban and Regional Research* 27: 1, 75–89.

Hunt, A. (2002), 'Regulating Heterosocial Space: Sexual Politics in the Early Twentieth Century', *Journal of Historical Sociology*, 15:1, 1–34.

Hunter, J. (1991), *Culture Wars* (New York: Basic Books).

Hunter, S., Shannon, C., Know, J. and Martin, J.I. (eds) (1998), *Lesbian, Gay and Bisexual Youths and Adults: Knowledge for Human Service Practice*, (London: Sage).

Illich, I. (1980), 'Vernacular Values', <http://www.preservenet.com/theory/Illich/Vernacular.html>, accessed 28 May 2006.

Ingram, G.B., Bouthillette, A-M. and Retter, Y. (eds) (1997), *Queers in Space: Communities/Public Places/Sites of Resistance*, (Seattle: Bay Press).

Irigary, L. (1993), *An ethics of sexual difference*, (Ithaca: Cornell University Press).

Islam, Y. (2004), 'Dissident Sexualities: Muslim and Gay in the UK', *Muslim Wakeup* <http://www.Muslimwakeup.com>, accessed 28 March 2005.

Iveson, K. (2003), 'Justifying Exclusion: The Politics of Public Space and the Dispute Over Access to McIvers Ladies' Baths, Sydney', *Gender, Place and Culture* 10, 215–228.

Jackson, S. (1999), *Heterosexuality in Question*, (London: Sage).

— (2003), 'Heterosexuality, Heteronormativity and Gender Hierarchy: Some Reflections on Recent Debates', in Weeks, J. et al. (eds).

Jacobs, J.M. and Nash, C. (2003), 'Too Little, Too Much: Cultural Feminist Geographies', *Gender, Place and Culture,* 10: 3, 265–279.

Jacobs, K. (2003), 'Queer Voyeurism and the Pussy-Matrix in Shu Lea Cheang's Japanese Pornography', in C. Berry et al. (eds).

Jacobs, K. (2004), 'Pornography in Small Places and Other Spaces', *Cultural Studies*, 18:1, 67–83.

Jagose, A. (1996a), *Queer Theory: An Introduction*, (Dunedin, NZ: Otago University Press).

Jagose, A. (1996b), 'Queer Theory', *Australian Humanities Review*, 4. <http://www.lib.latrobe.edu.au/AHR/archive/Issue-Dec-1996/jagose.html>.

Jay, E. (1997), 'Domestic Dykes: The Politics of "In-Difference"', in G.B. Ingram et al. (eds).

Jeffreys, S. (1984), 'The Queer Disappearance of Lesbians: Sexuality in the Academe', *Women's Studies International Forum* 17: 459–472.

— (2003), *Unpacking Queer Politics: A lesbian feminist perspective*, (Cambridge: Polity Press).

Johnston, L. (1996), 'Flexing Femininity: Female Body-Builders Refiguring "The Body"', *Gender, Place and Culture* 3, 327–340.

Johnston, L. and Valentine, G. (1995), 'Wherever I Lay My Girlfriend, That's My Home: The Performance and Surveillance of Lesbian Identities in Domestic Environments', in D. Bell and G. Valentine (eds).

Johnston, R. and Sidaway, D. (2004), *Geography and Geographers: Anglo-American Geography Since 1945*, (London: Hodder Arnold).

Johnston, R., Hepple, L., Hoare, T., Jones, K. and Plummer, P. (2003), 'Contemporary Fiddling in Human Geography while Rome Burns', *Geoforum*, 34, 157–161.

Johnston, R.J., Gregory, D., Pratt, G. and Watts, M.J. (eds) (2000), *Dictionary of Human Geography* (Oxford: Blackwell).

Jolly, S. (2000), '"Queering" Development: Exploring the Links Between same-sex Sexualities, Gender, and Development', *Gender and Development* 8: 1, 78–88.

— (2002), *Gender and Cultural Change. Overview Report* (Brighton: Institute of Development Studies).

Jones, C. and O'Doherty, D. (eds) (2005), *Manifestoes for the Business School of Tomorrow*, (Helsinki: Dvalin Press).

Jordan, J. (1998), 'The Art of Necessity: The Subversive Imagination of Anti-road Protest and Reclaim the Streets', in McKay (ed.).

Joshi, S. (2003), '"Watcha Gonna Do When They Cum All Over You?" What Police Themes in Male Erotic Video Reveal About (Leather)Sexual Subjectivity' *Sexualities* 6: 3–4, 325–342.

K., I. (2003), 'Ask Me Box', *Free! Magazine for the Irish Gay Scene* 13, 18.

Kalipeni, E., Craddock, S., Oppong, J. and Ghosh, J. (2004), *HIV/AIDS in Africa: Beyond Epidemiology* (Malden, MA and Oxford: Blackwell).

Kamikaze, I. (1995), 'I Used To Be An Activist, But I'm Alright Now', in I. O'Carroll and E. Collins (eds).

Kamel, G. (1995), 'Leathersex: Meaningful Aspects of Gay Sadomasochism', in T. Weinberg (ed.).

Kates, S. (1998), *Twenty Million New Customers: Understanding Gay Men's Consumer Behaviour*, (London: Harrington Park Press).

Katz, S. and Marshall, B. (2004), 'Is the Functional "Normal"? Aging, sexuality and the Bio-marking of Successful Living', *History of Human Sciences* 17:1, 53–75.

Kawale, R. (2004), 'Inequalities of the Heart: The Performance of Emotion Work by Lesbian and bisexual women in London, England', *Social and Cultural Geography*, 5: 4, 565–581.

Kearns, R. (1993), 'Place and Health: Towards a Reformed Medical Geography', *The Professional Geographer* 45:2, 139–147.

— (1994a), 'Putting Health and Health Care into Place: An Invitation Accepted and Declined', *The Professional Geographer* 46:1, 111–115.

— (1994b), 'To Reform is Not to Discard: A Reply to Paul', *The Professional Geographer*, 46:4, 505.

— (1998), '"Going it Alone": Place, Identity, and Community Resistance to Health Reforms in Hokianga, New Zealand', in Robin Kearns and Wilbert Gesler (eds).

Kearns, R. and Gesler, W. (eds) (1998), *Putting Health Into Place: Landscape, Identity and Well-Being* (Syracuse: Syracuse University Press).

Keith, M. (1992), 'Angry Writing: Representing the Unethical World of the Ethnographer', *Environment & Planning D: Society & Space* 10: 551–568.

Keith, M. and Pile, S. (eds) (1993a), *Place and Politics of Identity*, (New York: Routledge).

— (1993b), 'Conclusion: Towards New Radical Geographies', in M. Keith and S. Pile (eds).

Kennedy, J. (2000), 'Promises, Promises: Top Cop Insists Bathhouses Are Not Being Targeted', *Fab*, 10, 28 September.

Kesby, M. (2005), '"Let's Talk About Sex, Baby ..."': Conversing with Zimbabwean Children about HIV/AIDS', *Children's Geographies* 3:2, 201–218.

Ketteringham, W. (1979), 'Gay Public Space and the Urban Landscape: A Preliminary Assessment', paper delivered at the Annual Meeting of the Association of American Geographers.

— (1983), 'The Broadway Corridor: Gay Businesses as Agents of Revitalization in Long Beach, California', paper delivered at the Annual Meeting of the Association of American Geographers, Denver, CO, April 26.

Kich, G.K. (1996), 'In the Margins of Sex and Race', in M.P.P. Root (ed.).

Kinnaird, V. and Hall, D. (eds) (1994), *Tourism: A Gender Perspective*, (Chichester: John Wiley and Sons).

Kinsman, G. (1996), *Regulation of Desire: Sexuality in Canada* (Second Edition) (Montreal: Black Rose Books).

Kitchin, R. and Lysaght, K. (2003), 'Heterosexism and the Geographies of Everyday Life in Belfast, Northern Ireland', *Environment and Planning A*, 35:3, 489–510.

Kochems, L. (forthcoming), *Gay Culture: Assertive Symbolic Activity Among Gay Men in America*, unpublished PhD dissertation, Anthropology, University of Chicago, Chicago.

Kochems, L. and Del Casino Jr, V. (2004), 'Manipulating Multiple Life Identity-shifts: Points of HIV Risk in Overlapping Drug Cultures and Gay Cultures in Long Beach, CA', Oral Presentation at the American Public Health Association Meetings, Washington, D.C.

Kochems, L. and Patti, V. (1993), 'Gay Culture: Redefining Anthropology and Psychology', paper presented at *Lesbian and Gay Studies Forum*, Sefton Hall University, N.J.

Koumans, E., Sternberg, M., Motamed, C., Kohl, K., Schilinger, J. and Markowitz, L. (2005), 'Sexually Transmitted Disease Services at US Colleges and Universities', *Journal of American College Health* 53:5, 211–217.

Knopp, L. (1987), 'Social Theory, Social Movements and Public Policy: Recent Accomplishments of the Gay and Lesbian Mmovements in Minneapolis, Minnesota', *International Journal of Urban and Regional Research*, 11, 243–261.

— (1990), 'Some Theoretical Implications of Gay Involvement in an Urban Land Market', *Political Geography Quarterly*, 9: 337–352.

— (1992), 'Sexuality and the Spatial Dynamics of Capitalism', *Environment & Planning D: Society & Space* 10: 651–669.

— (1995), 'Sexuality and Urban Space: A Framework for Analysis', in D. Bell and G. Valentine (eds).

— (1997), 'Rings, Circles and Perverted Justice: Gay Judges and Moral Panic in Contemporary Scotland', in S. Pile and M. Keith (eds).

— (1998), 'Sexuality and Urban Space: Gay Male Identity Politics in the United States, the United Kingdom and Australia' in R. Fincher and J.M. Jacobs (eds).

— (2004), 'Ontologies of Place, Placelessness, and Movement: Queer Quests for Identity and their Impact on Contemporary Geographic Thought', *Gender, Place and Culture* 11:1, 121–134.

— (ed.) (1999), 'Queer Theory, Queer Pedagogy: New Spaces and New Challenges in Teaching Geography', *Journal of Geography in Higher Education* 23, 77–123.

Knopp, L. and Brown, M. (2003), 'Queer diffusions', *Environment & Planning D: Society & Space* 21, 409–424.

Knox, P. and Pinch, S. (2001), *Urban Social Geography* (Harlow: Longmans).

Kobayashi, A. (1994), 'Coloring the Field: Gender, Race, and the Politics of Fieldwork', *The Professional Geographer* 46, 73–80.

Kobayashi, A. and Peake, L. (1994), 'Unnatural Discourse: "Race" and Gender in Geography', *Gender, Place and Culture* 1, 225–243.

Kramer, J.L. (1995), 'Bachelor Farmers and Spinsters: Gay and Lesbian Identities and Communities in Rural North Dakota', in D. Bell and G. Valentine (eds).

Kremmler, K. (2001), *Blaubarts Handy* (Hamburg: Argument Verlag).

— (2003), *Die Sirenen von Coogee Beach* (Hamburg: Argument Verlag).

Kristeva, J. (1982), *Powers of Horror. An Essay on Abjection* (New York: Columbia University Press).

Kugle, S. (2003), 'Sexuality, Diversity, and Ethics in the Agenda of Progressive Muslims', in O. Safi (ed.).

Kulick, D. and Willson, M. (eds) (1995), *Taboo: Sex, Identity and Erotic Subjectivity in Anthropological Fieldwork*, (London: Routledge).

Kuntsman, A. (2003), 'Double Homecoming: Sexuality, Ethnicity, and Place in Immigration Stories of Russian Lesbians in Israel', *Women's Studies International Forum*, 26:4, 299–311.

Kushner, T. (1993), *Angels in America, Part Two: Perestroika* (New York: Theatre Communications Group).

Kwan, M. (2002), 'Quantitative Methods and Feminist Geographic Research', in P. Moss (ed.).

— (2004), 'Beyond Difference: From Canonical Geography to Hybrid Geographies', *Annals of the Association of American Geographers*, 94, 756–763.

Kylmika, W. (2001), *Politics in the Vernacular: Nationalism, Multiculturalism, and Citizenship*, (Oxford: Oxford University Press).

Lambevski, S. (1999), 'Suck My Nation: Masculinity, Ethnicity and (Homo)sex', *Sexualities* 2, 397–419.

Langdridge, D. and Butt, T. (2003),'A Hermeneutic Phenomenological Investigation of the Construction of Sadomasochistic Identities', *Sexualities* 7:1, 31–53.

Larsen, W. (1989), *Confessions of a Mail-Order Bride*, (Far Hills: New Horizon Press).

Latham, A. with Conradson, D. (2003), 'The Possibilities of Performance', *Environment and Planning A* 35, 1901–1906.

Lauria, M. and Knopp, L. (1985), 'Towards an Analysis of the Role of Gay Communities in the Urban Renaissance', *Urban Geography*, 6, 152–169.

Law, L. (1997), 'Dancing on the Bar: Sex, Money and the Uneasy Politics of Third Space', in S. Pile and M. Keith (eds).

— (2000), *Sex Work in Southeast Asia: The Place of Desire in a Time of AIDS*, (London: Routledge).

Lawson, V. (1995), 'The Politics of Difference: Examining the Quantitative/ Qualitative Dualism in Poststructuralist-feminist Research', *The Professional Geographer*, 47, 449–457.

Lawson, V.A. and Staeheli, L.A. (1991), 'On Critical Realism, Human Geography, and Arcane Sects!', *The Professional Geographer*, 43:2, 231–233.

Lazreg, M. (2002), 'Development: Feminist Theory's Cul-de-sac', in K. Saunders (ed.).

Leap, W. (ed.) (1999), *Public Sex: Gay Space* (New York: Columbia University Press).

Lechat, L. (2002), *Queeruption 4* [video] Cut & Paste Films.

Lee, J.Y. (1996), 'Why Suzy Wong is Not a Lesbian: Asian and Asian American Lesbian and Bisexual Women and Femme/Butch/Gender Identities', in B. Beemyn and M. Eliason (eds).

Lewin, E. and Leap, W. (eds) (1996), *Out in the Field: Reflections of Lesbian and Gay Anthropologists*, (Champaign: University of Illinois Press).

Lewis, M. (1994), 'A Sociological Pub Crawl Around Gay Newcastle', in S.Whittle (ed.).

Lim-Hing, S. (ed.) (1994), The Very Inside: An Anthology of Writing by Asian and Pacific Islander Lesbian and Bisexual Women, (Toronto: Sister Vision Press).

Little, J. (2003), 'Riding the Rural Love Train: Heterosexuality and the Rural Community', *Sociologica Ruralis* 43:4, 401–417.

Litva, A. and Eyles, J. (1995), 'Coming Out: Exposing Social Theory in Medical Geography', *Health and Place* 1:1, 5–14.

Lo, J. and Healey, T. (2000), 'Flagrantly Flaunting It? Contesting Perceptions of Locational Identity Among Urban Vancouver Lesbians', in G. Valentine (ed.).

Longhurst, R. (1995), 'The Body and Geography', *Gender, Place and Culture* 2, 97–105.

— (2001), *Bodies: Exploring Fluid Boundaries* (London and New York: Routledge).

Longino, H. (2002), *The Fate of Knowledge* (Princeton: Princeton University Press).

Lorber, J. (2000), 'Using Gender to Undo Gender: A Feminist Degendering Movement', *Feminist Theory* 1, 79–95.

Loreck, H. (2002), 'Körper, die ich Nicht Gewesen Sein Werde. Allegorische Konfigurationen in der Zeitgenössischen Kunst', in M-L. Angerer et al. (eds).

Lorimer, H. (2005), 'Cultural Geography: The Busyness of Being 'More-than-representational', *Progress in Human Geography* 29: 1, 83–94.

Lorway, R. (2003), 'Constructing Namibian Queer Selfhood in the Era of HIV/ AIDS', paper presented at the Sex & Secrecy Conference, Johannesburg, June.

— (forthcoming), 'Thinking Through the "Foreign Fetish": HIV-Vulnerability and "(Queer) Subjects of Desire"', *Anthropologica, Journal of the Canadian Anthropology Society*.

Luibhéid, E. (2005), 'Introduction: Queering Migration and Citizenship', in E. Luibhéid and L. Cantú Jr. (eds).

Luibhéid, E. and Cantú Jr., L. (eds) (2005), *Queer Migrations: Sexuality, U.S. Citizenship, and Border Crossings*, (Minneapolis and London: University of Minnesota Press).

Lyotard, J.F. (1984), *The Postmodern Condition*, (Minneapolis: University of Minnesota Press).

Lyttleton, C. (1994a), 'The Good People of Isan: Commerical Sex Work in Northeast Thailand', *The Australian Journal of Anthropology* 5:3, 257–259.

— (1994b), 'Knowledge and Meaning: The AIDS Reduction Campaign in Rural Northeast Thailand', *Social Science and Medicine* 38:1, 135–136.

— (1996), 'Messages of Distinction: The HIV/AIDS Media Campaign in Thailand', *Medical Anthropology* 16, 363–389.

— (2000), *Endangered Relations: Negotiating Sex and AIDS in Thailand* (Singapore: Harwood Academic Publishers).

MacDougal, B. (2000), *Queer Judgments: Homosexuality, Expression and the Courts in Canada*, (Toronto: University of Toronto Press).

Mackenzie, S. and Rose, D. (1983), 'Industrial Change, the Domestic Economy and Home Life', in J. Anderson and S. Duncan (eds).

Mackie, V. (2001), 'The Trans-sexual Citizen: Queering Sameness and Difference', *Australian Feminist Studies* 16, 185–192.

Marcos, Subcomandante (2001),*Our Word is Our Weapon: Selected Writings of Subcomandante Insurgente Marcos*, (London: Serpent's Tail).

Markusen, A. (2005), *Urban Development and the Politics of a Creative Class: Evidence from the Study of Artists*, <www.hhh.umn.edu/projects/prie/>, accessed 9 November 2006.

Marshall, B. (2002), '"Hard Science": Gendered Constructions of Sexual Dysfunction in the "Vaigra age"', *Sexualities* 5:2, 131–158.

Marston, S.A. (2004), 'Space, Culture, State: Uneven Developments in Political Geography', *Political Geography* 23, 1–16.

Martin, A. (1993), *The Lesbian and Gay Parenting Handbook*, (New York and London: Harper Perennial).

Martin, E. (1994), *Flexible Bodies: Tracking Immunity in American Culture from the Days of Polio to the Age of AIDS*, (Boston: Beacon Press).

Massad, J. (2002), 'Re-orienting Desire: The Gay International and the Arab world', *Public Culture* 14:2, 361–385.

— (2003), 'The Intransigence of Orientalist Desires: A Reply to Arno Schmitt', *Public Culture* 15:3, 593–594.

Massey, D. (1993), 'Politics and Space/Time', in M. Keith and S. Pile (eds).

Massey, D., Allen, J. and Sarre, P. (eds) (1999), *Human Geography Today*, (Cambridge: Polity Press).

Massumi B. (2002), *Parables for the Virtual: Movement, Affect, Sensation*, (Durham: Duke University Press).

May, J. (1958), *The Ecology of Human Disease* (New York: MD Publications).

— (ed.) (1961), *Studies in Disease Ecology* (New York: Hafner Publishing Company).

May, T. (ed.) (2002), *Qualitative Research in Action*, (London: Sage).

Mayer, J. and Meade, M. (1994), 'A Reformed Medical Geography Reconsidered', *The Professional Geographer* 46:1, 103–106.

McClintock, A. (1995), *Imperial Leather. Race, Gender and Sexuality in the Colonial Contest*, (New York: Routledge).

McCloskey, D. (1999), *Crossing: A Memoir*, (Chicago: University of Chicago Press).

McCormack D.P. (2002), 'A Paper with an Interest in Rhythm', *Geoforum* 33, 469–485.

— (2003), 'An Event of Geographical Ethics in Spaces of Affect', *Transactions of the Institute of British Geographers NS* 28, 488–507.

— (2005), 'Diagramming Practice and Performance', *Environment and Planning D: Society and Space* 23, 119–147.

McDowell, L. (1995), 'Body Work: Heterosexual Gender Performances in City Workplaces', in D. Bell and G. Valentine (eds).

— (1997), *Capital Culture: Gender at Work in the City*, (Oxford: Blackwell).

McKay, G. (ed.) (1998), *DiY Culture: Party & Protest in Nineties Britain*, (London: Verso).

McLafferty, S. (1995), 'Counting for Women', *The Professional Geographer*, 47, 436–441.

McNee, B. (1984), 'If You Are Squeamish', *East Lakes Geographer*, 19, 16–27.

Meade, M. and Earickson, R. (2000), *Medical Geography,* 2nd Edition (New York: Guilford).

Medhurst, A. and Munt, S.R. (eds) (1997), *Lesbian and Gay Studies*, (London: Cassell).

Metcalf, B. (1996a), 'Introduction: Sacred Words, Sanctioned Practice, New Communities', in B. Metcalf (ed.).

Metcalf, B. (ed.) (1996b), *Making Muslim Space in Europe and North America*, (Berkeley: University of California Press).

Michaels, E. (1994), *Unbecoming*, (Durham, NC: Duke University Press).

Miller, D. (1998), *A Theory of Shopping*, (Cambridge: Polity Press).

Minwalla, O., Rosser, B., Feldman, J. and Varga, C. (2005), 'Identity Experience Among Progressive Gay Muslims in North America: A Qualitative Study Within al-Fatiha', *Culture, Health, and Sexuality* 7:2, 113–128.

Mirza, H. (ed.) (1997), *Black British Feminism*, (London: Routledge).

Mistress Steel (no date), 'Aftercare', Albany Power eXchange (APeX) Inc. <http://www.albanypowerexchange.com/BDSMinfo/aftercare.htm>, accessed 9 November 2006.

Mitchell, D. (2000), *Cultural Geography: A Critical Introduction*, (Oxford: Blackwell).

Mohanty, C.T. and Alexander, M.J. (eds) (1997), *Feminist Genealogies, Colonial Legacies, Democratic Futures*, (New York and London: Routledge).

Momsen, J. (ed.) (1999), *Gender, Migration and Domestic Service*, (London and New York: Routledge).

Moran, L., Skeggs, B., Tyrer, P. and Corteen, K. (2001), 'Property, Boundary, Exclusion: Making Sense of Heteroviolence in Safer Spaces', *Social and Cultural Geography* 2, 407–420.

Moran, L. et al. (2003), 'The Formation of Fear in Gay Space: The "Straights" Story', *Capital & Class* 80, 173–198.

Morrill, R. (1984), 'Recollections of the Quantitative Revolution's Early Years: The University of Washington 1955–1965', in M. Billinge and D. Gregory (eds).

— (2002), 'Pausing for Breath', in P. Gould and F. Pitts (eds) (2002) *Geographical Voices* (Syracuse, Syracuse University Press).

Morris, G. (2005), 'Little Stabs of Happiness (and Horror). Film Reviews', *Bright Lights Film Journal* <http://www.brightlightsfilm.com/47/stabs.htm#iexist>, accessed 28 March 2005.

Morris, J. (1991), *Pride Against Prejudice: Transforming Attitudes to Disability*, (London: The Women's Press).

Moser, C. and Levitt, E. (1987), 'An Exploratory–Descriptive Study of a Sadomasochistically Oriented Sample', *The Journal of Sex Research*, 23, 322–337.

Moser, C. and Madeson, J. (2005), *Bound to be Free: The SM Experience*, (New York & London: Continuum).

Mosher, W., Chandra, A. and Jones, T. (2005), *Sexual Behaviour and Selected Health Measures: Men and Women 15-44 Years of Age, United States, 2002*, (Atlanta, Georgia: Centers for Disease Control and Prevention).

Moss, P. (ed.) (2002), *Feminist Geography In Practice*, (Oxford: Blackwell).

Moss, P. and Dyck, I. (2002), *Women, Body, Illness: Space and Identity in the Everyday Lives of Women with Chronic Illness*, (Oxford: Rowman and Littlefield).

Mouffe, C. (1992), *Dimensions of Radical Democracy*, (London: Verso).

Moya, P.L. (1997), 'Postmodernism, "Realism", and the Politics of Identity', in C.T. Mohanty and M.J. Alexander (eds).

Muller, T. (2007a), 'Liberty for All? Contested Spaces of Lesbian Resistance', *Gender, Place and Culture*.

Muller, T. (2007b), 'Performing "Community"? "Lesbian Community" in WNBA Spaces', *Social & Cultural Geography*.

Munoz, J.E. (1999), *Disidentifications: Queers of Color and the Performance of Politics*, (Minneapolis: University of Minnesota Press).

Murray, A. (2001), *Pink Fits: Sex, Subcultures and Discourses in the Asian-Pacific* (Clayton, Australia: Monash Asia Institute).

Murray, S. and Roscoe, W. (1997), *Islamic Homosexualities: Culture, History, and Literature* (New York: New York University Press).

Myslik, W. (1996), 'Renegotiating the Social/Sexual Iidentities of Places', in N. Duncan (ed.).

Nagel, C. and Staeheli, L. (2004), 'Citizenship, Identity, and Transnational Migration: Arab Immigrants to the United States', *Space and Polity* 8:1, 3–23.

Nagle, J. (1997), *Whores and Other Feminists*, (London: Routledge).

Nairn, K. (2003), 'What has the Geography of Sleeping Arrangements got to do with the Geography of our Teaching Spaces?', *Gender, Place and Culture*, 10:1, 67–81.

Nakashima, C.L. (1992), 'An Invisible Monster,' in M. Root (ed.).

Namaste, K. (1996a), 'Genderbashing: Sexuality, Gender, and the Regulation of Public Space', *Environment and Planning D: Society and Space* 14, 221–240.

Namaste, K. (1996b), 'Tragic Misreadings: Queer Theory's Erasure of Transgender Subjectivity', in B. Beemyn and M. Eliason (eds).

Nanda, S. (1986), 'The Hijras of India: Cultural and Individual Dimensions of an Institutionalised Third Gender Role', *Journal of Homosexuality* 11, 35–54.

Nardi, P.N. and Schneider, B.E. (eds) (1998), *Social Perspectives in Lesbian and Gay Studies: A Reader*, (London: Routledge).

Nash, C.J. (2005), 'Gay Politics and Ethnic Minorities: The Struggle for Gay Identity in Toronto in the Late 1970s', *Gender, Place and Culture* 12, 113–135.

— (2006), 'Toronto's Gay Village (1969 to 1982) Plotting the Politics of Gay Identity', *Canadian Geographer* (March) 50:1, 1–16.

Nash, C.J. and Bain, A.L. (forthcoming), 'Reclaiming Raunch'? Spatializing Queer Identities at Toronto Women's Bathhouse Events', *Social and Cultural Geography*.

Nast, H.J. (1998), 'Unsexy Geographies', *Gender, Place and Culture* 5, 191–206.

— (2002), 'Queer Patriarchies, Queer Racisms, International', *Antipode*, 34:5, 877–909.

Nataf, Z.I. (2001), 'Whatever I feel ...', < http://www.newint.org/issue300/trans.html> accessed 10 November 2006.

National Coalition for Sexual Freedom (2005), <http://www.ncsfreedom.org/>, accessed 10 November 2006.

Nealon, J. (1998), *Alterity Politics: Ethics and Performative Subjectivity*, (Durham, NC: Duke University Press).

Negroni, F. de (1992), *Afrique Fantasmes*, (Paris: Plon).

New Zealand AIDS Foundation (2004), *Negative Role Model*, press release, 14 June 2004, Auckland. <http://www.nzaf.org.nz/> (home page), accessed 10 November 2006.

News From Nowhere (eds) (2003), *We Are Everywhere: The Irresistible Rise of Global Anticapitalism* (London: Verso).

Newton, E. (1993), 'My Best Informant's Dress: The Erotic Equation in Fieldwork', *Cultural Anthropology*, 8, 3–23.

Ng, V. (1997), 'Race Matters', in A. Medhurst and S.R. Munt (eds).

Nguyen, V.K. (2002), 'Sida, ONG et la Politique du Témoignage en Afrique de l'Ouest', *Anthropologie et Sociétés* 26:1, 69–87.

— (2005), 'Uses and Pleasures: Sexual Modernity, HIV/AIDS, and Confessional Technologies in a West African Metropolis', in V. Adams and S.L. Pigg (eds).

Norris, A. (2005a), 'Introduction: Giorgio Agamben and the Politics of the Living Dead', in A. Norris (ed.).

— (2005b), *Politics, Metaphysics, and Death: Essays on Giorgio Agamben's Homo Sacer*, (Durham, NC: Duke University Press).

North, P. (1999), 'Explorations in Heterotopia: Local Exchange Trading Schemes (LETS) and the Micropolitics of Money and Livelihood', *Environment and Planning D* 17, 69–86.

Northwest Lesbian and Gay History Museum Project (2004), 'Claiming Space: Seattle's Lesbian and Gay Historical Geography', collaborative product consisting of glossy fold-out map with annotations and illustrations, (Seattle: Northwest Lesbian and Gay History Museum Project). Available from NWLGHMP, PO Box 797, 1122 E. Pike St, Seattle, WA 98122.

O'Carroll, I. and Collins, E. (eds) (1995), *Lesbian and Gay Visions of Ireland*, (London: Cassell).

O'Neil, J. (2001), '*Horror Autotoxicux*: The Dual Economy of AIDS', in R. Holiday and J. Hassard (eds).

Oelsen, T. (2005), *International Zapatismo: the construction of solidarity in the age of globalization*, (London: Zed Books).

Ogborn, M. (1997), 'Locating the Macaroni: Luxury, Sexuality and Vision in Vauxhall Gardens', *Textual Practice* 11, 445–461.

Oppermann, M. (ed.) (1998), *Sex Tourism and Prostitution. Aspects of Leisure, Recreation and Work* (New York: Cognizant Communication).

Ordover, N. (2003), *American Eugenics: Race, Queer Anatomy and the Science of Nationalism*, (Minneapolis: University of Minnesota Press).

Osborne, P. (ed.) (1996), *A Critical Sense: Interviews with Intellectuals*, (London: Routledge).

Oswin, N. (2005), 'Towards Radical Geographies of Complicit Queer Futures', *ACME: An International E-Journal for Critical Geographers* < http://www.acme-journal.org/vol3/Oswin.pdf>, 3:2, 79–86.

Padva, G. (2002), 'Heavenly Monsters: The Politics of the Male Body in the Naked Issue of *Attitude* Magazine', *International Journal of Sexuality and Gender Studies*, 7:4, 281–292.

Papayanis, M. (2000), 'Sex and the Revanchist City: Zoning Out Pornography in New York', *Environment and Planning D: Society and Space*, 18, 341–353.

Parker, D. and Song, M. (eds) (2001), *Rethinking 'Mixed Race'*, (London: Pluto Press).

Parker, M. (2001), 'Fucking Management: Queer Theory and Reflexivity', *Ephemera*, 1, 36–53.

— (2002), 'Queering Management and Organization', *Gender, Work & Organization*, 9, 146–166.

— (2005), 'Fucking', in C. Jones and D. O'Doherty (eds).

Parr, H. (2002), 'Medical Geography: Diagnosing the Body in Medical and Health Geography, 1999-2000', *Progress in Human Geography*, 26:2, 240–251.

Parr, H. and Butler, R. (1999), 'New Geographies of Illness, Impairment and Disability', in Ruth Butler and Hester Parr (eds).

Patton, C. (1990), *Inventing AIDS* (New York: Routledge).

— (1994), *Last Served? Gendering the HIV Pandemic* (London: Taylor and Francis).

Patton, C. and Sánchez-Eppler, B. (eds.) (2000), *Queer Diasporas* (Durham: Duke University Press).

Paul, B. (1994), 'Comments on Kearn's "Place and Health: Towards and Reformed Medical Geography"', *The Professional Geographer* 46:4, 504–505.

Peace, R. (2001), 'Producing Lesbians: Canonical Proprieties' in D. Bell et al. (eds).

Peake, L. (1993), '"Race" and Sexuality: Challenging the Patriarchal Structuring of Urban Social Space', *Environment and Planning D: Society and Space*, 11, 415–432.

Peters, W. (2001), 'Queer Identities: Rupturing Identity Categories and Negotiating Meanings of Queer', unpublished MA thesis, University of Toronto.

Phelan, S. (ed.) (1997), *Playing With Fire: Queer Politics, Queer Theories*, (London and New York: Routledge).

Phillips, J. (2002), 'The Beach Boys of Barbados: Post-colonial Entrepreneurs', in S. Thorbek and B. Pattanaik (eds).

Phillips, R., Watt, D. and Shittleton, D. (eds) (2000), *De-Centering Sexualities: Politics and Representations Beyond the Metropolis*, (New York and London: Routledge).

Philo, C. (1992), 'Neglected Rural Geographies: A Review', *Journal of Rural Studies*, 8, 193–207.

— (1996), 'Staying In? Invited Comment on "Coming Out: Exposing Social Theory in Medical Geography"', *Health and Place*, 2:1, 25–40.

— (2005), 'Sex, Life, and Death: Fragmentary Remarks Inspired by "Foucault's Population Geographies"', *Population, Space, and Place*, 11, 325–333.

Pile, S. and Keith, M. (eds) (1997), *Geographies of Resistance*. (New York: Routledge).

Pinder, D. (2005), *Visions of the City*, (Edinburgh: Edinburgh University Press).

Pitman, B. (2002), 'Re-mediating the Spaces of Reality Television: America's Most Wanted and the Case of Vancouver's Missing Women', *Environment and Planning A*, 34, 167–184.

Plows, A. (1998), 'Earth First! Defending Mother Earth, Direct-style', in G. McKay (ed.).

Plummer, K. (ed.) (1992), *Modern Homosexualities: Fragments of Lesbian and Gay Experience*, (London: Routledge).

— (2003), 'Intimate Citizenship and the Culture of Sexual Story Telling', in J. Weeks et al. (eds.).

Plummer, P. and Sheppard, E. (2001), 'Must Emancipatory Geography be Qualitative?', *Antipode*, 30, 758–763.

Podmore, J.A. (2001), 'Lesbians in the Crowd: Gender, Sexuality and Visibility along Montreal's Boul St-Laurent', *Gender, Place and Culture*, 8:4, 333–355.

Pointon, S. (1997), 'Transcultural Orgasm as Apocalypse. Urosukidoji: The Legend of the Overfiend', *Wide Angle*, 19: 3, 41–63.

Poore, G. (1996), 'Three Movements in A Minor: Lesbians and Immigration', *Off Our Backs*, August-September 1996, 12.

Popkin, R. and Stroll, A. (1993), *Philosophy Made Simple*, (New York: Doubleday).

Potts, A. (2004), '"Viagra Stories": Challenging "Erectile Dysfunction"', *Social Science & Medicine,* 59, 489–499.

Potts, A., Gavey, N., Grace, V. and Vares, T. (2003), 'The Downside of Viagra: Women's Experiences and Concerns', *Sociology of Health & Illness*, 11, 325–333.

Pratt, G. (1998), 'Grids of Difference: Place and Identity Formation', in R. Fincher and J. Jacobs (eds).

Pratt, G. and Hanson, S. (1994), 'Geography and the Construction of Difference', *Gender, Place and Culture*, 1, 5–30.

Pred, A. (1984), 'Place as Historically-Contingent Process: Structuration and Time-Geography of Becoming Places', *Annals of the Association of American Geographers* 74, 279–297.

Preecha, P. (1989), *Dichan Mai Chai Sophenii*, (Bangkok: Samnakphim).

Pritchard, A., Morgan, N.J. and Sedgely, D. (2002), 'In Search of Lesbian Space? The Experience of Manchester's Gay Village', *Leisure Studies*, 21, 105–123.

Prosser, J. (1998), *Second Skins: The Body Narratives of Transsexuality*, (New York: Columbia University Press).

Puar, J. (2002a) 'A Transnational Feminist Critique of Queer Tourism', in *Antipode*, 34:5, 935–945.

— (ed.) (2002b), 'Queer Tourism: Geographies of Globalization', *GLQ: A Journal of Lesbian and Gay Studies,* 8, 1–2.

— (2002c), 'Circuits of Queer Mobility: Tourism, Travel, and Globalization', *GLQ: A Journal of Lesbian and Gay Studies*, 8, 101–137.

— (2006), 'On Terror: Queerness, Secularism, and Affect', keynote lecture at the *Out of Place: Interrogating Silences in Queerness and Raciality* conference, Lancaster University, March 24–25.

Purkis, J. and Bowen, J. (ed.) (2004), *Changing Anarchism: Anarchist Theory and Practice in a Global Age* (Manchester: Manchester University Press).

Queen, c. and Schimel, L. (eds) (1997), *Pomosexuals: Challenging Assumptions about Gender and Sexuality*, (San Francisco: Cleis Press).

Queeruption London (2003), *Queerewind* (London: Queeruption Collective). Available from Infoshop, 56a Crampton Street, London SE17 3AE.

Quilley, S. (1995), 'Manchester's "Village in the City": The Gay Vernacular in a Post-Industrial Landscape of Power', *Transgressions*, 1:1, 36–50.

— (1997), 'Manchester's "New Urban Village": Gay Space in the Entrepreneurial City', in G.B. Ingram et al. (eds).

R. v. Hornick and Aitcheson (31 January 2002), No. G 0087 (12/94) (Ont. Gen. Div.).

Rand, E. (2003), 'Breeders on a Golf Ball: Normalizing Sex at Ellis Island', *Environment and Planning D: Society and Space*, 21, 441–460.

Retter, Y. (1997), 'Lesbian Spaces in Los Angeles', in G.B. Ingram et al. (eds).

Retzloff, T. (1997), 'Cars and Bars: Assembling Gay Men in Postwar Flint, Michigan', in B. Beemyn (ed.).

Rich, A. (1986), *Blood, Bread and Poetry: Selected Prose 1979-1985*, (New York: Norton).

Richardson, D. (2005), 'Desiring Sameness? The Rise of a Neoliberal Politics of Normalisation', *Antipode*, 37:3, 515–535.

Richardson, D. and Seidman, S. (eds) (2003), *Handbook of Lesbian and Gay Studies*, (London: Sage).

Richardson, D., McLaughlin, J. and Casey, M. (eds) (2006), *Intersections Between Feminist and Queer Theory*, (Basingstoke, Palgrave Macmillan).

Robinson, K. (2003), 'The Passion and the Pleasure: Foucault's Art of Not Being Oneself', *Theory, Culture and Society*, 20: 2, 119-144.

Robinson, V., Hockey, J. and Meoh, A. (2004), '"What I Used to Do On My Mother's Settee": Spatial and Emotional Aspects of Heterosexuality in England', *Gender, Place and Culture*, 11:3, 417–435.

Roen, K. (2001), 'Transgender Theory and Embodiment: The Risk of Racial Marginalisation', *Journal of Gender Studies*, 10, 253–263.

Rofes, E. (2001), 'Imperial New York: Destruction and Disneyfication under Emperor Giuliani', *GLQ: A Journal of Lesbian and Gay Studies*, 7, 101–109.

Root, M.P.P. (1996a), 'The Multiracial Experience', in M.P.P. Root (ed.).

Root, M.P.P. (ed.) (1996b), *The Multiracial Experience: Racial Borders as the New Frontier*, (London: Sage).

Rose, G. (1999), 'Performing Space', in D. Massey et al. (eds).

Rose, N. (1999), *Powers of Freedom: Reframing Political Thought*, (Cambridge: Cambridge University Press).

Roseneil, S. (2000), *Common Women, Uncommon Practices: The Queer Feminisms of Greenham*, (London: Cassell).

Rosneau, P. (1991), *Postmodernism and the Social Sciences*, (Princeton, NJ: Princeton University Press).

Rotello, G. (1997), *Sexual Ecology: AIDS and the Destiny of Gay Men* (New York: Dutton).

Rothenburg, T. (1995), '"And She Told Two Friends": Lesbians Creating Urban Social Space', in D. Bell and G. Valentine (eds).

Routledge, P. (1997), 'The Imagineering of Resistance: Pollock Free State and the Practices of Postmodern Politics', *Transactions of the Institute of British Geographers* 22, 359–376.

— (2003), 'Convergence Space: Process Geographies of Grassroots Globalization Networks', *Transactions of the Institute of British Geographers*, 28, 333–349.

Rubin, G. (1998), 'The Miracle Mile: South of Market and Gay Male Leather 1962–1997', in J.Brook et al. (eds).

Rushbrook, D. (2002), 'Cities, Queer Space, and the Cosmopolitan Tourist', *GLQ: A Journal of Lesbian and Gay Studies*, 8, 183–206.

Sachs, W. (ed.) (1992), *The Development Dictionary. A Guide to Knowledge and Power*, (London: Zed Books).

Safi, O. (2003a), 'Introduction', in O. Safi, (ed.).

Safi, O. (ed.) (2003b), *Progressive Muslims*, (Oxford: Oneworld Publications).

Saint-Blancat, C. (2002), 'Islam in Diaspora: Between Reterritorialization and Extraterritoriality', *International Journal of Urban and Regional Research*, 26:1, 138–152.

Saldanha, A. (2005), 'Trance and Visibility at Dawn: Racial Dynamics in Goa's Rave Scene', *Social and Cultural Geography*, 6:5, 707–721.

Salecl, R. (1997), 'The Postsocialist Moral Majority', in J.W. Scott et al. (eds).

Sanchez, L. (2004), 'The Global Erotic Subject, the Ban, the Prostitute-free Zone', *Environment and Planning D: Society and Space*, 22, 861–883.

Sanchez-Eppler, B. and Patton, C. (2000), 'Introduction: With a Passport Out of Eden', in C. Patton and B. Sánchez-Eppler (eds).

Sandliands, C. (2002), 'Lesbian Separatist Communities and the Experience of Nature: Toward a Queer Ecology', *Organization & Environment*, 15, 131–163.

Sant, M. and Waitt, G. (2000), 'Sydney: All Day Long, All Night Long', in J. Connell (ed.).

Saunders, K. (ed.) (2002), *Feminist Post-Development Thought. Rethinking Modernity, Post-Colonialism and Representation*, (London: Zed Books).

Schmitt, A. (2003), 'Gay Rights Versus Human Rights: A Response to Joseph Massad', *Public Culture*, 15:3, 587–591.

Schmitt, A. and Sofer, J. (1992), *Sexuality and Eroticism among Males in Moslem Societies*, (Binghamton, NY: Harrington Park Press).

Schueller, M.J. (2006), 'Analogy and (White) Feminist Theory: Thinking Race and the Color of the Cyborg Body', *Signs* 31:1, 63–92.

Schurman, N. (2004), *GIS: A Short Introduction*, (Oxford: Blackwell).

Scott, J. (1992), 'The Evidence of Experience', in H. Abelove et al. (eds).

Scott, J.W., Kaplan, C. and Keates, D. (eds) (1997), Transitions, Environments, Translations. Feminism and International Politics, (New York: Routledge).

Scott, T.D. (1997), 'Le Freak, C'est Chic, Le Fag, Quelle Drag!', in C. Queen and L. Schimel, (eds).

Sedgwick, E.K. (1990), *Epistemology of the Closet*, (Berkley: University of California Press).

— (1993), *Tendencies*, (Durham, NC: Duke University Press).

— (2003), *Touching Feeling: Affect, Pedagogy, Performativity*, (Durham, NC: Duke University Press).

Serlin, D. (1996), 'The Twilight (Zone) of Commercial Sex', in Dangerous Bedfellows (eds.).

Shakespeare, T. (1996), 'Power and Prejudice: Issues of Gender, Sexuality and Disability', in L. Barton (ed.).

— (2003), 'I Haven't Seen That in the Kama Sutra: The Sexual Stories of Disabled People', in J. Weeks et al. (eds).

Shakespeare, T., Gillespie-Sells, K. and Davies, D. (1996), *The Sexual Politics of Disability*, (London: Cassell).

Shannon, G. and Dever, G. (1974), *Health Care Delivery: Spatial Perspectives*, (New York: McGraw-Hill).

Shannon, G., Pyle, G. and Baskur, R. (1991), *The Geography of AIDS*, (New York and London: Guilford).

Sharp, J.P., Routledge, P., Philo, C. and Paddison, R. (eds) (2000), *Entanglements of Power: Geographies of Domination/Resistance*, (London: Routledge).

Sheppard, E. (2001), 'Quantitative Geography: Representations, Practices, and Possibilities', *Environment and Planning D: Society and Space*, 19, 535–554.

— (2004), 'Practicing Geography', *Annals of the Association of American Geographers*, 94, 744–747.

Shildrick, M. (2002), *Embodying the Monster: Encounters with the Vulnerable Self*, (London: Sage).

Shildrick, M. (2004), 'Queering Performativity: Disability after Deleuze', *SCAN: Journal of Media Arts Studies* 1:3 [website] <http://scan.net.au/scan/journal/display.php?journal_id=36>, accessed 11 November 2006.

Shrage, L. (1997), 'Passing Beyond the Other Race or Sex,' in N. Zack (ed.).

Signorile, M. (1997), *Life Outside: The Signorile Report on Gay Men: Sex, Drugs, Muscles and the Passages of Life*, (New York: Harper Collins).

Simmons, G. (2003), 'Are We Up to the Challenge? The Need for a Radical Re-ordering of the Islamic Discourse on Women', in O. Safi, (ed.).

Simon, S. (2006a), 'Christians Sue for Right not to Tolerate Policies", *Los Angeles Times*, 10 April, A-1.

— (2006b), 'Hundreds of Teenagers Plan to Make Valentine's a "Day of Purity"', *Los Angeles* Times,13 February, A-9.

Simpson, M. (1999), *It's a Queer World: Deviant Adventures in Pop Culture*, (New York: Harrington Park Press).

Singhanetra-Renard, A., Chongsatitmun, C. and Wibulswasdi, P. (1996), *Household and Community Responses to HIV/AIDS in Thailand*, (New York: WHO/GOA/UNAIDS).

Skeggs, B. (1997), *Formations of Class & Gender*, (London: Routledge).

— (1999), 'Matter Out of Place: Visibility and Sexualities in Leisure Spaces', *Leisure Studies*, 18, 213–232.

— (2001), 'The Toilet Paper', *Women's Studies International Forum* 24:2–3, 295–307.

— (2002), 'Techniques for Telling the Reflexive Self', in T. May (ed.).

— (2004), 'Uneasy Alignments, Resourcing Respectable Subjectivity', *GLQ: A Journal of Lesbian and Gay* Studies, 10:2, 291–298.

Skeggs, B., Moran, L., Tyrer, P. and Binnie, J. (2004), 'Queer as Folk: Producing the Real of Urban Space', *Urban Studies*, 41:9, 1839–1856.

Sleigh, A., Chee Heng Leng, Yeoh, B.S.A., Phua Kai Hong and Safman, R. (eds) (2006), *Population Dynamics and Infectious Diseases in Asia*, (London and Singapore: World Scientific Publishers, Springer).

Smith, A. (2003), 'S&M–event Organizers Whipped in Maryland: Outcry from Public Causes Cancellation of "World's Largest" Bondage Conference', *WorldNetDaily* [website], (published online 17 October 2003) <http://worldnetdaily.com/news/article.asp?ARTICLE_ID=35121>, accessed 11 November 2006.

Smith, J. (1999), *Islam in America*, (New York: Columbia University Press).

Smith, M. (1986), 'Physician's Specialities and Medical Trade Areas: An Application of Central Place Theory', paper presented at *Applied Geography Conference*, Kansas State University.

Smyth, F. (2005), 'Medical Geography: Therapeutic Places, Spaces, and Networks', *Progress in Human Geography*, 29:4, 488–495.

Sothern, M. (2004), '(Un)queer Patriarchies, or "What Do We Think When We Fuck?"', *Antipode*, 36, 183–190.

— (2006), 'On Not Living With AIDS: Or, AIDS-As-Post-Crisis', *ACME: An International Journal for Critical Geographies*, 5:2, 144–162. <http://www.acme-journal.org/vol5/MSo.pdf>, accessed 12 November 2006.

Spencer, A (2005), *DIY: The Rise of Lo-fi Culture*, (London: Marion Boyars Publishers).

Spickard, P. (2001), 'The Subject is Mixed Race', in D. Parker and M. Song (eds).

Squire, S. (2004), *Liminal Lives: Imagining The Human at the Frontiers of Biomedicine*, (Durham, NC: Duke University Press).

Squires, J. (ed.) (1993), *Perversity,* Special Issue of *New Formations*, 19.

Stacey, J. (1988), 'Can There be a Feminist Ethnography', *Women's Studies International Forum*, 11, 21–47.

Staeheli, L. (1996), 'Publicity, Privacy, and Women's Political Action', *Environment and Planning D: Society and Space* 14, 601–619.

Staeheli, L. and Lawson, V. (1995), 'Feminism, Praxis, and Human Geography', *Geographical Analysis*, 27, 321–338.

Stein, A. (1997), 'Sleeping with the Enemy?', *Sex and Sensibility*, (Berkeley: Berkeley University Press).

Stewart, A. (1995), 'The Early Modern Closet Discovered', *Representations* 50,

76–100.

Stiell, B. and England, K. (1999), 'Jamaican Domestics, Filipina Housekeepers, and English Nannies: Representations of Toronto's Foreign Domestic Workers', in J. Momsen (ed.).

Stratham, P. (1999), 'Political Mobilization by Minorities in Britain', *Journal of Ethnic and Migration Studies*, 25, 597–626.

Streeter, C. (1996), 'Ambiguous Bodies', in M.P.P. Root (ed.).

Stryker, S. (2004), 'Transgender Studies: Queer Theory's Evil Twin', *GLQ: Journal of Lesbian and Gay Studies*, 10, 212–215.

Stychin, C. (2003), *Governing Sexuality: The Changing Politics of Citizenship and Law Reform*, (Oxford: Hart).

Sullivan, A. (1995), *Virtually Normal: An Argument about Homosexuality*, (New York: Alfred A Knopf).

Sullivan, N. (2003), *A Critical Introduction to Queer Theory*, (New York: New York University Press).

Sweat, M., Nopkesorn, T., Mastro, T., Sangkharomya, S., MacQueen, K., Pokapanichwong, W., Sawaengdee, Y. and Weniger, B. (1995), 'AIDS Awareness Among a Cohort of Young Thai Men: Exposure to Information, Level of Knowledge, and Perception of Risk', *AIDS Care*, 7:5, 573–591.

Tani, S. (2002), 'Whose Place is This Space? Life in the Street Prostitution Area of Helsinki, Finland', *International Journal of Urban and Regional Research*, 26:2, 343–359.

Tapinc, H. (1992), 'Masculinity, Femininity, and Turkish Male Homosexuality', in K. Plummer, (ed.).

Tattelman, I. (2000), 'Presenting a Queer (Bath)House', in J.A. Boone et al. (eds).

Taylor, G. and Ussher, J. (2001), 'Making Sense of S&M: A Discourse Analytic Account', *Sexualities*, 4:3, 293–314.

Taylor, V., Kaminski, E. and Dugan, K. (2003), 'From the Bowery to the Castro: Communities, Identities and Movements', in D. Richardson and S. Seidman (eds).

Taylor, Y. (2004), 'Negotiation and Navigation – An Exploration of the Spaces/ Places of Working-class Lesbians', *Sociological Research Online*, 9:1, <http:// www.socresonline.org.uk/9/1/taylor.html>.

Thien, D. (2005), 'After, or Beyond Feeling? A Consideration of Affect and Emotion in Geography', *Area*, 37: 4, 450–456.

Thomas, M.E. (2004), 'Pleasure and Propriety: Teen Girls and the Practice of Straight Space', *Environment and Planning D: Society and Space*, 22, 773–789.

Thompson, M. (ed.) (1987), *Gay Spirit: Myth and Meaning*, (New York: St Martin's Press).

Thongchai, W. (1994), *Siam Mapped: A History of the Geo-Body of a Nation*, (Honolulu: University of Hawaii Press).

Thorbek, S. and Pattanaik, B. (eds) (2002), *Transnational Prostitution. Changing Global Patterns*, (London: Zed Books).

Thornton, S. (1995), *Club Cultures: Music, Media and the Subcultural Capital*, (Cambridge: Polity Press).

Thrift, N. (1996), *Spatial Formations*, (London: Sage).

— (1999), 'Steps to an Ecology of Place', in D. Massey et al. (eds.).

— (2000), 'Entanglements of Power: Shadows?', in J. Sharp et al. (eds.).

— (2003), 'Performance and…', *Environment and Planning A*, 35, 2019–2024.

— (2004), 'Intensities of Feeling: Towards a Spatial Politics of Affect', *Geografiska Annaler*, 86B: 1, 57-78.

Tolia-Kelly, D.P. (2006), 'Affect – An Ethnocentric Encounter? Exploring the "Universalist" Imperative of Emotional/Affectual Geographies', *Area*, 38:2, 213–217.

Treichler, P. (1999), *How to Have a Theory in an Epidemic: Cultural Chronicles of AIDS*, (Durham and London: Duke University Press).

Triantifillou, P. and Nielsen, M.R. (2001), 'Policing Empowerment: The Making of Capable Subjects', *History of Human Sciences*, 14: 2, 63–86.

Turner, L. (2001), 'The Wrong Body of/and Knowledge', paper presented at *Researching under the Rainbow: The Social Relations of Research with Lesbian and Gay Men,* Lancaster University, 27 September.

Valentine, G. (1989), 'The Geography of Women's Fear', *Area* 21, 385–390.

— (1993a), '(Hetero)sexing Space: Lesbian Pperceptions and Experiences of Everyday Spaces', *Environment and Planning D: Society and Space*, 11, 395–413.

— (1993b), 'Negotiating and Managing Multiple Sexual Identities: Lesbian Time-Space Strategies', *Transactions of the Institute of British Geographers NS*, 18: 2, 237–248.

— (1993c), 'Desperately Seeking Susan: A Geography of Lesbian Friendships", *Area*, 25, 109–116.

— (1995a), 'Creating Transgressive Space: The Music of kd lang', *Transactions of the Institute of British Geographers NS*, 20:4, 474–485.

— (1995b), 'Out and About: Geographies of Lesbian Landscapes', *International Journal of Urban and Regional Research*, 19:1, 96–111.

— (1996), '(Re)negotiating the Heterosexual Street: Lesbian Productions of Space', in N. Duncan (ed.).

— (1997), 'Making Space: Lesbian Separatist Communities in the United States', in P. Cloke and J. Little (eds).

— (1999a), 'Ode to a Geography Teacher: Sexuality and the Classroom', *Journal of Geography in Higher Education*, 23, 417-423.

— (1999b), 'What it Means to be a Man: The Body, Masculinities, Disability', in R. Butler and H. Parr (eds).

— (ed.) (2000a), *From Nowhere to Everywhere: Lesbian Geographies*, (Cambridge: Polity Press).

— (2000b), '*Sexuality, Geography and*', in R.J. Johnston et al. (eds).

— (2002), 'Queer Bodies and the Production of Space', in D. Richardson and S. Seidman (eds).

Valentine, G. and Skelton, T. (2003), 'Finding Oneself, Losing Oneself: The Lesbian and Gay "Scene" as a Paradoxical Space', *International Journal of Urban and Regional Research*, 27, 849–866.

Valocchi, S. (2005), 'Not Yet Queer Enough: The Lessons of Queer Theory for the Sociology of Gender and Sexuality', *Gender and Society*, 19, 750–770.

Valverde, M. and Cirak, M. (2003), 'Governing Bodies, Creating Gay Spaces: Policing and Security Issues in "Gay" Downtown Toronto', *British Journal of*

Criminology, 43, 102–121.

Vanderbeck, R. (2005), 'Masculinities and Fieldwork: Widening the Discussion', *Gender, Place and Culture* 12: 4, 387–402.

Van Deusen, R. (2004), 'The State, Culture and Rights: A Response to Sallie Marston's Space, Culture, State: Uneven Developments in Political Geography', *Political Geography*, 23, 27–34.

Van Hoven, B. and Horschelmann, K. (eds) (2005), *Spaces of Masculinities*, (London: Routledge).

VanLandingham, M. Supraset, S., Sittitrai, W., Vadhanaphuti, C. and Grandjean, N. (1993), 'Sexual Activity Among Never-married Men in Northern Thailand', *Demography*, 30:3, 297–313.

Volcano, D.L.G. and Halberstam, J. (1999), *The Drag King* Book, (London: Serpent's Tail).

Wadud, A. (2003), 'American Muslim Identity: Race and Ethnicity in Progressive Islam', in O. Safi, (ed.).

Walby, S. (1989), 'Theorising Patriarchy', *Sociology* 23:2, 213–234.

Ward, A. (1997), 'Which Side Are You On?', *Amerasia Journal* 19:2, 109–112.

Warner, M. (ed.) (1993), *Fear of a Queer Planet: Queer Politics and Social Theory*, (Minneapolis: University of Minnesota Press).

Warner, M. (2002), *Publics and Counterpublics*, (New York: Zone Books).

Warner, M. and Berlant, L. (2002), 'Sex in Public', in M. Warner.

Warner, T. (2002), *Never Going Back: A History of Queer Activism in Canada*, (Toronto: University of Toronto Press).

Warren, C. (1974), 'Space and Time', reprinted in P. Nardi and B. Schneider, (eds) (1998).

Watney, S. (1988), 'AIDS: The Cultural Agenda', *Radical America*, 21:4, 49–53.

— (1994), *Practices of Freedom: Selected Writings on HIV/AIDS*, (Durham, NC: Duke University Press).

Weeks, J. (1985), *Sexuality and its Discontents: Myths, Meanings and Modern Sexualities*, (London: Routledge and Kegan Paul).

— (2003), 'Necessary Fictions: Sexual Identities and the Politics of Diversity', in J. Weeks et al. (eds).

Weeks, J., Holland, J. and Waites, M. (eds) (2003), *Sexuality and Society: A Reader*, (Cambridge: Polity Press).

Weightman, B. (1980), 'Gay Bars as Private Places', *Landscape Research*, 23, 9–16.

— (1981), 'Commentary: Towards a Geography of the Gay Community', *Journal of Cultural Geography*, 1, 106–112.

Weinberg, T. (ed.) (1995), *S&M: Studies in Dominance & Submission*, (Amherst, NY: Prometheus Books).

Weinberg, T. and Magill, M. (1995), 'Sadomasochistic Themes in Mainstream Culture', in T. Weinberg (ed.).

Weissman, D. and Weber, B. (dir.) (2002), *The Cockettes*, (London: Tartan Video).

West, P. (2004), *Report into the Medical and Related Needs for Trans People in Brighton and Hove: The Case for an Integrated Service*, (Brighton: Spectrum). Available from <http://www.pfc.org.uk/medical/spectrum.pdf>.

WGSG (1997), *Feminist Geographies*, (London: Longman).

WGSG (2004) (eds) (2004), *Gender and Geography: 20 Years On* (Women in Geography Study Group, Glasgow).

Whatmore, S. (1999), 'Hybrid Geographies: Rethinking the "Human" in Human Geography', in D. Massey et al. (eds).

— (2003), *Hybrid Geographies: Natures, Cultures, Spaces* (London: Sage Publications).

Whippen, D. (1987), 'Science Fictions: The Making of a Medical Model for AIDS', *Radical America*, 20:6, 39–53.

Whittle, S. (ed.) (1994a), *The Margins of the City: Gay Men's Urban Lives*, (Manchester: Arena).

— (1994b), 'Consuming Differences: The Collaboration of the Gay Body with the Cultural State' in S. Whittle (ed.).

— (2000), *The Transgender Debate: The Crisis Surrounding Gender Identities*, (Reading: South Street Press).

Wilchins, R. (1997), *Read My Lips: Sexual Subversion and the End of Gender*, (Ann Arbor, MA: Firebrand Books).

Williams, T.K. (1992), 'Prism Lives', in M.P.P. Root (ed.).

Williams-León, T. (2001), 'The Convergence of Passing Zones,' in T. Williams-León and C.L. Nakashima (eds).

Williams-León, T. and Nakashima, C.L. (eds) (2001), *The Sum of Our Parts: Mixed Heritage Asian Americans*, (Philadelphia: Temple University Press).

Wilton, R. (1996), 'Diminishing Worlds? The Impact of HIV/AIDS on the Geography of Daily Life', *Health and Place*, 2:2, 1–16.

Wilton, T. (2000), 'Out/performing Ourselves: Sex, Gender and Cartesian Dualism', *Sexualities*, 3, 237–254.

Wincapaw, C. (2000), 'The Virtual Spaces of Lesbian and Bisexual Women's Electronic Mailing Lists', *Journal of Lesbian Studies*, 4, 45-59.

Winchester, H. and White, P. (1988), 'The Location of Marginalised Groups in the Inner City', *Environment and Planning D: Society and Space*, 6, 37–54.

Wiseman, J. (1996), *SM101: A Realistic Introduction*, 2nd edition (San Francisco: Greenery Press).

Witz, A. (2000), 'Whose Body Matters? Feminist Sociology and the Corporeal Turn in Sociology and Feminism', *Body and Society*, 6:2, 1–24.

Woodhead, D. (1995), 'Surveillant Gays': HIV, Space and the Constitution of Identities', in D. Bell and G. Valentine (eds).

World Health Organization (2001), *Global Prevalence and Incidence of Selected Curable Sexually Transmitted Infections: Overviews and Estimates*, (Geneva: World Health Organization).

Wright, J. and Rowson, E. (eds) (1993), *Homoeroticism in Classical Arabic Literature*, (New York: Columbia University Press).

Yingling, T.E. (1997), *AIDS and the National Body*, (ed.) R. Wiegman, (Durham, NC: Duke University Press).

Yip, A. (2004), 'Negotiating Space with Family and Kin in Identity Construction: The Narratives of British Non-heterosexual Muslims', *The Sociological Review*, 52:3, 336.

— (2005), 'Queering Religious Texts: An Exploration of British Non-heterosexual Christians' and Muslims' Strategy of Constructing Sexuality-affirming Hermeneutics', *Sociology*, 39:1, 47–65.

Young, P. (2003), 'Infamous Gay Spot Cleared Up', *Evening Chronicle*, 18 June, Newcastle, UK.

Young, R. (1995), *Colonial Desire. Hybridity in Theory, Culture and Race*, (London: Routledge).

Zack, N. (ed.) (1997), *Race/Sex*, (New York: Routledge).

Žižek, S. (1991), *Looking Awry. An Introduction to Jacques Lacan through Popular Culture*, (Cambridge, MA: MIT Press).

Index

by Margaret Binns